Chemical Processing of Dielectrics, Insulators and Electronic Ceramics

MATERIALS RESEARCH SOCIETY
SYMPOSIUM PROCEEDINGS VOLUME 606

Chemical Processing of Dielectrics, Insulators and Electronic Ceramics

Symposium held November 29–December 1, 1999, Boston, Massachusetts, U.S.A.

EDITORS:

Anthony C. Jones
Inorgtech Limited
Mildenhall, Suffolk, United Kingdom
and
The University of Liverpool
United Kingdom

Janice Veteran
Advanced Micro Devices
Austin, Texas, U.S.A.

Donald Mullin
SPAWAR System Center
San Diego, California, U.S.A.

Reid Cooper
University of Wisconsin-Madison
Madison, Wisconsin, U.S.A.

Sanjeev Kaushal
Tokyo Electron America
Austin, Texas, U.S.A.

Materials Research Society
Warrendale, Pennsylvania

CAMBRIDGE UNIVERSITY PRESS

Cambridge, New York, Melbourne, Madrid, Cape Town,
Singapore, São Paulo, Delhi, Mexico City

Cambridge University Press
32 Avenue of the Americas, New York NY 10013-2473, USA

Published in the United States of America by Cambridge University Press, New York

www.cambridge.org
Information on this title: www.cambridge.org/9781107413207

Materials Research Society
506 Keystone Drive, Warrendale, PA 15086
http://www.mrs.org

First published 2002
First paperback edition 2013

Single article reprints from this publication are available through
University Microfilms Inc., 300 North Zeeb Road, Ann Arbor, MI 48106

CODEN: MRSPDH

ISBN 978-1-107-41320-7 Paperback

CONTENTS

CHEMICAL VAPOR DEPOSITION
OF OXIDE CERAMICS

*Invited Paper

CHEMICAL VAPOR DEPOSITION OF
NON-OXIDE CERAMICS

SOLUTION DEPOSITION OF
ELECTRONIC CERAMICS

*Invited Paper

ALTERNATIVE CHEMICAL PROCESSING
METHODS AND CHARACTERIZATION
OF ELECTRONIC CERAMICS

PREFACE

This volume contains papers from Symposium NN, "Chemical Processing of Dielectrics, Insulators and Electronic Ceramics," held November 29–December 1 at the 1999 MRS Fall Meeting in Boston, Massachusetts. This symposium continues the theme of a previous symposium (Mater. Res. Soc. Symp. Proc. **495** (1998)) on the creative use of chemistry in the fabrication of advanced electronic ceramics. The symposium focused on the chemical fabrication of a variety of oxide and non-oxide materials which are likely to play a crucial role in the development of the next generation of microelectronics devices.

The symposium consisted of eight oral and four poster sessions with a total of 68 papers being presented, 44 of which are included in this volume. These clearly demonstrate the multidisciplinary nature of the field, involving inorganic precursor chemistry, gas-phase and solid state chemistry, materials science, chemical physics, and chemical engineering.

A number of particularly "hot" areas of research were featured in the symposium, including the deposition of high-k dielectric gate oxides, ferroelectric oxide films for infrared and memory applications, low-k dielectrics, TiN and TaN diffusion barriers, and new precursors for III-V nitrides.

The emphasis throughout is on chemical methods for the controlled deposition of thin films, for which chemical vapor deposition (CVD) has proven to be a useful and versatile technique. A particularly noteworthy development is the use of liquid injection MOCVD for the deposition of oxide multilayers and superlattices. Despite the increasing use of CVD, solution deposition techniques such as sol-gel, metalorganic decomposition (MOD), hydrothermal processing, and chemical bath techniques were also prominently featured.

These proceedings overlap to some extent with a number of other symposia in the 1999 MRS Fall Meeting, including "Ferroelectric Thin Films VIII," "GaN and Related Alloys," and "Structure and Properties of Ultrathin Dielectric Thin Films on Silicon and Related Materials." It is intended that the current volume complements and forms a valuable supplement to these related symposia.

It is the sincere hope of the symposium organizers that this volume will prove to be a useful overview of current research trends in a dynamic and exciting area of solid state technology.

Anthony C. Jones
Janice Veteran
Donald Mullin
Reid Cooper
Sanjeev Kaushal

January 2000

ACKNOWLEDGMENTS

The success of the symposium is due to the efforts of many people to whom we are very grateful. We are grateful to all of the speakers, poster presenters, and authors whose contributions are represented in these proceedings. We thank the MRS staff and the Meeting Chairs whose patience and efforts made our tasks much easier. We are also very grateful to the organizations who provided generous financial support.

Invited Speakers and Session Chairs

William S. Rees Jr., Georgia Tech, Atlanta, Georgia
Yoshihide Senzaki, Schumacher Inc., Carlsbad, California
S.A. Campbell, University of Minnesota, Minneapolis, Minnesota
Jean-Pierre Senateur, LMGP, ENS de Physique de Grenoble, France
M. Yoshimura, Tokyo Institute of Technology, Japan
R.P. Raffaelle, Rochester Institute of Technology, New York, New York
Roy Gordon, Harvard University, Boston, Massachusetts
Paul O'Brien, University of Manchester, UK
Janice Veteran, Advanced Micro Devices, Austin, Texas
Anthony C. Jones, Inorgtech Ltd. and Liverpool University, UK

Financial Support

MKS Instruments
Strem Chemicals Inc.

MATERIALS RESEARCH SOCIETY SYMPOSIUM PROCEEDINGS

MATERIALS RESEARCH SOCIETY SYMPOSIUM PROCEEDINGS

Prior Materials Research Society Symposium Proceedings available by contacting Materials Research Society

Chemical Vapor Deposition
of Oxide Ceramics

DESIGN, SYNTHESIS AND CHARACTERIZATION OF PRECURSORS FOR CHEMICAL VAPOR DEPOSITION OF OXIDE-BASED ELECTRONIC MATERIALS

OLIVER JUST[*], BETTIE OBI-JOHNSON[*], JASON MATTHEWS[*], DIANNE LEVERMORE[*], TONY JONES[**], AND WILLIAM S. REES, JR[*].
[*]School of Chemistry and Biochemistry and School of Materials Science and Engineering, and the Molecular Design Institute, Georgia Institute of Technology, Atlanta, GA 30332-0400
[**]InorgTech, 25 James Carter Road, Mildenhall, Suffolk, IP28 7 DE, United Kingdom

ABSTRACT

Ferroelectric and other high dielectric constant metal oxides currently are sought-after for a variety of applications in the electronics industry. To meet the demand of preparation of these interesting materials in a manner compatible with traditional silicon-based fabrication procedures, chemical vapor deposition routes are desired for film growth. Compounds displaying high vapor phase stability are necessary as precursors for these applications. Additionally, in general, it is preferred to utilize compounds in a liquid state, due to the more rapid re-establishment of equilibrium at a liquid-vapor interface, compared to that present at a solid-vapor interface. This combination of desired molecular properties, in turn, presents a great challenge to the coordination chemist. Several of the metals of interest for these uses reside in groups 2–5. Common design features are emerging for the ligands best suited for attachment to these metals for subsequent utilization in the deposition of metal oxides. In order to achieve coordinative saturation of the relatively high ionic radii exhibited by most of these elements, multidentate, monoanionic ligands are relied upon. In the past, most often, homoleptic ligand sets have been employed, thereby reducing the chance for ligand scrambling to occur during the growth process. Such disproportionation processes have been credited, in previous work, with the observation of a temporal decay in vapor pressure of heteroleptic compounds. In some interesting new developments, it has been found that heteroleptic compounds possess sufficient vapor phase integrity to permit their evaluation as CVD precursors. These, and related, results are presented herein.

INTRODUCTION

One of the greatest challenges in materials chemistry is to close the loop between evaluation of final device performance and the design of precursors, which enter into processes, utilized in device manufacture. This Holy Grail of "post-mortem" detection of failure devices, and its integration into the "pre-embryonic" design of molecular precursors, has attracted substantial interest from researchers in recent years. As shown in Figure 1, it is incumbent on researchers in the area of precursor development to take a broad view of what are considered as inputs and outputs to the overall area of precursor design. Frequently, it is viewed that the singular input is design and the only output is a CVD precursor. Design includes the components of cost, technical specifications, equipment limits, and process parameters, each of which must be independently considered and weighed against one another in decisions regarding precursor design. The output is not only the compound itself, but also additionally equipment, and process recommendations to accompany all chemistries, which have been developed. In this vein, one may have discovered a compound which is not amenable to delivery by traditional (vapor phase) modes. An example of this is the emergence of liquid delivery systems to accommodate precursors, which are not useable in processes relying exclusively on traditional vapor delivery schemes.

3

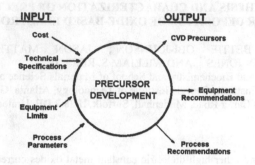

INPUT

Cost

Technical
Specifications

PRECURSOR
DEVELOPMENT

Equipment
Limits

Process
Parameters

OUTPUT

CVD Precursors

Equipment
Recommendations

Process
Recommendations

Figure 1: Inputs and Outputs for MOCVD precursors.

This manuscript follows the two themes of design and characterization in the following sections.

DESIGN

Statement of the Challenge

Many desired dielectric, insulating, and other electronic materials contain elements residing in groups 1-5 of the periodic chart. The heaviest representatives among these elements have the smallest known charge/size ratio among the entire periodic chart. Thus, this problem is among the most difficult for a coordination chemist to tackle. Additionally, there are substantial chemistry knowledge gaps present in these s block and early d block transition elements. Therefore, the wide pyramid base which was present in p block chemistry, and contributed to the early development of alternative precursors in III-V compound semiconductors, is absent in this region. Furthermore, organometallic chemistry (which is known for the early transition elements) is often not directly applicable to the growth of metal oxides, which are necessary for most modern electronic materials. Therefore, substantial basic research effort must be invested to compensate for these fundamental chemistry knowledge gaps among the elements, which are vital to the preparation of the next generation of electronic devices.

RECENT RESULTS

As shown in Figure 2, magnesium with a *bis(β-diketonate)* ligand is four coordinate. The magnesium being divalent, binds two monovalent ligands, each ligand being bidentate, with the net result being a coordination number two less than the optimum number of six for magnesium. As shown in the Figure, the material picks up two additional diethyl ether molecules in axial positions to become octahedral, and, therefore, six coordinate. The diethyl ether ligands are intermolecular in nature, and their weak coordination is capable of becoming disassociated in vapor phase transport.

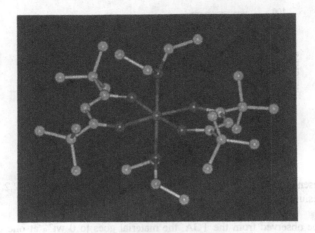

Figure 2: Ball-and-stick representation of Mg(tmhd)₂(Et₂O)₂.

In order to compensate for this loss, recently intramolecularly coordinating ligands have been designed. In the specific case of magnesium, these were tridentate monoanionic ligands. A structure of one of these resultant products is depicted in Figure 3,

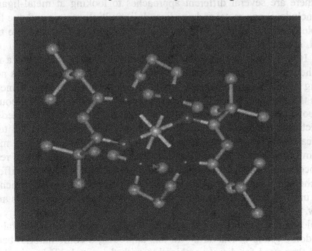

Figure 3: Ball-and-stick representation of bis(5-N-dimethylamino-2,2,7-trimethyl-3-octanato) magnesium.

and the thermogravimetric analysis plot is presented in Figure 4.

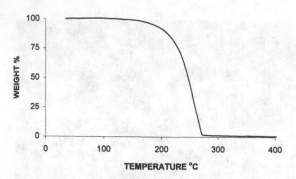

Figure 4: Representation of the thermal behavior of bis(5-N-dimethylamino-2,2,7-trimethyl-3-octanato) magnesium.

As can be observed from the TGA, the material goes to 0 wt% at one atmosphere of pressure at approximately 260°C in a single step; therefore there is neither solid nor vapor state decomposition prior to sublimation and during transport of this material, respectively. This example of intramolecular coordination satisfying the high coordination number of low-valent metal cations has been successfully employed in several research groups, notably those of Rees and Marks for the group 2 elements.[1]

Overall, there are several different approaches to looking at metal-ligand interaction systems. These include: i) one metal and one ligand; ii) multiple metals and one ligand; iii) one metal and multiple ligands; and iv) multiple metals and multiple ligands. The traditional one metal/one ligand approach has been used for decades in the preparation of compound semiconductors. It is the one, which is most frequently the entry point into a new materials system. Thus, when an initial result appears in chemical vapor deposition of a new material it is inevitably using off-the-shelf precursors, which are well known to be one metal/one ligand compositions. In the chemical sense, these are referred to as homoleptic compounds. The last one mentioned above (multiple metals and multiple ligands) is generally utilized in the sol-gel processing of electronic materials and, to the best of our knowledge, has yet to be met with success in the area of chemical vapor deposition of processing of electronic materials. The notion of having multiple metals and one ligand is also primarily (at this stage) reserved for the use of sol-gel processing. The final remaining one (one metal and multiple different ligands), referred to in a chemist's vocabulary as a heteroleptic compound, is one which has received limited attention, until recently. There are several recent success stories in this area, which are highlighted below.

One early example of this heteroleptic approach was the use of a single metal mixed ligand system from Gordon's group at Harvard.[2] In this approach, a combinatorial batch of ligands was prepared and then used for direct combination with the metal. This mixed ligand, single metal system was an ambient condition liquid. Although not highlighted by the authors in their original work, a key advantage of this approach is that the ligand purification step has been circumvented. Therefore, the frequently time-consuming and difficult process of purifying an organic compound to absolute homogeneity prior to being mixed with a metal has been skipped. This avoids the inevitable loss of material that occurs during purification, as well as the concomitant increased cost of the final product.

Another recent example in this area is the dimeric heteroleptic zirconium isopropoxide β-diketonate, which has been prepared at InorgTech, and was used in a liquid injection system. It has a higher vapor pressure than the homoleptic *tetrakis*-tmhd zirconium complex, as indicated in Figure 5.

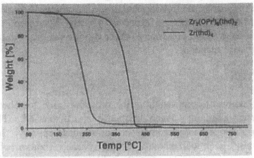

Figure 5: Comparative thermal decomposition data for $Zr_2(O^iPr)_6(thd)_2$ and $Zr(thd)_4$.

The deposition rate of this heteroleptic compound is very similar to that of lead *bis*-tmhd as shown in Figure 6.

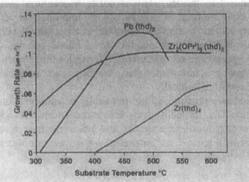

Figure 6: Metal oxides growth rates as a function both of substrate temperature and utilized precursor.

The evaporator is reported to remain residue-free after long-term use of this material.[3] The design of this particular material is such that the bridging ligands between the two zirconiums are both alkoxides. Purely from consideration of a traditional coordination chemistry point of view, and looking at the electrostatics present within the various anionic ligands, it is predicted that alkoxides would bridge, and the β-diketonates would function as terminal ligands in this case.

Despite the above highlighted successes in one metal/mixed ligand heteroleptic systems for chemical vapor deposition precursors, there are nevertheless numerous concerns, which remain unsolved with these approaches. Ligand scrambling may lead to a temporal instability in vapor pressure. If such scrambling is present, then congruent evaporation of the resulting temporally changing evaporator contents cannot be compensated for. There may be inconsistent thermal decomposition profiles resulting from the differing bond association energies, which are present amongst the ligand set. In some respects, this may be turned to the advantage of the chemist, because they can then have a selective loss of ligand attachment of metal to the

7

surface and subsequent thermal decomposition of the remaining ligands on the metal in a different regime. Additionally, there are challenges to *in-situ* feedback control monitors when multiple ligands are present in the flowing reaction co-products. Frequently, these monitors are optimized for single point detection of species. Lastly, heteroleptic compounds may have different kinetic stabilities in the presence of other vapor reactants, which lead to the possibility of premature vapor phase reactions.

CHARACTERIZATION

In characterization of compounds in the precursor area for chemical vapor deposition, there are two different domains, which frequently are discussed. The first one is directly intended towards application in organometallic vapor phase epitaxy. Separate from that, there is the issue of gaining chemical insights through characterization of compounds, which is entered back into the feed cycle to generate a better design for subsequent generations of precursors. In this particular approach, the motive may be summed up as follows. In order to achieve improvements on existing source compounds, one must understand the beneficial and detrimental components of their decomposition mechanisms. This may be regarded as studying failure cases to examine the mode(s) of failure, thereby eliminating the identified failure mode(s) from subsequent molecules. This is not to be confused with designing the perfect precursor. In the same vein, it is elimination of unproductive growth pathways, which practitioners are after, not the creation of exclusively perfect growth pathways.

The characterization of compounds should be focused on the critical properties for organometallic vapor phase epitaxy precursors. There are three such properties. Precursors must have suitable vapor pressure, vapor phase stability, and condensed phase stability. Each of these issues is addressed individually.

In the case of suitable vapor pressure, all too frequently it is simply a one atmosphere TGA plot that is relied on for this data. In reality, very few manufacturing processes are operated at one atmosphere, thus it is at reduced pressure that the most important measurement of usefulness can occur. As shown in Figure 7, an automated vapor pressure determination apparatus has been constructed at Georgia Tech.

a.)

b.)

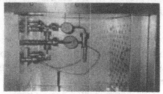

Figure 7: Photographs of the VPDA apparatus
a.) outside the oven (turbo pump and cold cathode
gauge are not shown), b.) inside the oven.

8

Figure 8 presents a CAD view of the system.

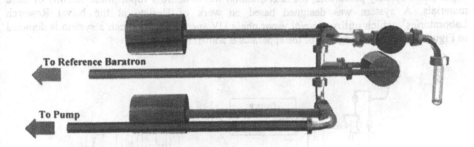

To Reference Baratron

To Pump

Figure 8: The CAD view of the VPDA system.

Representative data obtained for yttrium, copper and barium tmhd compounds are shown in Figure 9a and for barium and strontium tmhd compounds in Figure 9b.

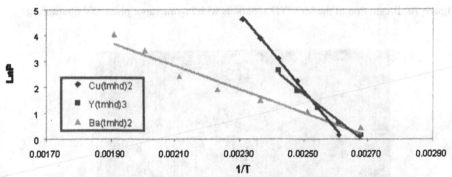

Figure 9a: Graphical depiction of 1/T versus lnP for Cu(thd)$_2$, Y(thd)$_3$ and Ba(thd)$_2$.

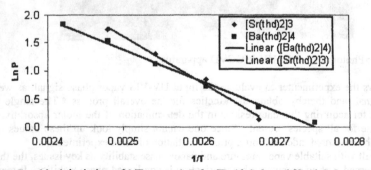

Figure 9b: Graphical depiction of 1/T versus lnP for [Ba(thd)$_2$]$_4$ and [Sr(thd)$_2$]$_3$.

Once having constructed an instrument specifically designed for accomplishing vapor pressure determination, which addressed the key issue for suitable vapor pressure for organometallic vapor phase epitaxy precursors, the next question was to address vapor phase stability of these materials. A system was designed based on work of Desisto at the Naval Research Laboratories[4], which utilizes *in-situ* vapor phase UV spectroscopy. Such a system is depicted in Figure 10, and a photograph of the apparatus is shown in Figure 11.

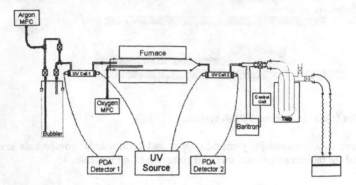

Figure 10: Illustration of the MOCVD apparatus with *in-situ* vapor phase UV monitoring.

Figure 11: Photograph of the UV-MOCVD apparatus.

This allows the experimenter to evaluate the input UV-VIS vapor phase signal, as well as the output signal, and thereby obtain the kinetics for the overall process. The single constant demanded for acquiring this data resides in the determination of the molar absorbtivity in the vapor phase for all species present. Since one cannot simply look up these values in tables, they must be measured independently, prior to initiation of these experiments.

Having dealt with suitable vapor pressure and vapor phase stability as key issues, the third issue to be concerned with in precursor characterization is condensed phase stability. It generally is regarded that changes in the condensed phase are irreversible processes occurring by ligand loss or fragmentation of materials in condensed phases. However, there are also reversible condensed phase interactions, which are typified by phase changes that may occur during cycling of bubbler lifetime. The result of a solid state ^{13}C NMR experiment demonstrating this both for variable temperature, as well as temporally over 24 hours, is presented in Figure 12 for yttrium tris(tmod).

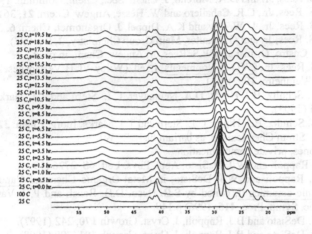

Figure 12: Variable temperature ^{13}C solid state NMR of [Y(tmod)$_3$]$_2$.

This compound, a dimer at ambient conditions when loaded into an evaporator, it converts upon heating to 60° C to a monomeric material (still in the solid state). Upon re-cooling to 25° C, it initially retains its monomeric structure to subsequently reverting back to a dimer within 24 hours. Thus, to sum up, the material is loaded into the evaporator as a dimer, at use conditions it is a monomer, when it is cooled back down it slowly reverts back to a dimer; however, the conversion is not complete for almost one day. This gives rise to a detailed understanding of the condensed phase stability and the thermal cycling which is demanded in solid state bubbler transport for organometallic vapor phase epitaxy employment in the preparation of electronic materials by such processes.

SUMMARY

Overall, several key criteria in design and characterization of compounds must be addressed. It is not just the synthesis of precursors, which will permit the next generation of electronic materials to be prepared, but chemical vapor deposition in a more efficient and rigorous manner. There must be a careful balance between exhaustive characterization for applications and building the knowledge base of chemical insight by studying the failure products of unsuccessful precursors. When designing new compounds, one must pay attention to the fundamental roles of coordination chemistry, and the gaps which are present today in our knowledge related to the s block and early d block elements. Specifically, with regard to critical properties for precursors in general, each of the three issues discussed herein must be answered with a resounding "yes". It must have suitable vapor pressure, it must have suitable vapor phase stability, and it must have suitable condensed phase stability prior to being employed in a CVD reactor.

REFERENCES

1.   a.) W.S. Rees, Jr. and D.A. Moreno, J. Chem. Soc., Chem. Commun. 1991, 1759.

b.) W.S. Rees, Jr., C.R. Caballero and W. Hesse, Angew. Chem. 21, 361 (1992).

c.) W.S. Rees, Jr., U.W. Lay and K.A. Dippel, J. Organomet. Chem. 6, 27 (1994).

d.) W.S. Rees, Jr. and G. Krauter, Main Group Chemistry, 2, 9 (1997).

e.) S.L. Castro, O. Just and W.S. Rees, Jr. Angew. Chem. in print.

f.) W.S. Rees, Jr., CVD of Nonmetals, 1st ed. (Wiley-VCH Verlagsgesellschaft mbH, D-69469 Weinheim, 1997), p. 1.

g.) D.L. Schultz, B.J. Hinds, D.A. Neumayer, C.L. Stern and T.J. Marks, Chem. Mater. 5, 1605 (1993).

h.) D.L. Schultz, B.J. Hinds, C.L. Stern and T.J. Marks, Inorg. Chem. 32, 249 (1993).

2.   Sold by Strem Chemical Co. as barium bis[BREW].

3.   A.C. Jones, private communication.

4.   a.) B.J. Rappoli and W.J. DeSisto, Appl. Phys. Lett. 68, 2726 (1996).

b.) B.J. Rappoli and W.J. DeSisto, in Metal-Organic Chemical Vapor Deposition of Electronic ceramics II, edited by S.B. Desu, D.B. Beach and P.C. Van Buskirk (Mat. Res. Soc. Symp. Proc. 415, 1996) p.149.

c.) W.J. DeSisto and B.J. Rappoli, J. Cryst. Growth 170, 242 (1997).

d.) W.J. DeSisto and B.J. Rappoli, J. Cryst. Growth 191, 290 (1998).

MOCVD of High-K Dielectrics and Conductive Metal Nitride Thin Films

Yoshihide Senzaki,[a,*] Richard F. Hamilton,[b] Kimberly G. Reid,[c] Christopher C. Hobbs,[c]
Rama I. Hegde[c], and Mike J. Tiner[c]
[a] Schumacher, Carlsbad, California, 92009
[b] Air Products and Chemicals, Inc., Allentown, Pennsylvania, 18195
[c] Motorola, Austin, Texas, 78721
* corresponding author (e-mail: senzaky@apci.com)

Abstract

A known liquid mixture of $[(CH_3CH_2)_2N]_3Ta=NCH_2CH_3$ and $[(CH_3CH_2)_2N]_3Ta[\eta^2-CH_3CH_2N=CH(CH_3)]$ was studied to deposit Ta_2O_5 and TaN thin films by CVD. Films were deposited at temperatures below 400°C using oxygen for oxide and ammonia for nitride, respectively. XRD analysis revealed that as-deposited amorphous tantalum oxide films were converted to hexagonal Ta_2O_5 after annealing under oxygen, while tantalum nitride thin films contained cubic TaN as deposited. The low viscosity, thermal stability, and sufficient volatility of the precursor allows direct liquid injection to deliver the precursor, which results in high deposition rate and uniformity of the deposited films.

Introduction

As the size of integrated circuit devices decreases, chemical vapor deposition (CVD) shows a unique advantage over physical vapor deposition (PVD) for device fabrication in terms of its excellent step coverage for trench, via, and stacked cell structures for applications in microelectronics devices. In general, liquid precursors are preferred to solid precursors for CVD applications due to the ease and reproducibility of the precursor delivery.

Ta_2O_5 is considered a promising material as a gate oxide and as a DRAM storage capacitor due to its relatively high dielectric constant.[1] For the CVD of Ta_2O_5 thin films, various precursors have been studied thus far. Tantalum halides (TaX_5, $X = F^2$, Cl^3) are solid and suffer from low volatility and difficulty in delivery. A solid metal-organic complex $Ta(NMe_2)_5$ has been reported to provide Ta_2O_5 films with significant levels of carbon and nitrogen impurities.[4] A dinuclear solid complex $[Ta(OMe)_5]_2$ has also been studied.[5] The most commonly studied liquid precursor $[Ta(OEt)_5]_2$ has a marginal vapor pressure and deposits films containing carbon impurities.[1] It has recently been reported that mononuclear complexes of the type $Ta(OEt)_4(L)$, where L = dimethylaminoethoxide[6] and β-diketonate,[7] have improved volatility for CVD applications.

As copper emerges as a substitute for aluminum in multilevel metallization of semiconductor devices, CVD precursors for conductive TaN are also actively being sought due to the excellent barrier properties of TaN against the copper diffusion into silicon substrates at high temperatures.[8] So far, a number of precursors have been evaluated for TaN CVD. None have shown the combination of being a stable liquid, ease of synthesis and suitability for low temperature CVD processing to provide pure TaN. Solid tantalum halides (TaX_5, $X = Br^9$, Cl^{10}) are not desirable as mentioned above. Reaction of solid $Ta(NMe_2)_5$ with ammonia yields non-conductive Ta_3N_5.[11] Liquid alkylamide type complexes have also been studied. An unstable liquid $Ta(NEt_2)_5$ has been used as a single-source to produce tantalum carbonitride.[12] The quality of TaN films deposited from $Ta(NEt_2)_5$ was improved by using ammonia as a reactant gas.[13] Tsai, et. al. reported the deposition of low resistivity TaN films from t-$BuN=Ta(NEt_2)_3$ at 650°C.[14] This high temperature process is incompatible with integrated circuit process integration. Chiu, et. al. reported the use of a liquid mixture of the compounds $[(CH_3CH_2)_2N]_3Ta=NCH_2CH_3$ and $[(CH_3CH_2)_2N]_3Ta[\eta^2\text{-}CH_3CH_2N=CH(CH_3)]$ (the structure is shown below). The films deposited at 500-650°C without any reactant gas contained a significant amount of carbons.[15]

$$[(CH_3CH_2)_2N]_3Ta \underset{C(H)CH_3}{\overset{NCH_2CH_3}{<}}$$

In this study, this liquid mixture $[(CH_3CH_2)_2N]_3Ta=NCH_2CH_3$ and $[(CH_3CH_2)_2N]_3Ta[\eta^2\text{-}CH_3CH_2N=CH(CH_3)]$ in the molar ratio of 7:3 was used in CVD applications in combination with oxygen or ammonia reactants at low temperatures (below 400°C) to provide tantalum oxide or nitride thin films. The precursor was delivered by direct liquid injection (DLI) and bubbling. DLI is a preferred method because it delivers the same ratio of constituents to the reactor as are in the source container. DLI has the added advantage of storing the precursor at room temperature and heating only the amount required to be delivered, and therefore, improving precursor shelf life.

Experimental

The precursor mixture $[(CH_3CH_2)_2N]_3Ta=NCH_2CH_3$ and $[(CH_3CH_2)_2N]_3Ta[\eta^2\text{-}CH_3CH_2N=CH(CH_3)]$ is an air sensitive yellow liquid and was prepared according to the literature method.[16, 17] Both compounds have similar volatility.[17]

Films were deposited on 4-inch diam. Si(100) wafers using a warm-wall single wafer LPCVD system at a chamber pressure of 0.5-1.5Torr. For precursor delivery by bubbling, the precursor container and delivery lines were heated to 85°C and 90°C, respectively. Silicon substrates were heated to 270-384°C. 40sccm of O_2 or 30sccm of NH_3 gas was introduced into the chamber. The precursor was delivered to the chamber with a helium carrier gas flow of 150-200sccm.

The DLI system(liquid flow controller/vaporizer/injector) is shown in Figure 1. The precursor was delivered into the vaporizer at 25°C at a flow rate of 0.08mL/min using a helium push gas (6psi) and vaporized at 90°C. The precursor vapor was delivered to the chamber with a helium sweep gas flow of 100-130sccm. A flow of 40-80sccm of O_2 gas was introduced into the chamber for oxide deposition. For nitride deposition, a flow of 50sccm of $(NH_3 + He)$gas was introduced into the chamber at volume ratios of NH_3: precursor vapor = 2:1 to10:1.

Thin films were characterized by ellipsometry, thin film X-ray diffraction (XRD), Auger electron spectroscopy (AES), X-ray photoelectron spectroscopy (XPS), transmission electron microscopy (TEM), and energy dispersive X-ray (EDX) analysis. Capacitance-voltage and leakage current (I-V) characteristics of the Ta oxide films were measured using a mercury probe. The Hg-probe area was calibrated using a 1500Å thick SiO_2 standard and was determined to be $0.229mm^2$ with a repeatability of 0.142%. A capacitance-voltage frequency of 100kHz was used

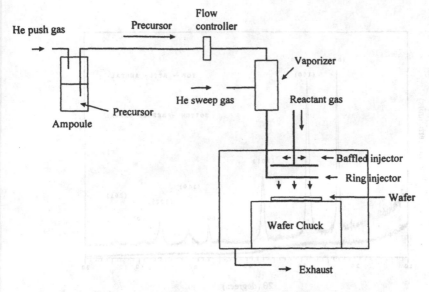

Figure 1   CVD reactor configuration with DLI system

and the equivalent oxide thickness (EOT) was calculated from the maximum accumulation capacitance.

Results and Discussion

<u>CVD of tantalum oxide</u>: The precursor delivery by bubbling provided an approximately 1400Å blue thin film after 10min of deposition at 380°C. The refractive index of the film 2.3 measured by ellipsometry was comparable to the literature value of 2.0-2.2.[1] XRD analysis revealed that films were amorphous as deposited. While XRD identified the film annealed at 800°C under nitrogen for 30min as hexagonal tantalum oxide (Figure 2), AES analysis demonstrated that the ratio of O/Ta is less than 2.5. Thus, the film is hexagonal δ-(Ta-O), or presumably oxygen deficient form of Ta_2O_5 as previously suggested by Terao.[18] The XRD of the sample annealed under O_2 for 30min closely resembled that of the film annealed under N_2. AES analysis demonstrated that the atomic ratio of O/Ta was 2.5, and therefore, the film annealed under O_2 was identified as hexagonal δ-Ta_2O_5. Based on the AES analysis, less than 5 atom.% of carbon was incorporated in the film while nitrogen was below the detection limit (<1 atm.%).

Using DLI delivery, thin films were grown at deposition rates reaching 170Å/min at 340°C. From the Arrhenius plot of the deposition rate against the reciprocal of the deposition temperature, the apparent activation energy is estimated to be 22kcal/mol (Figure 3). The refractive indexes, 2.1-2.2 of the films were similar to those of the films deposited using a precursor bubbling method. The film uniformity degraded with increasing deposition temperature. The standard deviations were about 1% at 270°C and increased to 6% at 340°C.

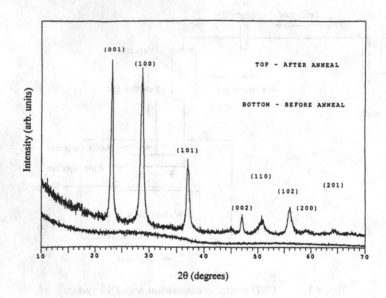

Figure 2. XRD of tantalum oxide before and after anneal under N_2 at 800°C for 30min.

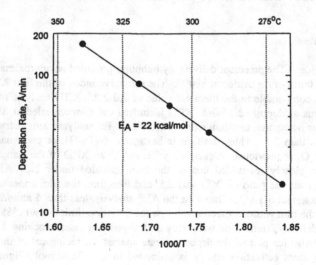

Figure 3. Arrhenius plot of TaOx deposition rate against $1/T(K^{-1})$.

For electrical characterization, Ta oxide thin films were deposited on HF etched Si substrates at 360°C using DLI method. Films with three different thickness (140, 190, and 580Å as-deposited) were annealed at 800°C for 10min under N_2 and O_2. The effects of thermal anneal on dielectric constants and EOT of Ta oxide films are shown in Figures 4 and 5, respectively. The as-deposited film with thickness of 580Å had a low dielectric constant of 16 while dielectric constants of 22-23, which are comparable to the literature value of $Ta_2O_5(\sim25)$,[1] were obtained after the anneal. The low dielectric constants of the as-deposited films could be attributed to the oxygen deficiency(O/Ta<2.5). The influence of the interfacial SiO_2 layer is reflected in the thickness dependence of the dielectric constant, namely, the measured dielectric constant decreases as the film thickness decreases. Independent of the film thickness, the leakage current of O_2 annealed samples consistently ranged from 10^{-7} to $10^{-6}A/cm^2$ as shown in Figures 6 and 7. The leakage current data points obtained from the as-deposited and N_2 annealed samples hovered in wider ranges. This may indicate that O_2 anneal is efficient in decreasing oxygen vacancies in the films. The improvement of leakage current by post-deposition treatment is commonly known.[19] In Figure 8, TEM cross section image of the 140Å (as-deposited) sample is shown. 9 ± 1Å of silicon oxide layer was grown between TaOx and Si during the deposition. The thickness of this interface layer increased to approximately 30-40Å after O_2 anneal as shown in Figure 9.

CVD of tantalum nitride: Using precursor delivery by bubbling, a gray thin film with thickness of 1500Å was grown at 384°C over a 10min deposition period. The XRD analysis in Figure 10 shows that the crystallinity of cubic TaN in the as-deposited film was slightly enhanced after annealing at 500°C under N_2 for 30 min. AES analysis after surface cleaning by sputtering revealed that the as-deposited film contained N/Ta =1.3, and the impurity levels of 7 atom.% C and 5 atom.% O. The oxygen impurity is due likely to the residual oxygen in the reactor.

Using DLI delivery, conductive thin films with high uniformity (standard deviation less than 1%) were obtained. The refractive index of the films were in the range of 2.5-2.6. EDX demonstrated that films consist of Ta and N, while no C or O was detected. Higher deposition rates, as compared to those obtained from a precursor bubbling method, ranged from 200 to 530Å/min at 340°C as the partial pressure of NH_3 increased.

Conclusions

The precursor used in this study is a thermally stable liquid with low viscosity and sufficient volatility. It was demonstrated for the first time that when it is delivered by direct liquid injection and used in the presence of O_2 or NH_3, the same precursor provides tantalum oxide or nitride films at temperatures below 400°C at high deposition rates.

From a chemical perspective, the high oxophilicity of the metal in this metal amide/imido complex renders it to be a useful precursor for metal oxide deposition in the presence of oxygen at relatively low temperatures. Its oxygen free character also enables deposition of tantalum nitride using ammonia as a reactant. It is postulated that a transamination reaction, similar to that of the TiN CVD using $Ti(NR_2)_4$ and NH_3,[20] contributes to the low deposition temperature for tantalum nitride deposition.

At Schumacher, an analogous liquid compound t-BuN=Ta(NEt$_2$)$_3$,[14] has also been studied to deposit tantalum oxide and nitride by CVD at low temperatures using oxygen and ammonia as reactant gases. The results will be reported elsewhere in due course.

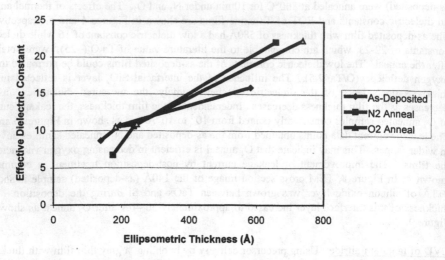

Figure 4. Effective dielectric constant of TaOx films before and after anneal

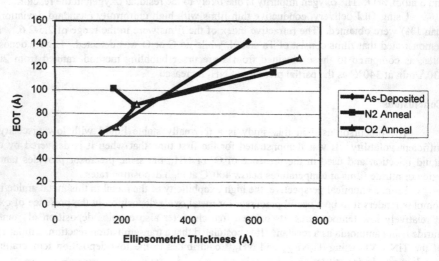

Figure 5. EOT vs ellipsometric thickness of TaOx films before and after anneal

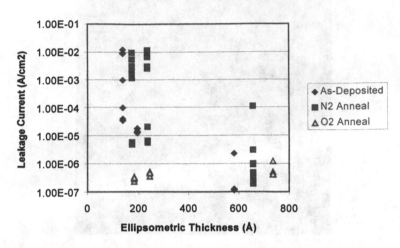

Figure 6. Leakage current vs thickness of TaOx films before and after anneal

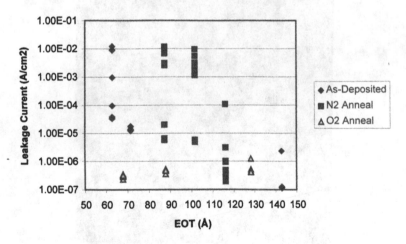

Figure 7. Leakage current vs EOT of TaOx films before and after anneal

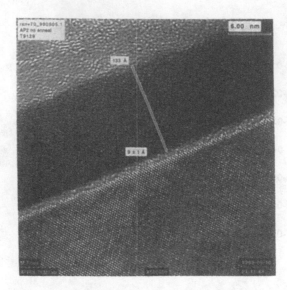

Figure 8. TEM cross section image of TaOx as-deposited.

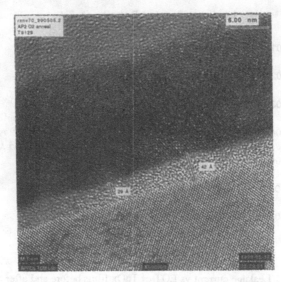

Figure 9. TEM cross section image of Ta_2O_5 after annealed at 800°C for 10 min under O_2.

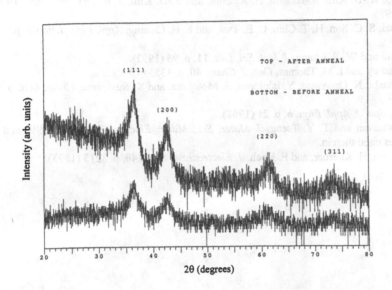

Figure 10. XRD of tantalum nitride before and after anneal under N_2 at 500°C for 30min.

References

1) For example, see H. Treichel, A. Mitwalsky, N. P. Sandler, D. Tribula, W. Kern, and A. P. Lane, *Adv. Mat. Opt. Elec.* **1**, p. 299 (1992).
2) R. A. B. Devine, L. Vallier, J. L. Autran, P. Paillet, and J. L. Leray, *Appl. Phys. Lett.* **68**, p. 1775 (1996).
3) S. R. Jeon, S. W. Han, and J. W. Park, *J. Appl. Phys.* **77**, p. 5978 (1995).
4) T. Tabuchi, Y. Sawado, K. Uematsu, and S. Koshiba, *Jpn. J. Appl. Phys.* **30**, p. L1974 (1991).
5) C. H. An and K. Sugimoto, *J. Electrochem. Soc.* **141**, p. 853 (1994).
6) A. C. Jones, T. J. Leedham, P. J. Wright, M. J. crosbie, D. J. Williams, P. A. Lane, and P. O'Brien, *Mat. Res. Soc. Symp. Proc.* **495**, p. 11, Pittsburgh, PA, 1998.
7) K. D. Pollard and R. J. Puddephatt, *Chem. Mater.* **11**, p. 1069 (1999).
8) M. T. Wang, Y. C. Lin, and M. C. Chen, *J. Electrochem. Soc.* **145**, p. 2538 (1998).
9) A. E. Kaloyeros, X. Chen, T. Stark, K. Kumar, S.-C. Seo, G. G. Peterson, H. L. Frisch, B. Arkles, and J. Sullivan, *J. Electrochem. Soc.* **146**, p. 170 (1999).
10) K. Hieber, *Thin Solid Films* **24**, p. 157 (1974).
11) R. Fix, R. G. Gordon, and D. M. Hoffman, *Chem. Mater.* **5**, p. 614 (1993).
12) G.-C. Jun, S.-L. Cho, K.-B. Kim, H.-K. Shin, and D.-H. Kim, *Jpn. J. Appl. Phys.* **37**, p. L30 (1998).

13) S.-L. Cho, K.-B. Kim, S.-H. Min, H.-K. Shin, and S.-D. Kim, *J. Electrochem. Soc.* **146**, p. 3724 (1999).
14) M. H. Tsai, S. C. Sun, H. T. Chiu, C. E. Tsai, and S. H. Chuang, *Appl. Phys. Lett.* **67**, p. 1128 (1995).
15) H.-T. Chiu and W.-P. Chang, *J. Mat. Sci. Lett.* **11**, p. 96 (1992).
16) D. C. Bradley and I. M. Thomas, *Can. J. Chem.* **40**, p. 1355 (1962).
17) Y. Takahashi, N. Onoyama, Y. Ishikawa, S. Motojima, and K. Sugiyama, *Chem. Lett.* p. 525 (1978).
18) N. Terao, *Jpn. J. Appl. Phys.* **6**, p. 21 (1967).
19) S. Ezhilvalavan and T. Y. Tseng, *J. Mater. Sci., Mater. Electron.* **10**, p. 9 (1999) and the references cited therein.
20) A. Intemann, H. Koerner, and F. Koch, *J. Electrochem. Soc.* **140**, p. 3215 (1993).

Group IVB Oxides as High Permittivity Gate Insulators

S. A. Campbell[*], B. He[*], R. Smith[**], T. Ma[*], N. Hoilien[*], C. Taylor[**], and W. L. Gladfelter[**]
[*]Department of Electrical and Computer Engineering, Campbell@ece.umn.edu
[**]Department of Chemistry
University of Minnesota
Minneapolis, Minnesota 55455

ABSTRACT

Increasing MOSFET performance requires scaling, the systematic reduction in device dimensions. Tunneling leakage, however, provides an absolute scaling limit for SiO_2 of about 1.5 nm. Power limitations and device reliability are likely to pose softer limits slightly above 2 nm. We have investigated the use of high permittivity materials such as TiO_2, ZrO_2, and their silicates as potential replacements for SiO_2. We have synthesized titanium nitrate ($Ti(NO_3)_4$ or TN), zirconium nitrate ($Zr(NO_3)_4$ or ZrN), and hafnium nitrate ($Hf(NO_3)_4$ or HfN) as hydrogen and carbon free deposition precursors. Several problems arise in the use of these films including the formation of an amorphous low permittivity interfacial layer. For TiO_2 this layer is formed by silicon up diffusion. Surface nitridation retards the formation of the interfacial layer. We discuss the effects of both thermal and remote plasma surface nitridation treatments on the properties of the film stack. ZrO_2 and HfO_2 appear to form a thermal layer of silicon oxide between the high permittivity film and the silicon and have excess oxygen in the bulk of the film.

INTRODUCTION

As field effect transistors have scaled deep into the submicron, the gate oxide thickness has scaled as well. Soon to be announced large scale integrated circuits have physical oxide thicknesses as low as 2.0 nm[1]. The national technology roadmap for semiconductors projects this scaling to continue, decreasing to approximately 1.5 nm when the devices reach the 0.1 μm node[2]. Such thin gate oxides display direct tunneling between the gate and substrate, and the gate and source/drain diffusions. The leakage current increases by approximately 100x for every 0.5 nm in thickness reduction. 2.0 nm gate oxides leak at about 200 mA/cm^2, while 1.5 nm gate oxides leak at about 50 A/cm^2. Power consumption constraints are expected to limit scaling of the gate oxide to approximately this thickness, although reliability concerns may limit the ultimate thickness of SiO_2 to about 2.0 nm.

Continuation of device scaling will therefore require the replacement of SiO_2 with a higher permittivity material, which will allow thicker layers at the same equivalent oxide thickness. In this paper we report on the deposition and characterization of the group IVB binary oxides TiO_2, ZrO_2, and HfO_2. The chemical vapor deposition of these films, especially TiO_2, has been practiced for over four decades[3 4 5 6]. The most commonly used precursors, the isopropoxides, can be directly thermolyzed or hydrolyzed. The problem with these reactions is that residual carbon is left in the films, leading to excessive leakage[7]. Often a post deposition oxygen anneal is required to

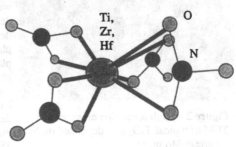

Figure 1 – Structure of nitrato precursor

obtain low leakage films. This oxidation tends to increase the thickness of the low permittivity interfacial layer that forms when high permittivity materials are deposited directly on silicon, making the formation of low oxide equivalent thickness films extremely difficult. The column IVB oxides, however, can all be deposited using nitrato precursors (Figure 1), a class of compounds of the form $M(O_3N)_4$, where M is the metal atom (Ti, Zr, or Hf)[8][9]. We have found that the use of these compounds avoids carbon contamination of the films, obviating the need for any post deposition oxygen anneals.

EXPERIMENTAL

The metal oxide films were deposited by chemical vapor deposition in three different low pressure CVD reactors. Comparable results were seen in all three systems. Precursors were heated to approximately 50°C (TN), 80°C (ZrN), or 90°C (HfN) to increase their volatility. The transfer lines from the bubbler to the chamber were heated 5 to 10 °C above the bubbler temperature to prevent precursor condensation in the lines. Controlling the precursor temperature and the flow of the carrier gas controlled the precursor flux. UHP Ar was used as a carrier for the TN and ZrN; UHP N_2 was used as the HfN carrier. The growth substrate for all experiments was (100) silicon. The TiO_2 and ZrO_2 substrate samples were RCA cleaned using an HF last process. HfO_2 samples were RCA cleaned and banked. Immediately before deposition the samples were first dipped in 4:1 H_2SO_4: H_2O_2 to remove any organic contaminants, rinsed with de-ionized water and blown dry with N_2, then dipped in 5% HF for several seconds to prepare the H-terminated surface, rinsed with de-ionized water and blown dry with N_2. Deposition experiments were performed in the 1 to 10 torr range with temperatures between 400 and 500 °C. Deposition rates were between 1 and 10 nm/min. Films were characterized by X-ray diffraction (Siemans D-5005), Rutherford backscattering (RBS), and scanning electron microscopy (SEM) (Hitachi S-900). Selected samples were also examined with high-resolution transmission electron microscopy (TEM) and medium energy ion scattering (MEIS). Film thicknesses were measured by cross-sectional SEM and ellipsometry.

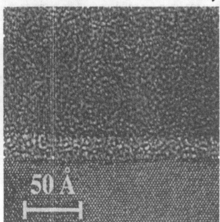

Electrical measurements were made using capacitors on both n- and p- type substrates. All of the wafers received a high dose backside implant for eventual ohmic contact formation. A 1000 °C furnace anneal was used to activate the implant. After deposition, some of the wafers had an inert ambient anneal. These anneals will be noted in the subsequent discussion. A gate electrode was then applied. Several electrodes were tested including RF sputtered Pt, CVD deposited polycrystalline silicon, and CVD deposited TiN. The gate was then patterned using photolithography and ion milling (Pt) or SF_6 plasma etching (poly and TiN). A layer of photoresist was then spun on the front of the wafer for protection and the backside was etched back to the silicon. A layer of Al was then deposited on the backside by DC sputtering. Finally the wafers were cleaned and annealed at 450 °C in forming gas.

Figure 2 – High resolution cross section TEM of typical TiO_2 film deposited on Si (courtesy Motorola.

TIO₂ RESULTS AND DISCUSSION

The deposited TiO₂ was found to be polycrystalline anatase for all films deposited between 400 and 600 °C. The grain size varied with film thickness. Secondary electron microscopy indicated that 50 nm thick films showed columnar growth with equi-dimensional grains extending the entire depth of the film and exhibiting fully developed facets at the air interface. The grain size were 20 to 40 nm in diameter at the deposition temperatures used in this study. Transmission electron micrographs showed that films that were grown for making high capacitance structures, which were 3 to 6

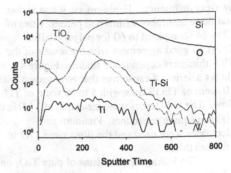

Figure 3 – Time of flight SIMS profile of TiO₂ deposited on Si.

nm thick, had a simple poly crystalline structure with grains 2 to 5 nm in diameter (Figure 2).

Figure 2 also illustrates a pervasive component of the structure produced when TiO₂ is deposited directly on top of silicon. An amorphous layer, 1.0 to 2.5 nm thick, is formed between the silicon and the high permittivity film. Other researchers have reported the formation of an oxide layer of similar thickness resulting from the deposition of metal oxides on silicon.[10][11] The thickness of this layer increases with increasing film deposition temperature and with increasing post deposition anneal time and/or temperature. The largest increases are seen when the films are post annealed in an oxidizing ambient. Even annealing at 450 °C in an oxygen atmosphere causes the thickness of the interfacial layer to increase[12]. By varying the thickness of the high permittivity layer and measuring the capacitance as a function of TiO₂ thickness, one can estimate the capacitance of this amorphous interfacial layer. Using the thickness measured by TEM, one can then assign an average permittivity to the layer. We find that the average permittivity of these layers is always in the range of 10 to 14.

This amorphous layer was originally thought to be SiO₂. Careful study of these films by time of flight secondary ion mass spectroscopy (TOFSIMS) analysis was carried out to investigate this hypothesis. A time of flight SIMS profile is shown in Figure 3. In the transition region between TiO₂ and the substrate silicon, the profile shows a reduction of the TiO₂ signal relative to TiO (not shown), indicating a reduction in the number of Ti atoms coordinated to two oxygen atoms. Also seen is a peak in the Ti-Si signal at or near the interfacial layer. This suggests that silicon up diffuses from the substrate, and replaces oxygen, forming a TiSi$_x$O$_y$ layer.

The effect of introducing silicon into titanium dioxide was studied by depositing films using both titanium nitrate and tetraethylorthosilicate (TEOS). By varying the ratios of the flows and/or the bubbler temperatures, we were able to vary the composition of the film. Over the range of compositions studied ($1 < $ Si/Ti < 6), all of the films were amorphous, as determined

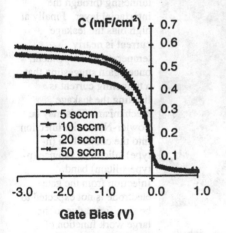

Figure 4 – 1 MHz C-V measurements of TiSi$_x$O$_y$ with varying TN flow and 10 sccm of TEOS.

25

by x-ray diffraction. Furthermore, it was shown that these amorphous films had permittivities of 12 to 14 (compared to 60 for polycrystalline TiO₂), in good agreement with the results of the TiO₂ thickness dependence studies. Figure 4 shows a series of capacitors that were grown with 10 sccm of TEOS flow with 5 to 50 sccm of TN flow. The C-V traces were measured at 1 MHz on 100 μm square capacitors. Platinum gate electrodes were used and the films were 11.1 to 16.1 nm thick.

The leakage mechanisms of pure TiO₂ on silicon were determined by measuring the voltage and temperature dependence of the leakage current. Both n- and p-type substrates were used to avoid minority carrier lifetime effects. Figure 5 summarizes the room temperature results in a

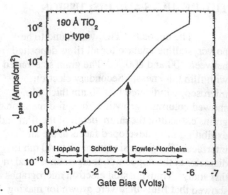

Figure 5 – Leakage current for Pt/TiO₂/Si structure showing mechanisms

relatively thick (19 nm) film on p-type silicon where all three regions are easily visible. At the lowest fields the leakage current increases linearly with the applied field and is only weakly temperature dependent. The leakage current in this region depends strongly on anneal condition. At moderate bias, the leakage current follows conventional Arrhenius behavior, suggesting Schottky barrier emission. By measuring the temperature dependence of both n- and p- type substrates we have found conduction and valence band offsets for TiO₂ relative to silicon of 0.95 +/- 0.1 and 1.0 +/- 0.1 eV, respectively. Combined with the silicon bandgap of 1.1 eV, this is in good agreement with the generally accepted anatase TiO₂ bandgap of 3.2 eV. In this regime the leakage current is not limited by tunneling through the interfacial layer. Finally at high bias the leakage current is nearly temperature independent, once again suggesting that a tunneling current is limiting the leakage mechanism. This may be Fowler-Nordheim injection into the conduction (n-type) silicon or valence (p-type silicon) band. Injection from the gate electrode is not expected to be significant due to the large work function of platinum.

Another problem illustrated by Figure 4 is

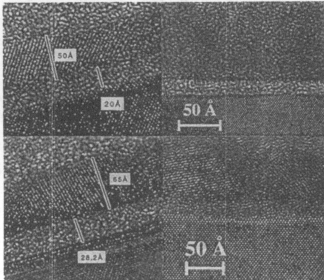

Figure 6 – Effect of an 850 °C thermal NH₃ nitridation (right side micrographs) before TiO₂ deposition, both as deposited (top row) and after a 750 °C N₂ anneal. The ammonia pretreatment reduces the amorphous layer thickness by 0.7 nm

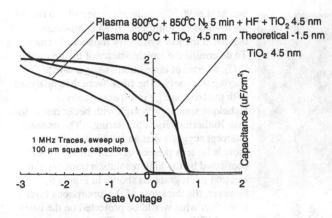

Plasma 800°C + 850°C N₂ 5 min + HF + TiO₂ 4.5 nm
Plasma 800°C + TiO₂ 4.5 nm Theoretical -1.5 nm
TiO₂ 4.5 nm

1 MHz Traces, sweep up
100 μm square capacitors

Capacitance (uF/cm²)

Gate Voltage

Figure 7 – A wet HF etch after surface nitridation almost completely recovers the observed fixed charge.

the value of the flat band voltage. A Pt gate capacitor on a lightly doped wafer should have a flat band voltage of +0.5 volts. The flat band of the $TiSi_xO_y$ layers were –0.1 V, while pure TiO_2 deposited directly on Si produces a flatband voltage of –0.6 V. This latter figure corresponds to a positive fixed charge density of 10^{13} cm⁻².

One possible explanation for this positive charge is the formation of traps

associated with the silicon up diffusion and replacement of the oxygen atoms bonded to titanium. The silicon up diffusion should be restricted by the formation of a diffusion barrier at the silicon surface before TiO_2 deposition. To test this, the surface of silicon wafers were nitrided using NH_3. The ammonia was either cracked thermally, or through the use of a 200 W remote plasma system. The wafers were nitrided in the same low pressure reactor used for TiO_2 deposition. Figure 6 shows the result of this work. The treatment succeeded in reducing the amorphous interfacial layer thickness by approximately 0.7 nm, both as deposited and after a 750 °C N₂ anneal.

The use of a surface nitridation was not found to suppress the flat band shift described above, however. Instead the flat band shift increased to –0.9 V. The use of the remote plasma, along with lower substrate temperatures produced results that were intermediate between that of the TiO_2 deposited directly on silicon and the 850 °C thermal nitridation. Figure 7 shows the effect of treating the surface by wet etching the nitrided surface using a 100:1 solution of HF in water. The curve shifts to the right and is close to the ideal C-V, however an extremely large concentration of interface states remains. As will be discussed later, these interface states are largely related to the gate electrode process.

ZrO_2 and HfO_2 RESULTS AND DISCUSSION

In an effort to overcome some of the difficulties in the thermodynamically unstable Ti-Si-O system, we have also begun to investigate the use of ZrO_2 and HfO_2, both of which are predicted to be thermodynamically stable on silicon. As with TiO_2, these compounds were deposited from their respective nitrato precursors. For ZrO_2, the precursor vessel

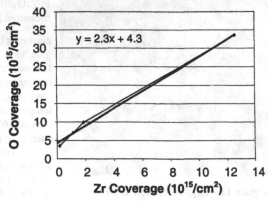

$y = 2.3x + 4.3$

O Coverage (10^{15}/cm²)

Zr Coverage (10^{15}/cm²)

Figure 8 – MEIS analysis of ZrO_2 films showing the oxygen coverage as a function of the zirconium

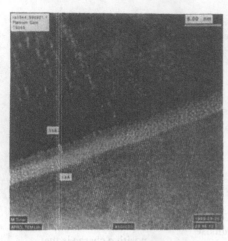

Figure 9 – TEM of Pt/ZrO₂/Si capacitor showing 1.8 nm thick amorphous interfacial layer.

two layer model, suggesting little of the inter layer mixing that was observed for TiO_2.

Capacitors were made by depositing 6.5 nm of ZrO_2 on both n- and p- type wafers, RF sputtered platinum as a gate electrode, then patterning the films using ion milling. As shown in Figure 10, the leakage

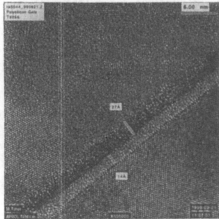

Figure 11 – High resolution TEM (courtesy Motorola) of ZrO_2 film with polysilicon gate electrode.

was held at 90 °C, producing a deposition rate of approximately 3 nm/min. Both compounds deposit in the low symmetry monoclinic phase. To determine the stoichiometry of the deposited ZrO_2, a series of depositions were run with varying thickness. The films were then analyzed with medium energy ion spectroscopy, a technique similar, although with better resolution, than Rutherford Back Scattering. The nonzero intercept suggests that an interfacial silicon dioxide layer is present in these films. This is confirmed by the high resolution transmission electron micrograph as shown in Figure 9, however, the thickness of the amorphous layer is more than what would be projected on the basis of the MEIS extrapolation. The energy dependence of the MEIS signal is well fit by a

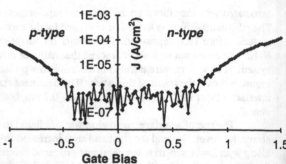

Figure 10 – J-V characteristics of 6.5 nm thick Pt/ZrO₂/Si films. Both substrate types are measured in accumulation at room temperature.

current is between 10 and 100 mA/cm² at +/- 1 volt gate bias. The leakage current was measured as a function of temperature and once again fit to a barrier emission model. The slope of the curve gave approximately 0.1 eV. This weak temperature dependence (only a factor of 10 between 60 and 160 °C), suggests that the leakage current in this regime may be due to trap hopping.

Due to the thermodynamic instability of silicon and TiO_2, low leakage polysilicon gates could not be achieved. Pt gate capacitors showed a flat band shift corresponding to a charge of 5×10^{12} cm⁻². Since MEIS results indicates a pure SiO_2 interfacial layer for ZrO_2, one would expect a

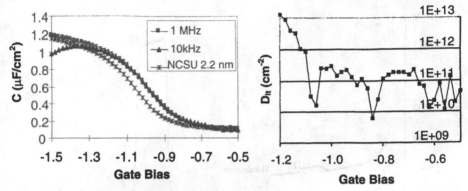

Figure 12 – Low and high frequency C-V traces of N+ poly / ZrO_2 / Si capacitors. The interface state density was extracted from the C-V data. Also shown is the ideal C-V curve.

better interface. To investigate the effect of the gate electrode, one would like to use a material deposited by CVD and etched in such a way as to minimize any residual damage. Since ZrO_2 is expected to be thermodynamically stable in the presence of silicon, it should be compatible with polysilicon gate electrode. Poly gate capacitors were therefore made by depositing in-situ phosphorus doped polysilicon on top of ZrO_2 deposited films. The poly was patterned photolithographically and etched in a low power SF_6 plasma system. The result is shown in the TEM of Figure 11. No evidence of a second (i.e. upper) interfacial layer is seen, confirming the stability of this interface. Of course, this is not proof that the interface is stable, merely that no amorphous region can be seen.

To investigate the electrical properties of this structure, capacitors were measured electrically. The results can be seen in Figure 12. Ten kilohertz was used to approximate the low frequency curve since the leakage current makes very low frequency measurements difficult. The traces show approximately 60 mV of shift from the ideal curve, however, the flat band of the ideal C-V depends on the doping concentration of the polysilicon. More importantly, little distortion of the curve is seen between the two frequencies. The difference in the high and low frequency data was then used to extract the interface state density. The data shows a broad minimum throughout depletion with a value of about 10^{11} cm^{-2}. The data is well fit by a 2.2 nm gate oxide equivalent thickness film. This compares to a 1.8 nm gate oxide equivalent thickness for the platinum gate.

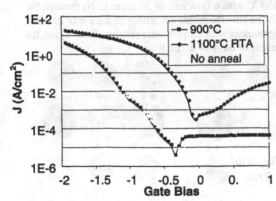

Figure 13 – Leakage current density of N+ poly / ZrO_2 / Si capacitors after two typical dopant activations.

To investigate the thermal stability of the poly / ZrO_2 / Si capacitor, the samples were heated and the leakage current and capacitance were remeasured after heating. Two anneals were used that represent

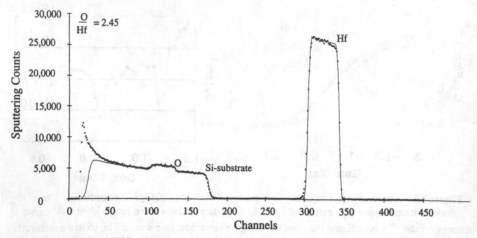

Figure 14 – RBS of HfO_2 deposited from hafnium nitrate shows excess oxygen in the film.

typical dopant activation steps. Both anneals were done in UHP nitrogen. As shown in Figure 13, the 30 minute, 900 °C furnace anneal produced only small changes in the leakage current, except at large accumulation biases. Substantially increased leakage, however, is seen for the 30 second 1100 °C RTP anneal. Similarly, the accumulation layer capacitance decreased slightly (~10%) after the 900 °C anneal. The leakage current was too large to measure the leakage current after the 1100 °C anneal. The decrease in the capacitance may be due to a thickening of the amorphous interfacial layer, but this has not yet been confirmed by TEM measurements.

It is known that the stability of the various phases of HfO_2 is significantly better than ZrO_2. The transition temperatures between the monoclinic, tetragonal, and cubic phases are approximately 300 °C higher in HfO_2. Once again, films have been deposited from hafnium nitrate in a low pressure reactor at temperatures between 400 and 500 °C. Figure 14 shows an RBS measurement of a film deposited at 450 °C with a flow rate of 20 sccm of N_2 through the precursor vessel, which was held at 82 °C. The film thickness according to RBS was 293 nm. The thickness measurement was in good agreement with ellipsometer results. Once again, the ion scattering measurements indicate an excess of oxygen in the film.

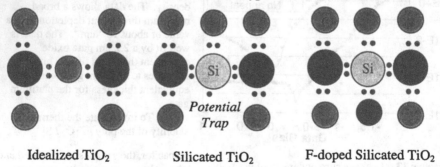

Idealized TiO_2 Silicated TiO_2 F-doped Silicated TiO_2

Figure 15 – Simple 2D model shows the effect of Si replacement of a bridging oxygen in TiO_2 and the potential passivating effect of fluorine.

DISCUSSION

TiO_2 has been deposited both directly on silicon and on nitrided surfaces. Films with an acceptably low leakage (< 1 mA/cm^2) and a 1.5 nm gate oxide equivalent thickness have been produced, but all films show signs of significant positive charge. This is reflected in the poor mobility of FETs built with these films. SIMS analysis shows that no impurity is present in sufficient quantity to account for the charge. These results also indicate the presence of a high concentration of Si atoms replacing O, forming Ti-Si bonds, at the interface between the silicon and the high permittivity film. Since the valence of Si is 4, compared to 6 for oxygen, it is likely that at least one unpaired electron results from the silicon up diffusion unless there is a significant local rearrangement in the physical structure. Furthermore, the loss of this electron could account for the observed positive charge (Figure 15). The addition of fluorine due to the HF pretreatment is likely to passivate these unsatisfied bonds, reducing the fixed charge at the interface. The fact that surface nitridation introduces a shift corresponding to additional positive charge in the insulator is consistent with observations of the effects of SiO_2 nitridation. Due to the relatively poor vacuum system in the low pressure reactor, it is expected that the nitridation step used here actually produced an oxynitride. Time of flight SIMS measurements after the high permittivity oxide was deposited never found a ratio of nitrogen to oxygen greater than 0.2.

The interface state density (and perhaps some of the fixed charge density) may be related to the use of sputtered Pt as the gate electrode. We have observed that the interface state density improves when a lower sputtering power is used and appears to be dramatically improved when CVD TiN is used as the gate material instead of Pt.

When ZrO_2 and HfO_2 are deposited on silicon, the films have a considerable amount of excess oxygen. This oxygen may be interstitial, however, the oxygen may also be incorporated as peroxide. The holes in which the anions sit in the monoclinic structure are large enough to accommodate the O_2 molecule in much the same way it does for sulfur molecules (S_2) occupying atomic sulfur sites in iron pyrite (FeS_2). X-ray diffraction studies indicate that the films are preferentially oriented in the (002) direction. This suggests that the film begins growth before the formation of the amorphous interfacial layer. This is also consistent with growth experiments that we have done under UHV conditions which show no difference in the interfacial layer thickness compared to conventional low pressure deposition. Thus interfacial oxidation may be occurring during the high permittivity film deposition. This agrees with the measurements of oxygen diffusion in ZrO_2 made recently by Garfunkel[13].

ACKNOWLEDGMENTS

This work was supported by Motorola and Texas Instruments through a Semiconductor Research Corporation Research Customization Award (Contract #BJ479) and by IBM through a University Partnership Award. The authors also wish to acknowledge the contributions of Chris Hobbs (Motorola), Glen Wilk (Texas Instruments), and Matt Coppell (IBM) who provided extremely valuable discussion and thin film analysis.

REFERENCES

[1] 1999 International Electron Devices Meeting

[2] Semiconductor Industry Association, "The National Technology Roadmap for Semiconductors Technology Needs", SIA, San Jose, California (1997).

[3] Yan, J.; Gilmer, D. C.; Campbell, S. A.; Gladfelter, W. L.; Schmid, P. G. *J. Vac. Sci. Technol. B* **1996**, *14*, 1706 .

[4] Campbell, S. A.; Gilmer, D. C.; Wang, X.-C.; Hsieh, M.-T.; Kim, H.-S.; Gladfelter, W. L.; Yan, J. *IEEE Trans. Electron Devices* **1997**, *44*, 104.

[5] Zhang, Q.; Griffin, G. L. *Thin Solid Films* **1995**, *263*, 65.

[6] Aarik, J.; Aidla, A.; Uustare, T.; Sammelselg, V. *J. Crystal Growth* **1995**, *148*, 268.

[7] Fictorie, C. P.; Evans, J. F.; Gladfelter, W. L. *J. Vac. Sci. Technol. A* **1994**, *12*, 1108.

[8] Garner, C. D.; Wallwork, S. C. *J. Chem. Soc. (A)* **1966**, 1496.

[9] D. G. Columbo, D. C. Gilmer, V. G. Young, S. A. Campbell, and W. L. Gladfelter, Chem. Vap. Deposition **4**, p. 220 (1998).

[10] Rausch, N.; Burte, E. P. *J. Electrochem. Soc.* **1993**, *140*, 145.

[11] Yoon, Y. S.; Kang, W. N.; Yom, S. S.; Kim, T. W.; Jung, M.; Park, T. H.; Seo, K. Y.; Lee, J. Y. *Thin Solid Films* **1994**, *238*, 12.

[12] Kim, H.-S.; Gilmer, D. C.; Campbell, S. A.; Polla, D. L. *Appl. Phys. Lett.* **1996**, *69*, 3860

[13] Private discussion, E. Garfunkel.

APPLICATION OF PULSED INJECTION MOCVD TO THE DEPOSITION OF DIELECTRIC AND FERROELECTRIC OXIDE LAYERS AND SUPERLATTICES

JEAN-PIERRE SENATEUR*, JOHANNES LINDNER*, FRANÇOIS WEISS*, CATHERINE DUBOURDIEU*, CARMEN JIMENEZ*, ADOLFAS ABRUTIS.**

* LMGP, ENS de Physique de Grenoble, INPG, UMR CNRS 5628, BP 46, 38402 Saint Martin d'Hères Cedex, FRANCE.
* Department of general and inorganic chemistry - 24 Naugarduko -Vilnius University, 2006 Vilnius, LITHUANIA.

ABSTRACT

The technique used for the control of the injection of fuel in thermal motors has been applied to the generation of active gases for MOCVD layers deposition. A wide variety of compounds and multilayers have already been grown using this new CVD source (from simple oxides, nitrides or metals to complex ternary oxides). Using two injection sources working sequentially, stackings like YBa2Cu3O7/CeO2 or SrTiO3/YBa2Cu3O7 double layers, Ta2O5/SiO2 amorphous or crystallized multilayers and SrTiO3/BaTiO3 or YBa2Cu3O7/PrBa2Cu3O7 epitaxial superlattices have been synthesized. The injection MOCVD seems to be a very promising technique for the stacking of multicomponents oxides having different electronic properties. This opens the way for the synthesis of a wide set of artificial materials, exhibiting completely new properties that cover a very large range of potential applications. After a description of the technique, we will demonstrate its performances on some results obtained on the synthesis of dielectric and ferroelectric layers and multilayers.

INTRODUCTION

The discovery of very exciting new properties in advanced oxides (high temperature superconductivity, colossal magnetoresistance, very high dielectric constants...) has aroused during the last ten years a considerable interest for the synthesis of thin layers of these materials. Electronic devices like microwave components obtained by the association of HTS and insulating, ferroelectric or magnetic oxides can now be realized (Josephson effect devices, spin-polarized quasi-particle injection devices, tunable filters..). New protective, thermal barriers or very efficient buffer layers can be obtained with the stacking of different oxides. Transparent thin layers or multilayers with different refractive index have numerous applications for optical amplifiers, narrow or large band filters, supermirrors or glass art. The recent development of MOCVD for the deposition of these complex materials is strongly related to the discovery of high temperature superconductivity.

Twelve years ago, the high temperature superconductivity observed in compounds like REBa2Cu3O7, (where RE = Y or a rare earth), has impulsed a fantastic research effort, in relation with their potential applications. These materials being very hard and brittle, it is evident that the best route for many industrial use is to realize thin layers deposited on a convenient substrate (single crystal for electronics, flexible metallic tape for high power applications). A wide number of experiments have been done by physicists, who have generally no experience within chemical techniques : this explain why most of the results already reported are related to layers obtained by physical deposition techniques (sputtering, MBE or co-evaporation, laser ablation). The research effort developed for chemical deposition techniques, like sol-gel, pyrosol gel or chemical vapor deposition (CVD) was lower by some orders of magnitude : in Europ for instance, less than ten laboratories are involved in the synthesis of complex oxide layers or multilayers by CVD. The first successful deposition of high quality high-temperature superconducting layers were obtained by CVD technique two years after the first results obtained by PVD. This long delay was mainly related to the non-existence of convenient volatile chemical precursors for barium. The main problem for the CVD deposition of complex oxide containing barium or strontium is due to the unavailability of precursors having a long term stability and reproducibility of the vapor pressure. The best precursors were (and are always) beta-diketonates (pure tetramethyl heptane dionates or fluorinated derivates), which are solids at room temperature. The sublimation rate of a solid is related to the free surface of the precursor, which depends on the grain size of the powder and to the amount of precursor remaining inside the evaporator : this leads to a drastic irreproducibility.

The barium precursors of the first generation were thermally unstable at their sublimation temperature, moisture sensitive, and had a strong tendency to oligomerization. In addition, these barium precursors were subjected to "aging" phenomena during their storage, even at room temperature, which increased the irreproducibility. After each deposition experiment, a lot of brown residue of barium precursor remained in the source : only twenty to fifty percent of the commercially available precursors were volatile. A strong effort has been made by the chemists to improve the purity of this precursor : now, high quality pure barium beta diketonate precursor is commercially available, and some adducts (o-phenantroline, tetraglyme, triglyme..) may be handled and stored without drastic precautions. But these adduct compounds are thermally unstable : they dissociate below the volatilization temperature, which leads to a drift in the vapor pressure, if conventional CVD sublimators are used. An other handicap for the conventional CVD processing of these compounds is that the barium beta-diketonate has a low vapor pressure at the temperature at which its decomposition is negligible. The source temperature has to be stabilized below 200 °C, which leads to growth rates of $YBa_2Cu_3O_7$ far below .2 microns/hours [1], too low for large scale applications. The quality and the reproducibility of the barium beta-diketonate precursor (pure or with adducts) have now reached their upper limit; it is evident that only the discovery of a completely new precursor will increase this growth rate using conventional CVD sources.

An other way to obtain convenient growth rates is to investigate new CVD sources generating high and stable vapor pressures over a long time, even with thermally unstable precursors. Many groups have developed different systems, which all are based on the same principle : the main part of the precursor is maintained in a close vessel at room temperature under inert gas. Only the amount needed for the vapor pressure generation is continuously, or sequentially introduced into an evaporator, held at "high" temperature, where a flash volatilization occurs. As compared to conventional CVD bubblers or sublimators, these systems have several advantages :

- The precursor(s) is maintained at room temperature under inert gas : there is no "aging" if pure precursors, commercially available now, are used.

- The time of transfer between the evaporator and the substrate being very short (< 1 sec), very high vapor pressure may be reached by overheating.

The main difference with conventional CVD bubblers (or sublimators) is that the vapor flow of the active species is not controlled by the source temperature and by the vector gas flow, but only by the injection flow rate of the solid/liquid precursor(s) inside the evaporator. There is no need for accurate temperature stabilization of the evaporator. The condition for a long term stability of the vapor pressure production is that the flow rate of the precursor(s) outside the evaporator is at least equal to its injection rate. If this condition is fulfilled, the accuracy of the control of the active vapor flow rate depends only on the accuracy of the injection of the precursor. Two different ways have been explored :

- Direct introduction of solid precursor(s) inside the evaporator.
- Injection of a solution of the precursor(s) in a convenient solvent, which is selected so that it does not change the chemistry of the precursor and is easily volatilized.

Direct injection of solid precursor inside the evaporator.

Different systems have been tested :

A mixture of powder of precursors is introduced inside an evaporator held at high temperature by a vibrating powder feeder [2].

The accurate handling of hygroscopic powders being very difficult, a variation of this technique has been recently proposed : little balls of inert material (alumina, stainless steel) are covered by the powder. These balls are introduced inside the evaporator and recycled after evaporation of the precursor [3].

A third possibility is to slowly introduce in a strong thermal gradient inside the evaporator, a solid rod made with a compacted mixture of the precursor(s) [4,5].

These techniques lead to very interesting results at laboratory scale, but the powders being hygroscopic or electrostatic (if dried), the realization of a homogeneous mixture of large amounts for the synthesis of multicomponents layers is a challenge : it seems difficult to extrapolate these techniques to industrial scale.

Injection of a solution of precursors in an organic solvent

The precursors are dissolved in an organic solvent (generally THF, monoglyme, diglyme, hexane...). The injection techniques which have been used are aerosols, micro pumps, syringes or liquid mass flow controllers [6 to 12]. In all these systems, the liquid is continuously introduced inside the evaporator, heated under low pressure. For the aerosol technique, low volatility solvents are needed (diglyme) to avoid their volatilization at room temperature inside the aerosol generator, which is at the same pressure as the reactor. For the other techniques, the injection of the solution is made through a capillary tube that limits the volatilization of the solvent in the cold zone before it reaches the evaporator. More sophisticated systems have been proposed to alleviate this problem : injection inside a three way valves, which allows the introduction of a mixture of vector gas and solution inside the capillary (ASM), or special design of the evaporator (MKS).

The vapors of the solvent in the reaction zone increase drastically the carbon concentration of the gas phase. Pure metal-organic precursors contain a large amount of carbon (for the deposition of one $YBa_2Cu_3O_{7-x}$ formula unit, 143 carbon atoms have to be expulsed in the exhaust of a CVD reactor when pure beta-diketonates precursors are used !). The presence of the solvent in the reaction zone increases the carbon concentration, but we have not found any significant difference in the carbon concentration of our layers deposited using eituer conventional or liquid-injection CVD sources (NRA, SIMS, SNMS, XPS or electronic properties). Nevertheless, even without chemical interactions between the solvent and the precursors, its presence in the reaction zone decreases the oxygen activity, which may have some drawbacks. For $YBa_2Cu_3O_{7-x}$ layers deposition, for instance, our thermodynamic calculations show that an increase of the solution feeding rate has to be compensated by an increase of the oxygen partial pressure, in relation with the decrease of the oxygen activity in the gas phase with the increase of its carbon-hydrogen content. If the oxygen activity is too low, $YBa_2Cu_3O_{7-x}$ is not stable.

The control of the homogeneity and of the flow rate being considerably easier for liquid than for solids, several solutions have been already proposed for the use of solution of precursors, but with elimination of the solvent in the reaction zone. The first technique, proposed by Kaul and Seleznev [13], is based on the ex-situ wetting of a fiber glass tape and the elimination of the solvent. Then the dried tape is mechanically carried inside an evaporator where only the remaining precursor(s) is flash volatilized. Other technique, with "in-situ" separation of the solvent, have been tested [14,15] : a travelling tape is wetted by the solution in a "cold" zone where the solvent is volatilized, then the dried precursor is carried toward a hot zone where the precursor is volatilized. Two vector gas flows in opposite direction are used to carry the vapors of the precursor(s) toward the substrate and those of the solvent directly toward a cold trap.

We have developed in LMGP, in collaboration with Vilnius University, a very simple liquid delivery system for CVD sources based on the principle used for the control of the fuel injection in recent thermal motors [16]. This system has been successfully applied for the deposition at laboratory scale of a wide variety of simple or complex compounds, and also for the deposition of multilayers or superlattices :

- Al_2O_3, Ta_2O_5, SiO_2, MgO, TiO_2, Y_2O_3, $ZrO_2(Y)$, CeO_2, Ta_2O_5-SiO_2 multilayers
- $YBa_2Cu_3O_{7-x}$ / $PrBa_2Cu_3O_{7-x}$ and $YBa_2Cu_3O_{7-x}$/$PrBa_2Cu_3O_{7-x}$ superlattices
- $BaTiO_3$, $SrTiO_3$, $Ba_{(1-x)}Sr_x TiO_3$ and $BaTiO_3$/$SrTiO_3$ superlattices on $LaAlO_3$
- $La_{(1-x)}MnO_3$, $La_{(1-x)}Sr_xMnO_3$, $Nd_{(1-x)}Sr_xMnO_3$
- Ti(C,N)
- Cu
- $YBa_2Cu_3O_{7-x}$ / $ZrO_2(Y)$//Metal
- $YBa_2Cu_3O_{7-x}$ / $SrTiO_3$//$LaAlO_3$

EXPERIMENT

Deposition technique

Our injection source is quite different from the other liquid delivery systems already used for CVD processing of thin layers. The principle is a sequential injection of micro amounts of solution of precursors (or of pure precursors, if they are liquid at room temperature), inside an evaporator held at high temperature. The active solution, maintained at room temperature in a close vessel pressurized by 2 Bar of inert gas (N_2 or Ar), is pushed inside the injector by the gas

pressure. The injector is a high speed electro valve driven by a computer or an electric pulse generator. The flow rate of the precursor inside the evaporator is controlled by :
- The electrical pulse width, which defines the volume injected at each injection.
- The injection frequency.
- The concentration of the solution.
- The differential pressure between the liquid inside the injector and the evaporator.
-The viscosity of the solution.

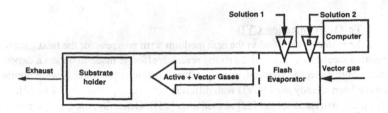

Fig 1 : Schematic representation of a direct injection CVD reactor. In this representation, two injectors (A and B) are connected to the same evaporator. The synthesis of binary oxides is obtained either by feeding each injector by the precursor of each element, or by injecting a mixture of the two precursors. The synthesis of multilayers is made by injecting alternatively different volumes of each solution of precursor(s) (defined by the number of injections and the electrical pulse width on A and B injectors).

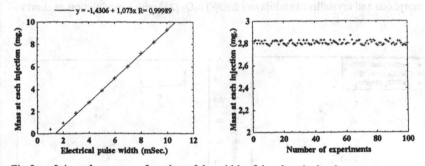

Fig 2-a : Injected mass as a function of the width of the electrical pulse.
Fig 2-b : Variation over 100 experiments of the total mass of hexane injected with 100 pulses (pulse width of 4 milliseconds, differential pressure = 1.5 Bar, liquid = hexane, frequency = 25 Hz, voltage = 17 V.)

The injectors used so far for our CVD experiments are commonly used for the fuel injection of recent thermal motors : they are low cost, very accurate high-speed electro-valves. Generally, there is no corrosion problems with many organic precursors, like for instance tetra methyl heptane dionates dissolved in monoglyme or hexane, but for special applications, some of them have been internally protected by a protective coating. There is a wide variety of commercially available injectors, but generally, the maximum working frequency is around 200

Hz, and the minimum width of the electric pulses which open the injector is close to 1 millisecond. This corresponds to a minimum mass of liquid injected (hexane) varying from .2 to 4 mg./injection., depending on the kind of injector used (working differential pressure : 150 KPa). The performances of one kind of injector is represented in Fig 2-a and 2-b.

We will present some examples of the application of this injection technique to the deposition of high-k dielectric thin layers of materials having a high probability for being the future generation of sub-0.1 micron ULSI technology.

RESULTS

<u>Deposition of Ta$_2$O$_5$ by injection CVD</u>

Ta$_2$O$_5$ is the material identified as the next medium term progress for the next generation of high integrated electronics. Presently, a strong research effort is made to obtain a deposition of layers with a high dielectric constant, at temperatures as low as possible. A wide number of techniques have been already used (CVD with different activation processes [17 to 20], sputtering [21], UV activated sol-gel [22] or oxidation [23]). Generally, the CVD deposition of Ta2O5 is made with tantalum ethoxide precursor. In a first set of experiments, we have deposited this oxide by injection CVD using tantalum ethoxide [24] diluted in hexane or THF. The growth rate is very low below 350 °C and reach a maximum value of 11 microns/ hour at 650 °C, far above the growth rates reported for conventional CVD sources. Below this temperature, the compounds are amorphous, above this temperature, the layers are crystallized with an orthorhombic structure. At 550 °C, the refractive index of the layers has a maximum close to 2.2, which is the bulk value. Using two injectors working sequentially, we have obtained amorphous and crystallized multilayers SiO$_2$-Ta$_2$O$_5$ [25] with a modulation as short as 8 nm.

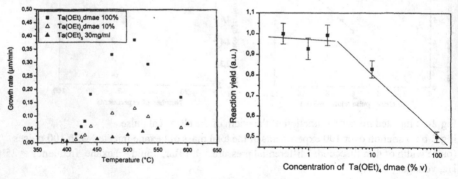

Fig 3-a : Growth rates of Ta$_2$O$_5$ obtained with Ta(OEt)$_5$ and Ta(OEt)$_4$DMAE : the difference in the maximum growth rate temperature is more than 100 °. The deposition conditions were : pressure = 6.7 hPa, oxygen partial pressure = 0.7 hPa.
Fig 3-b : Deposition yield of Ta$_2$O$_5$ as a function of the dilution of the precursor in hexane. Precursor : Ta(OEt)$_4$DMAE, deposition temperature : 450 °C, Solution feeding rate : 0.25 cc/min., injection frequency : 1 Hz, evaporator temperature : 120 °C

38

More recently, we have successfully deposited Ta_2O_5 using $Ta(OEt)_4(DMAE)$ (tantalum tetraethoxy dimethylaminoethoxide), a new precursor provided by Inorg-Tech. This precursor has a better volatility than $Ta(OET)_5$, and displays a lower thermal stability, which allows the deposition of tantalum oxide at lower temperature. The growth rates obtained for both precursors are reported in Fig 3a.

As compared to the conventional CVD sources, an interesting characteristic of the injection is that the layer's growth rate is determined by the injection flow rate of the precursor(s). This parameter can be controlled by the computer driving the injector(s) (electrical pulse width and injection frequency), or by the dilution of the precursors. Fig 3-b shows the results obtained with different dilution in hexane of pure $Ta(OEt)_4DMAE$. It appears that, with a too high feeding rate of the precursor, the reaction yield decreases (the reaction yield if defined by the ratio between the thickness and the mass of the precursor injected, normalized to 1 at its higher value). The higher reaction yield is obtained for a concentration lower than 10% of precursor in hexane. This explains why, when keeping constant the injection flow rate, the mean growth rate is divided by a factor of three if the dilution of the precursor (vol.%) is increased by a factor of ten (but these factors are qualitative, they depend on the deposition parameters and on the geometry of the reactor).

Recently, Ta_2O_5 films with breakdown field higher than 2MV/cm. and dielectric constant values 18-24 have been obtained by the association of this injection technique with the UV activation of the reaction [26, 27] (dilution of $Ta(OEt)_4DMAE$ in hexane = 10%, UV wavelength = 222 nm, substrate temperature << 400°C). The leakage currents are comparable to thOSE reported for unannealed films obtained by other techniques.

Deposition experiments on Ta_2O_5- TiO_2 mixed oxides

Cava et al. [28] have reported that ceramic of mixed Ta-Ti oxides may have a dielectric constant as high as 126, which is 6 times higher than the highest value obtained for pure Ta_2O_5. Thin films have been recently obtained by sputtering [29]. Preliminary experiments on the deposition of these mixed oxides have been carried out by injection CVD [30], using a single source containing a mixture of $Ta(OEt)_4DMAE$ and $Ti(OEt)_2(DMAE)_2$ (INORG TECH) diluted in hexane (10% vol.). The ligands being the same, there is no reaction between the precursors in the liquid source. The films have been grown at 500 °C, with an evaporator temperature of 250 °C, and feeding rate of the solution of 0.3 cc/min.. The Ti/Ta ratio of the films, determined by EDX, EPM and RBS is very close to that of the solution The films are amorphous, very transparent, with refractive index 2.2 - 2.3. We have not found in the as-grown films any increase of the dielectric constant, these films exhibit high leakage currents and further experiments are in progress. Nevertheless, these experiments show that mixed Ta-Ti oxides thin layers with various compositions are easily obtained by injection CVD.

Deposition of $Ba_{(1-x)}Sr_x TiO_3$ and $BaTiO_3 / SrTiO_3$ superlattices

The very high dielectric constant of these materials offers very promising possibilities of application for capacitors and DRAM of third generation. Thin layers are obtained by injection

CVD using a mixture of Ba or Sr(thd)$_2$, triglyme and Ti(Oipr)$_2$(thd)$_2$ dissolved in hexane (total concentration = 0,02 mol/l.). The deposition parameters are :
- Substrate temperature   800 °C (LaAlO$_3$, SrTiO$_3$, Al$_2$O$_3$)
- Pressure   6.7 hPa
- Gas flows   250 cc/min O$_2$ + 250 cc/min. Ar
- Evaporator temperature   280 °C
- Injection parameters   1 milliseconds, 1 Hz (solution flow rate = 0.1 cc/min.).

Using these conditions, the mean increase of the thickness is close to 0.1 nm at each injection. For SrTiO$_3$ deposited on LaAlO$_3$, the EDX analysis shows that the concentration of strontium in the film is lower than the concentration in the solution (Fig 4-a). For a Sr/Ti ratio in the solution varying between 1.2 and 1.5, no extra diffraction lines are visible on the x-ray diffraction pattern, but the line width of the 002 diffraction line varies drastically with the composition (Fig 4-b).

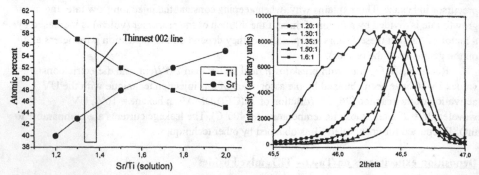

Fig 4-a : Composition of the Sr$_x$TiO$_3$ layers as a function of the source composition
Fig 4-b : evolution of the width of the 002 diffraction line (4-b)

From these analysis, we have chosen for the deposition of the mixed layers and multilayers the composition of the solution which give layers exhibiting the thinnest 002 diffraction line, although the corresponding Sr/Ba ratio in the film, determined by EDX, is not 1. The same procedure has been applied to BaTiO$_3$ single layers, which has shown that for both compounds, the optimum ratio Sr (or Ba)/Ti in the solution is 1.35. The layers obtained with this solution have a good surface morphology. The surface roughness, determined by AFM on 150 nm thick SrTiO$_3$ and BaTiO$_3$ layers are respectively 1 and 2 nm. (the surface roughness of the bare LaAlO$_3$ substrate was .5 nm.).

Mixed Ba$_{(1-x)}$Sr$_x$TiO$_3$ oxides are obtained by a injecting a simple mixture of different amounts of the solutions used for the deposition of each compounds. The X ray patterns of the samples over the full solid solution and the corresponding variation of unit cell parameters are reported in Fig 5-a and 5-b. Both on LaAlO$_3$ or SrTiO$_3$ single crystals substrates, X-rays pole figure shows a "cube-on cube" epitaxy of the layers. A striking result is that the tetragonal structure of BaTiO$_3$ is not observed, which may be related to a very little grain size. [31].

BaTiO$_3$ / SrTiO$_3$ superlattices are obtained by injecting alternatively the two solutions used for the deposition of each compounds. The period of the superlattice, deduced from the satellites lines observed on the X-ray patterns, is reported on Fig 6. The linearity observed

40

between the period and the number of injections at each cycle clearly shows the good performances of the injection system. XPS and high-resolution RBS measurements confirm the modulation of the composition, and are in good agreement with X-ray diffraction.

The electronic properties of these mixed samples deposited on $SrTiO_3(Nb)$ conducting single crystal substrates have been studied, both as solid solution or superlattices. The two kinds of samples show very different behavior of the dielectric constant : in the solid solution, the dielectric constant has a maximum at the Curie temperature, while it is quasi constant in the superlattices, with values 300-500, depending on the thicknesses. The dielectric constants of these layers are comparable to those of the layers obtained by MBE [32], or laser ablation [33].

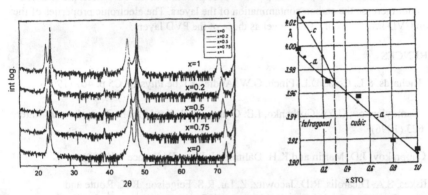

Fig 5-a : X ray diffraction patterns of mixed $Ba_{(1-x)}Sr_xTiO_3$ epitaxial oxides layers (log scale).
Fig 5-b : lattice parameter variation in the solid solution (Injection CVD layers   Bulk materials)

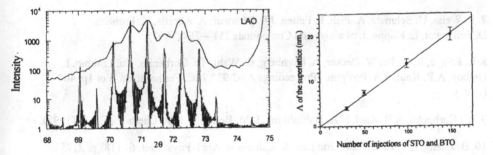

Fig 6-a : Satellites lines observed on a $BaTiO_3$ / $SrTiO_3$ superlattice together with the simulation. The parameters for the simulation are : no interdiffusion, 10 double layers containing 24 unit cell of $BaTiO_3$ (a=.399 nm.) + 27 unit cell of $BaTiO_3$ (a=.3905 nm.)).
Fig 6-b : variation the superlattices modulation as a function of the number of injections, which were the same for each source in this experiment..

CONCLUSIONS

These results shows that the sequential liquid injection technique allows the preparation of a wide number of good quality thin layers, multilayers or superlattices in a very simple way.

The layers already deposited are smooth when they are amorphous or "epitaxially" grown (surface roughness of the order of 1 nm). The main difference from conventional CVD is that only one source is needed for the deposition of thin layers of complex compounds, like for PVD techniques. But sources with different stoichiometries for injection CVD are considerably easier to realize than sources for laser ablation or sputtering : they are only mixtures of liquids. It appears that, for the layers deposited up to now, the presence of the solvent in the reaction zone did not increase the carbon contamination of the layers. The electronic properties of the injection CVD layers is at the same level as those of the PVD layers.

REFERENCES

1. B.C. Richards, S.L. Cook, D.L. Pinch, G.W. Andrews J. de Phys IV C5 p. 407 (1995)

2 . S.V. Samoylenkov, O.Yu. Gorbenko, I.E. Graboy, A.R. Kaul, Yu.D. Tretyakov, J. Mater. Chem. 623 (1996).

3. I.S. Chuprakov, J.D. Martin and K.H. Dahmen, J. Phys. IV France 9 (1999).

4. R. Hiskes, S.A. Dicarolis, R.D. Jacowitz, Z. Lu, R.S. Feigelson, R.K. Route and J.L. Young, Journal of Crystal Growth 128 781 – 787 (1993).

5. C. Dubourdieu, S.B. Kang, Y.Q. Li, G. Kulesha, B. Gallois, Thin solid films 339 165 (1999).

6. M. Jergel, (Review article). Supercond. Sci. And Technol. 867 (1995).

7. F. Weiss, U. Schmatz, A. Pish, F. Felten, J.P. Sénateur, A. Abrutis, K. Frohlich, D. Selbmann, L. Klippe, J. of alloys and Compounds 251 – 264 (1997).

8. L. Klippe, R. Stolle, W. Decker, A. Nurnberg, G. Wahl, Yu. Gorbenko, Yu. Erokhin, I. Graboy, A.R. Kaul, S.A. Pozigun, "Proceedings ASC 93", H.C. Freihardt Ed. Vol 1p. 407 (1993).

9. Yu Gorbenko, A.R. Kaul, N.A. Babushkina, L.M. Belova, J. Mat. Chem. 7 (5) 747 (1997).

10. B. Zheng, E.T. Eisenbraun, Jun Liu, A. Kaloyeros, Appl. Phys. Lett. 61 (18) p. 2175 (1992).

11. Jung-Hyun Lee and Shi Was Rhee, Electrochem. and solid state lett. 2 (10) p. 510 (1999).

12. J. Zhang, R.A. Gardiner, P.S. Kirling, R.W. Boerstler and J. Steinbeck, Appl. Phys. Lett. 61 (24) p. 2884 (1992).

13. A.R. Kaul, B. Seleznev, J. de Phys. C3 3 p. 375 (1993)

14; G. Wahl, M. Pulver, W. Decker. L. Klippe, Surface & Coatings Technology 101, 132 (1998).

15. S. Pignard, H. Vincent, J.P. Sénateur, P.H. Giaugne, Chem. Solid films 347 p. 161 (1999)

16. J.P. Sénateur, R. Madar, O. Thomas, F. Weiss, A. Abrutis Patent n° 93/08838 - Licensed J.I.P. Elec - 11, chemin du Vieux Chène - 38240 MEYLAN - France - Fax 33 4 76 04 81 40

17. C. Chaneliere, S. Four, J.L. Autran, R. Devine,N.P. Sandler, J. Appl. Phys. 83 9 p 4823 (1998)

18. G.B. Alers, D.J. Werder, Y. Chabal, H.C. Lu, E.P. Gusev, E. Garfunkel, T. Gustafsson, R.S. Urdahl, Appl. Phys. Lett. 73 11 p. 1517 (1998)

19. H. Shinriki, M. Sugiura, Y. Liu, K. Shimomura, J. Electrochem. Soc. 145 9 p. 3247 (1998)

20. J.Lin, N. Masaaki, A. Tsukune, M. Yamada, Appl. Phys. Letters 74, n°17, p. 2370 (1999)

21. S. Ehilvalavan and Tseung-Yuen Tseng, Appl. Phys. letters 74 n°17 p.77 (1999)

22. J.Y Zhang, L.J. Bie, I.W. Boyd, Jpn. J. of Appl. Phys 37 L 27 (1998)

23. K. Ohta, K. Yamada, R. Shimuzu,Y Tarui, IEEE Trans Eletr. Devices ED 29 368 (1982)

24. F. Felten, J.P. Sénateur, F. Weiss, A. Abrutis, J. de Phys. IV Colloque C5 5 p. 1079 (1995)

25. F. Felten, M. Labeau, K. Yu-Zhang, A. Abrutis, Thin Solid Films 296 p. 79 (1997)

26. J.Y Zhang, I. Boyd, M.B. Mooney, P.K. Hurley, B.J. O'Sullivan, J.T. Beechinor, P.V. Kelly, G.M. Crean, Presented at MRS Spring Meeting, San Francisco, USA, April 5-9 1999 - Mat. Res. Symp. Proc., 567 p.397-402

27. M.B. Mooney, P.K. Hurley, B.J. O'Sullivan, J.T. Beechinor, P.V. Kelly, G.M. Crean, J.Y. Zhang, I. Boyd, Proceedings of INFOS 99 – Microelectronic Engineering 48 p. 283 (1999)

28. R.F. Cava, W.F. Peck, J.J. Krajewski , Letters to Nature 377 - 215 - (1995)

29. J.Y Gan, Y.C. Chang, T.B. Wu, Appl. Phys. Lett. 72 3 332 (1998)

30. C. Jimenez, M. Paillous, R. Madar, J.P.Sénateur, A.C. Jones, J. Phys. IV - 9 p. Pr8-569 (1999)

31 R. Waser, O. Lohse, Int. Ferroel. 21 1-4 p. 27 (1998)

31. B.D. Qu, M. Estvignee, D.J. Johnson, R.H. Prince, Appl. Phys. Lett. 72 11 (1998)

33. H. Tabata, T. Kawai, Appl. Phys. Lett. 70 3 (1997)

GROWTH OF MgO BY METAL-ORGANIC MOLECULAR BEAM EPITAXY

Feng Niu, Brent.H.Hoerman and Bruce.W.Wessels
Department of Materials Science and Engineering and Materials Research Center, Northwestern University, Evanston, Illinois 60208
Draft, 15 November 1999

ABSTRACT

MgO thin films were deposited on (100) Si substrates by metal-organic molecular beam epitaxy (MOMBE). Magnesium acetylacetonate was used as the precursor and an oxygen RF plasma was used as the oxidant. The films were characterized by a combination of transmission electron microscopy, Auger spectrometry and atomic force microscopy. Analyses indicate that the films directly deposited on Si substrates are stoichiometric, phase-pure, polycrystalline MgO with a [100] texture. Carbon contamination of the films resulting from precursor decomposition was not observed within detection limits. Furthermore, the growth rate of MgO has been systematically investigated as a function of growth temperature.

INTRODUCTION

MgO thin dielectric films deposited on Si are potential of considerably technological importance. For example they can be used potentially as an optical isolation layer for ferroelectric oxide waveguide on Si. They can be used potentially as a gate oxide in MOS devices [1] or they can serve as a buffer between other oxides and Si wafer where the integration of these materials with Si wafer is becoming a more and more important issue [2]. To date, a variety of deposition techniques have been used to grow MgO thin films on Si including metal-organic chemical vapor deposition (MOCVD) [3,4], sputtering [5-8], laser deposition [9-12], e-beam-assisted molecular beam epitaxy (MBE) [13] and sol-gel method [14]. An alternative approach to deposit MgO that may have several advantages is metal-organic molecular beam epitaxy (MOMBE). This technique utilizes metal organic compounds as the metal source [15-19]. The ultra-high vacuum enables in situ surface diagnosis such as reflection high-energy electron diffraction (RHEED), which is not possible for MOCVD growth. Another major advantage of MOMBE is that the O_2 pressure is low (<10^{-6} Torr), which minimizes amorphous silicon dioxide formation at the Si substrate interface which precludes epitaxial growth. The technique should be quite versatile potentially enabling the deposition of both refractory and reactive metal oxides. One of major challenges concerning the use of MOMBE is to identify suitable metal-organic precursors. The ideal solid precursor for MOMBE deposition should have 1) Sufficiently volatile (> 10^{-5} Torr); 2) A single metal-containing volatile species in the molecular beam; 3) A high sticking coefficient; 4) High long-term stability at the sublimation temperature (No decomposition or oxidation of the source materials).

In the present work, we describe the growth of MgO thin films by MOMBE using Mg (acac)$_2$ (acac= acetylacetonate) as the Mg precursor and a RF O plasma as the oxidant. The growth conditions, structure, compositions and surface morphology of the deposited films characterized by transmission electron microscopy (TEM), Auger spectrometry and atomic force microscopy (AFM). Further more, in order to optimize the growth conditions, the effects of the growth temperature on growth rate and surface morphology have been systematically investigated.

EXPERIMENTAL

The films were deposited in a SVT Associates SN35 MBE system equipped with low temperature effusion cells and oxygen plasma as shown in Figure 1. The solid metal-organic precursor was contained in an effusion cells. The cell temperature can be precisely controlled to within 0.2 °C. Mg

45

precursor was contained in an effusion cell. The cell temperature can be precisely controlled to within 0.2 °C. Mg (acac)$_2$ was used as the precursor. The source pressure was controlled by a flux monitor. An oxygen plasma from an SVT RF Plasma O$_2$ gun with output energy of 300-400 W was used as an oxidant. The substrate temperature was measured by a thermocouple calibrated by an infrared optical pyrometer.

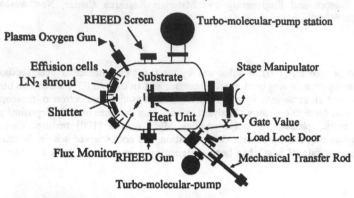

Figure 1 Schematic of a SVT Associate MBE system

N-type Si (100) wafers with resistivities of 1-10 Ωcm were used as substrates. Prior to growth, the Si surface was chemically etched in a HNO$_3$:HF:H$_2$O (5:3:92 by volume percentage) solution for several seconds to remove the surface oxide layer. It was then was subsequently cleaned in methanol and acetone for 15 minutes. The chemically etched Si wafer was immediately placed in the system load lock and then transferred to the growth chamber pre-pumped down to a base pressure of 10^{-9} to 10^{-10} Torr. The Si was subsequently degassed for 30 minutes at 200°C and desorped for 30 minutes at 850°C. The Si surface cleaned by this simple procedure showed a clear and streaky RHEED pattern with Kikuchi lines indicating the Si surface was atomically clean.

The film thickness was measured by a Tencor P-10 surface profiler. The film composition was measured by Auger electron spectroscopy (AES). The microstructure was determined by a Hitachi HF-2000, cold field emission TEM operated at 200kV. TEM samples were prepared by mechanical grinding and polishing followed by further mechanical dimpling to a thickness of less than 10 μm and final thinning in a Gatan Model 691 precision ion polishing system. Surface morphology and roughness of the MgO thin films were measured with a Nanoscope III AFM with a Si$_3$N$_4$ tip. Table I lists the typical growth conditions for preparing the MgO thin film.

Table I Typical growth conditions for deposition of MgO thin film

Metal-organic source	Mg(acac)$_2$
Substrate	(100) Si (n-type)
Growth temperature(°C)	500-850
Plasma O$_2$ pressure (Torr)	1.0×10^{-5}
Pressure of Source Mg(acac)$_2$ (Torr)	$(1.0\text{-}5.0) \times 10^{-6}$
Deposition time (hour)	1-4
Growth rate (nm/hr)	10-50

RESULTS

Figure 2 shows a TEM diffraction pattern from a 60 nm thick MgO thin film deposited at 700°C for 3 hours by MOMBE. A ring diffraction pattern was observed which is due to

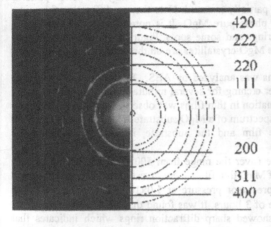

420
222
220
111

200

311
400

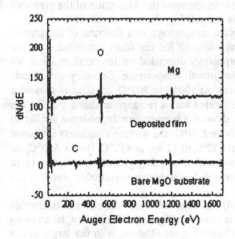

100nm

Figure 2. TEM diffraction pattern from a 60 nm
thick MgO film deposited at 700°C for 3 hours.

Figure 3. TEM bright field micrograph
from the MgO film deposited at 700°C
for 3 hours.

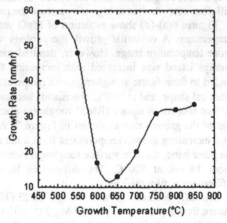

Figure 4. AES spectra from the MgO
film deposited at 700°C for 3 hours
and MgO substrate.

Figure 5. Growth rate as a function
of growth temperature.

47

electron diffraction from polycrystalline MgO. In addition, some extra diffraction spots due to MgO <001> zone axis were observed that were superimposed on the ring diffraction pattern, which indicated that the MgO films were partially textured. Note that no diffraction from any other phase is observed, thus the film is phase pure MgO. It is noted that RHEED patterns showed similar features with diffraction rings and some spots. Figure 3 shows a bright field TEM micrograph of MgO crystallites. The MgO crystallites are on the order of ten nanometers in size and had an irregular shape.

The chemical composition of the films was analyzed by AES. Figure 4 shows a typical Auger spectrum from the film after sputter etching for several minutes. Only Mg and O peaks are observed. No trace of carbon contamination in thin films was observed within the detection limit of AES. For comparison the Auger spectrum of a MgO substrate is also shown. The peak intensities are comparable for both the film and the substrate, indicating the films are stoichiometric.

The effects of the growth temperature (over the range from 500°C to 850°C) on growth rate, surface roughness and morphology of MgO thin films were investigated. The other growth conditions were kept constant with the precursor pressure at 1.0×10^{-6} Torr, plasma oxygen pressure at 1.0×10^{-5} Torr and growth time of 3 hours. It was found that RHEED patterns from the films deposited at all temperatures showed sharp diffraction rings which indicates that polycrystalline MgO films could be grown over the whole temperature range. However, RHEED patterns consisting of both rings and spots only appeared at 650°C and 700°C, indicating that textured MgO thin films were grown only in this temperature range. Figure 5 shows the growth rate as a function of the growth temperature. At growth temperatures above 650°C, the growth rate increased with increasing growth temperature from about 12nm/hr at 650°C to about 33 nm/hr at 850°C, while below 650°C the growth rate increased with decreasing growth temperature up to about 56 nm/hr at 500°. This complex temperature dependence indicates that there are two different kinetic regimes that dominate for the temperature ranges below and above 650°C. At low temperature the desorption of the precursor will dominate whereas at high temperature the rate is limited by the surface reaction.

Figures 6(a)-(d) show evolution of MgO surface morphology as a function of the growth temperature. A columnar growth morphology was formed for the films deposited over the entire temperature range. However, detailed morphology depended on temperature. First the average island size increased with increasing the growth temperature. Secondly the islands tended to form facets at higher growth temperatures. At 500°C to 700°C, the islands showed a spherical shape, and at 800°C, the islands became faceted with a rectangular shape. Finally, the surface root mean square (RMS) roughness also showed a temperature dependence similar to that of the growth rate as shown in Figure 7. Above 650°C, the surface roughness increased with increasing growth temperatures from 2 nm at 650°C to 11 nm at 850°C. Below 650°C, on the other hand, the film surface roughness increased with decreasing growth temperature up to about 14 nm at 500°C. The differences in measured roughness presumably result from differences in film thickness.

Recently epitaxial growth of MgO on Si (100) has been achieved. An electron diffraction pattern from a plan view epitaxial MgO thin film is shown in Figure 8. The pattern, taken along Si <001> zone axis, consisted of two sets of diffraction spots. The set with the larger lattice constant is from the Si (001) substrate. The other set is due to the MgO overlayer. It is noted that the MgO overlayer and underlying Si have an epitaxial relationship of MgO (001)//Si (001) and MgO [110]//Si [110].

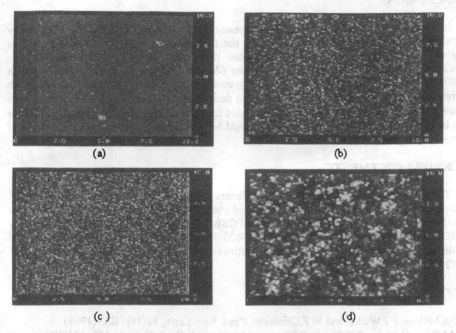

(a)

(b)

(c)

(d)

Figure 6. AFM images from MgO films deposited at different temperatures.
(a) 500°C; (b)600°C; (c)700°C; (d)800°C

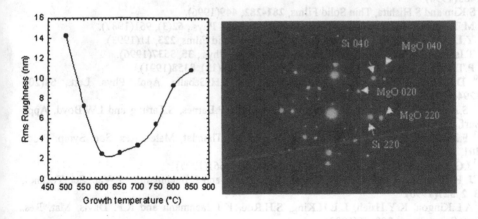

Si 040   MgO 040

MgO 020

MgO 220

Si 220

Figure 7. RMS surface roughness of
MgO films via growth temperatures.

Figure 8. TEM diffraction pattern from an
epitaxial MgO film taken along Si <100> zone.

CONCLUSIONS

In conclusion, we have demonstrated deposition of polycrystalline MgO thin films by metal-organic molecular beam epitaxy using the precursor Mg (acac)$_2$ and an RF oxygen plasma. AES analysis indicated the deposited films were stoichiometric MgO. Carbon contamination in the films was negligible. The films were phase pure and crystalline with <100> texture as determined by TEM. The growth rate of MgO thin films increased with increasing growth temperature above 650°C and decreased with increasing growth temperature below 650°C. From this study, it can be concluded that high quality, crystalline MgO thin films can be grown by MOMBE. The technique should be useful for deposition of a wide range of thin insulating oxide films.

ACKNOWLEDGEMENTS

The authors are grateful to Dr Antonio DiVenere for his technical assistance with MOMBE equipment installation. This work was supported under DARPA through the MURI program at Northwestern University monitored by the AFOSR under grant F49 620-96-1-0262 and by BMDO through the DURIP program under F49620-96-1-0460. Extensive use was also made of facilities supported by MRSEC program at Northwestern University under grant number DMR 9632472.

REFERENCES

[1] R.A.McKee, F.J.Walker and M.F.Chisholm, Phys. Rev. Letts., **81**(14), 3014(1998).

[2] J.M.Zeng, H.Wang, S.X.Shang, Z.Wang and M.Wang, J. Cryst. Growth, **169**, 474(1996).

[3] E.Fujii, A.Tomozawa, S.Fujii, H.Torii, M.Hattori and R.Takayama, Jpn. J. Appl. Phys., **32**, L1448 (1993).

[4] S.Fujii, A.Tomozawa, E.Fujii, H.Torii, R.Takayawa and T.Hirao, Appl. Phys. Lett., **65**(11), 1463(1994).

[5] S.Kim and S.Hishita, Thin Solid Films, **281-282**, 449(1996).

[6] M.Tonouchi, Y.Sakaguchi and T.Kobayashi, J. Appl. Phys., **62**(3), 961(1987).

[7] Y.Li, G.C.Xiong, G.J.Lian, J.Li and Z.Gan, Thin Solid Films, **223**, 11(1993).

[8] T.Ishiguro, Y.Hiroshima and T.Inoue, Jpn. J. Appl. Phys., **35**, 3537(1996).

[9] P.Tiwari, S.Sharan and J.Narayan, J. Appl. Phys., **69**(12), 8358(1991).

[10] D.K.Fork, F.A.Ponce , J.C.Tramontana and T.H.Geballe, Appl. Phys. Lett., **58**(20), 2294(1991).

[11] S.Amirhaghi, A.Archer, B.Taguiang, R.McMinn, P.Barnes, S.Tarling and I.W.Boyd, Appl. Surf. Sci., **54**, 205(1992).

[12] F.J.Walker, R.A.McKee, S.J.Pennycook and T.G.Thundat, Mater. Res. Soc. Symp. Proc., **401**, 13 (1996).

[13] J.G.Yoon and K. Kim, Appl. Phys. Lett., **66** (20), 2661(1995).

[14] J. P. Bade, E. A Baker, A. I. Kingon, R. F. Davis, and J. Bachmanm, J. Vac. Sci. Technol. B, **2**, 327(1990).

[15] A.L.Kingon, K.Y.Hsieh, L.L.H.King, S.H.Rou, K.J.Bachmann and R.F. Davis, Mat. Res. Soc. Symp. Proc., **200**, 49(1990).

[16] D.J.Lichtenwalner, and A.I.Kingon, Appl. Phys. Lett. **59**(23), 3045(1991).

[17] S.Ikegawa, and Y.Motoi, Thin Solid Films, **281-282**, 60(1996).

[18] K.Hayama, T.Togun, and M.Ishida, J. Cryst. Growth, **179**, 433(1997).

SELECTION AND DESIGN OF PRECURSORS FOR THE MOCVD OF LEAD SCANDIUM TANTALATE

A.C. JONES*, H.O. DAVIES*, T.J. LEEDHAM*, M.J. CROSBIE**, P.J. WRIGHT**,
P.O'. BRIEN***, AND K.A. FLEETING***
*Inorgtech Limited, 25 James Carter Road, Mildenhall, Suffolk, IP28 7DE, UK.
tony@inorgtech.co.uk
**Defence, Evaluation and Research Agency, St. Andrews Road, Malvern, Worcestershire,
WR14 3PS, UK.
***Department of Chemistry, Imperial College of Science, Technology and Medicine, London,
SW7 2BP, UK

ABSTRACT

Metalorganic chemical vapour deposition (MOCVD) is a promising technique for the deposition of the pyroelectric oxide lead scandium tantalate, $Pb(Sc_{0.5}Ta_{0.5})O_3$. In order to exploit the full potential of the method, it is important to identify the optimum combination of precursors so that process parameters and film properties are optimised. In this paper, the molecular design of new, more compatible Ta and Sc oxide precursors is described and it is shown how the use of carefully matched precursors allows the growth of $Pb(Sc_{0.5}Ta_{0.5})O_3$ in the required perovskite phase at low substrate temperatures.

INTRODUCTION

Lead scandium tantalate, $Pb(Sc_{0.5}Ta_{0.5})O_3$ (PST) is a sensitive pyroelectric oxide with a high figure of merit which has potential applications in uncooled thermal imaging, such as night sight technologies and fire detection[1]. However, the application of PST in pyroelectric devices has been restricted by the high temperatures (typically over 1000°C) required for the processing of bulk materials to the desired perovskite phase and to eliminate the pyrochlore phase (with its inferior electrical properties), which tends to form at low temperature.

Although PST in the perovskite phase has been grown at low temperature by sol-gel techniques[2], there have been few reports of successful MOCVD studies. Early work[3] involved a processing temperature of 800°C and a post-growth anneal step. More recent work[4] has shown that the perovskite phase of PST can be grown in one step at temperatures as low as 600°C. However, in both these studies, conventional MOCVD techniques were used, in which the Pb, Sc and Ta precursors were held in "bubblers" at relatively high source temperatures, leading to the possibility of source decomposition with time.

The low volatility of most oxide precursors has led to the increased use of some form of liquid injection MOCVD[5]. This is especially so for the growth of complex oxides such as PST where composition control is important. This variant of MOCVD alleviates the problem of "thermal ageing" and decomposition of the precursor in the bubbler which contributes to changing precursor fluxes with time from conventional bubblers. However, it requires that the precursors are soluble and stable in a suitable dry solvent. The precursors, especially those used for complex oxide deposition, need to have similar vaporisation and decomposition temperatures in order to achieve good layer uniformity. Therefore a key element in the selection of precursors is to maximize their compatibility.

Conventional precursors for the MOCVD of PST include $Pb(thd)_2$ (thd = 2,2,6,6-tetramethylheptane-3,5-dionate), $Sc(thd)_3$ and $Ta(OEt)_5$[4]. However, these are generally incompatible, with significantly different physical properties and decomposition characteristics. For instance, $Sc(thd)_3$ deposits oxide in the region 550 - 600°C compared to 400 - 450°C for the less thermally stable $Ta(OEt)_5$ source, which can lead to poor layer uniformity. $Sc(thd)_3$ is

51

also significantly less volatile than Ta(OEt)$_5$ and requires an evaporator temperature of at least 300°C to give acceptable oxide growth rates. Consequently, a more stable Ta source than Ta(OEt)$_5$ is required, together with a more volatile Sc precursor. In order to optimise PST layer uniformity, the Ta and Sc oxide sources should also deposit oxide in a similar temperature range to Pb(thd)$_2$ (i.e. 450 - 500°C[5]).

In this paper, the modification of Ta(OR)$_5$ precursors to give more thermally stable sources is described, together with chemical techniques for increasing the volatility of Sc β-diketonates. The use of these more compatible Pb, Sc and Ta precursors has allowed the deposition of Pb(Sc$_{0.5}$Ta$_{0.5}$)O$_3$ in the perovskite phase.

EXPERIMENTAL

The Pb, Sc and Ta oxide precursors used in the study were manufactured by Inorgtech Limited and were fully characterised by ^{1}H NMR, ICP-MS and elemental microanalysis before use.

Sc$_2$O$_3$ and Ta$_2$O$_5$ films were deposited over a range of temperatures on Si(100) substrates at atmospheric pressure using a liquid injection MOCVD reactor described elsewhere[5]. The films were grown from individual solutions of the Sc and Ta oxide precursors dissolved in tetrahydrofuran (THF).

Pb(Sc$_{0.5}$Ta$_{0.5}$)O$_3$ thin films were deposited on Pt(111) substrates at 450-600°C using a single solution of the precursors in THF. Growth conditions are given in Table I.

Table I. Growth conditions used to deposit single oxides and Pb(Sc$_{0.5}$Ta$_{0.5}$)O$_3$

Oxide	Precursor	THF solution concentration (molar)	Solution injection rate (cm^3 hr^{-1})	Evaporator temperature (°C)
Sc$_2$O$_3$	Sc(thd)$_3$ Sc(tmod)$_3$ Sc(mhd)$_3$	0.1	2	300
Ta$_2$O$_5$	Ta(OR)$_5$ Ta(OR)$_4$(L)	0.1	2	225
Pb(Sc,Ta)O$_3$	Pb(thd)$_2$ Sc(thd)$_3$ Ta(OEt)$_4$(thd)	0.066 0.033 0.033	2	300

Reactor pressure 1000 mbar, substrates Si(100) or Pt(111), argon flow rate 4000 cm^3 min^{-1}, oxygen flow rate 1000 cm^3 min^{-1}

RESULTS AND DISCUSSION

Modification of Ta Alkoxide Precursors

The selection of a suitable alternative to Ta(OEt)$_5$ presents a considerable challenge in the MOCVD of PST. There are no Ta β-diketonates available and therefore Ta(OMe)$_5$ was investigated initially, in the hope that substitution of [EtO] groups with [MeO] groups would impart increased thermal stability to the molecule. However, the optimum region of oxide deposition from Ta(OMe)$_5$ actually occurs at a lower substrate temperature than for Ta(OEt)$_5$

(see Fig. 1) and both precursors deposit oxide at significantly lower temperatures than Sc(thd)$_3$, which is likely to result in poor uniformity PST.

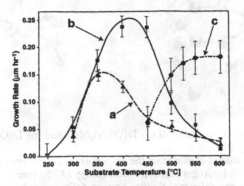

Figure 1. Variation in metal oxide growth rate with substrate temperature for (a) Ta(OMe)$_5$, (b) Ta(OEt)$_5$ and (c) Sc(thd)$_3$.

However, we have found that the thermal stability of tantalum alkoxide precursors can be strongly influenced by the insertion of β-diketonate groups such as thd or acac (2,4–pentanedionate). For instance, the optimum oxide deposition temperature is higher for Ta(OMe)$_4$(acac) and Ta(OEt)$_4$(thd) than the parent alkoxides, Ta(OMe)$_5$ and Ta(OEt)$_5$ and increases in the order Ta(OMe)$_4$(acac) < Ta(OPr^i)$_4$(thd) < Ta(OEt)$_4$(thd) < Ta(OMe)$_4$(thd), as shown by the growth rate data in Fig. 2.

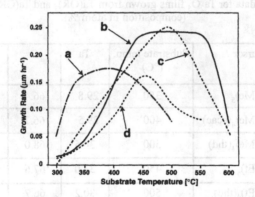

Figure 2. Variation in Ta$_2$O$_5$ growth rate with substrate temperature for (a) Ta(OMe)$_4$(acac), (b) Ta(OMe)$_4$(thd), (c) Ta(OEt)$_4$(thd) and (d) Ta(OPr^i)$_4$(thd)

These Ta(OR)$_4$(L) precursors are monomeric[6] (Fig. 3) and are more volatile and less air sensitive than the dimeric parent alkoxides[7]. They also undergo significantly less decomposition in the evaporator during MOCVD than the parent Ta(OR)$_5$ compound. For instance, at an evaporator temperature of 225°C, Ta(OEt)$_4$(thd) leads to a Ta$_2$O$_5$ growth rate of 0.25μm hr^{-1} compared with 0.06μm hr^{-1} achieved using Ta(OEt)$_5$, which further indicates that Ta(OR)$_4$(L) precursors have thermal stabilities which are more compatible with Sc(thd)$_3$.

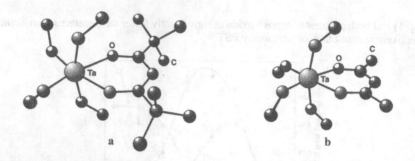

Figure 3. Crystal structures of (a) Ta(OMe)$_4$(thd) and (b) Ta(OMe)$_4$(acac) [6]

X-ray diffraction studies showed that the as-deposited films were predominantly amorphous, and analysis by Auger electron spectroscopy (AES) (see Table II) showed that the films were slightly oxygen deficient. Similar oxygen deficiencies have been observed in as-deposited films grown by a number of techniques. Residual carbon levels in the films varied from close to the detection limit (*ca.* 1 at%) up to 3.8 at% and showed no clear dependence on the nature of the alkoxide or β-diketonate group. These levels are similar to those reported in Ta$_2$O$_5$ grown by conventional MOCVD using Ta alkoxides[7], which indicates that the organic solvent THF used in liquid injection contributes no extra carbon to the films.

Annealing the tantalum oxide films in air at 1000°C produced films with the expected 2:5 atomic ratio of Ta:O, but had no measurable effect on the carbon levels in the films. On annealing the films crystallised in the δ-Ta$_2$O$_5$ phase.

Table II. AES data for Ta$_2$O$_5$ films grown from Ta(OR)$_5$ and Ta(OR)$_4$(L) precursors (composition in atom %).

Precursor	Substrate temp. (°C)	Ta	O	C
Ta(OMe)$_5$	350	29.8	66.4	3.8
Ta(OMe)$_4$(acac)	400	29.5	66.7	3.8
Ta(OMe)$_4$(thd)	500	30.5	68.0	1.5
Ta(OEt)$_5$	450	31.0	67.6	1.4
Ta(OEt)$_4$(thd)	500	30.2	66.7	3.1
Ta(OPr^i)$_4$(thd)	450	30.7	68.0	1.3

Reference to Figs. 1 and 2 indicates that, compared to the parent alkoxides, Ta(OMe)$_4$(thd), and Ta(OEt)$_4$(thd) deposit oxide at temperatures which are more compatible with Sc(thd)$_3$. These temperatures are also compatible with the oxide deposition temperature of Pb(thd)$_2$ (i.e. 450 - 500°C[5]). PST films were subsequently grown at substrate temperatures of between 400 and 600°C using a single solution of Pb(thd)$_2$, Sc(thd)$_3$ and Ta(OEt)$_4$(thd) dissolved in THF. At substrate temperatures of 500°C and below, the PST was found to be a mixture of the pyrochlore and perovskite phases. However, at 600°C the perovskite phase was predominant (Fig. 4), demonstrating the high potential of MOCVD for the growth of PST-based devices.

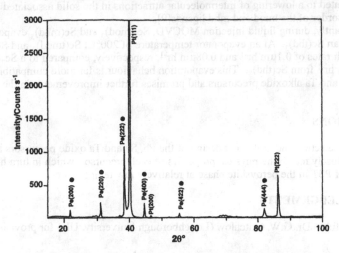

Figure 4. X-ray diffraction data for PST grown at 600°C from Pb(thd)$_2$, Sc(thd)$_3$ and Ta(OEt)$_4$(thd) [Pe = perovskite peaks (●), Pt = diffraction peaks from Pt substrate]

Modification of Sc Oxide Precursors

Despite the successful growth of PST using Sc(thd)$_3$, the use of high evaporator temperatures (300°C) is undesirable, potentially leading to decomposition of the Ta(OR)$_4$(L) and Pb(thd)$_2$ precursors and we have thus sought a more volatile Sc source. The insertion of alkoxide groups into six-coordinate Sc(thd)$_3$ is not an option as it would result in an unsaturated and unstable Sc centre. However, asymmetric β-diketone ligands such as tmod (2,2,7–trimethyloctane-3,5-dionate) and mhd (6-methylheptane-2,4-dionate) have been shown to significantly influence the volatility of yttrium β-diketonates[8]. We have thus investigated[9] the sterically hindered Sc β-diketonates Sc(tmod)$_3$ and Sc(mhd)$_3$ as alternatives to Sc(thd)$_3$ (see Fig. 5).

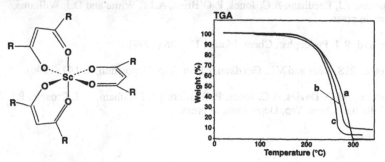

Figure 5. Effect of β-diketonate ligand symmetry on volatility of Sc β-diketonates
(a)Sc(thd)$_3$[R = Bu^t, Bu^t]   (b)Sc(mhd)$_3$[R = Me, Bu^i]   (c)Sc(tmod)$_3$[R = Bu^t, Bu^i]

Thermogravimetric analysis (TGA) on $Sc(thd)_3$, $Sc(tmod)_3$ and $Sc(mhd)_3$ (see Fig. 5), indicates that $Sc(tmod)_3$ and $Sc(mhd)_3$ evaporate at lower temperature than $Sc(thd)_3$ and this çan be attributed to a lowering of intermolecular attractions in the solid associated with increased disorder of the tmod and mhd ligands[9].

Significantly, during liquid injection MOCVD, $Sc(tmod)_3$ and $Sc(mhd)_3$ evaporate more efficiently than $Sc(thd)_3$. At an evaporator temperature of 200°C, $Sc(tmod)_3$ and $Sc(mhd)_3$ give Sc_2O_3 growth rates of $0.1\mu m\ hr^{-1}$ and $0.08\mu m\ hr^{-1}$, respectively, compared to a Sc_2O_3 growth rate of $<5nm\ hr^{-1}$ from $Sc(thd)_3$. This evaporation behaviour is far more compatible with the available Pb and Ta alkoxide precursors and promises further improvements in the MOCVD of PST.

CONCLUSIONS

Careful selection and molecular design of the Pb, Sc and Ta oxide precursors has allowed us to more closely match the physical properties of each precursor, which in turn has facilitated the growth of PST in the perovskite phase at relatively low temperatures.

ACKNOWLEDGEMENTS

We are grateful to Dr. G.W. Critchlow (Loughborough University, UK) for provision of the AES data.

REFERENCES

1. R.W. Whatmore, S.B. Stringfellow and N.M. Shorrocks, Infrared Technology, **XIX**, 391 (1993).

2. D. Liu, D.A. Payne and D.D. Viehland, Mater. Lett., **17**, 319 (1993).

3. F.W. Ainger, K. Bass, C.J. Brierly, M.D. Hudson, C. Trundle and R.W. Whatmore, Progr. Cryst. Growth Charact., **22**, 183 (1991).

4. D. Liu and H. Chen, Mater. Lett., **28**, 17 (1996).

5. A.C. Jones, T.J. Leedham, P.J. Wright, M.J. Crosbie, P.A. Lane, D.J. Williams, K.A. Fleeting, D.J. Otway and P. O'Brien, Chem. Vap. Deposition, **4**, 46 (1998).

6. H.O. Davies, T.J. Leedham, A.C. Jones, P. O'Brien, A.J.P. White and D.J. Williams, Polyhedron, in press.

7. K.D. Pollard, R.J. Puddephat, Chem. Mater., **11**, 106 (1999).

8. H.A. Luten, W.S. Rees and V.L. Goedken, Chem. Vap. Deposition, **2**, 149 (1996).

9. K.A. Fleeting, H.O. Davies, A.C. Jones, P. O'Brien, T.J. Leedham, M.J. Crosbie, P.J. Wright and D.J. Williams, Chem. Vap. Deposition, in press.

DEPOSITION OF SiO$_2$:F:C WITH LOW DIELECTRIC CONSTANT AND WITH HIGH RESISTANCE TO ANNEALING

J. Lubguban, Jr., Y. Kurata, T. Inokuma and S. Hasegawa
Department of Electronics, Faculty of Engineering, Kanazawa University,
920-8667 Kanazawa, JAPAN

ABSTRACT

SiO$_2$:F:C films were deposited using a plasma-enhanced chemical vapor deposition (PECVD) technique from SiH$_4$/O$_2$/CF$_4$/CH$_4$ mixtures. Increasing amount of carbon were introduced to the SiO$_2$:F films by changing the CH$_4$ flow rate, [CH$_4$], while keeping constant other conditions such as rf power and deposition temperature, T_D. It was found that the addition of CH$_4$ decreases the dielectric constant, k, from 3.36 for [CH$_4$] = 0 sccm to 2.95 for [CH$_4$] = 8 sccm. For the [CH$_4$] condition where the film has a lowest k, the deposition temperature and rf power were optimized by depositing films using different values of T_D and rf power. The k for films in the new series as well as the stress and water absorption was investigated. Results show that the dielectric constant further decrease up to 2.85. Some films were then annealed from 400 - 800 °C and it was found that the k for films deposited with higher [CH$_4$] has a better stability with respect to annealing up to a temperature of 600 °C.

INTRODUCTION

Continuing demands in ultra large-scale integrated circuit, are in shrinking device dimensions to less than 0.25 μ m. When device size decreases, problems due to wiring capacitance like propagation delay, crosstalk noise and power dissipation occur [1]. The most effective way to solve these problems is to use a suitable material with a low dielectric constant, low k. SiO$_2$:F films with a dielectric constant of 3.3 has been widely investigated. Other materials like PTFE, a-C:F, porous materials and organic silica with low k were also studied [2-4]. However, problems related to thermal stability, water absorption, step coverage and poor adhesion arise for these films.

In this study, carbon doping to SiO$_2$:F films were investigated. The optimal condition for the deposition of these films was investigated by changing deposition parameters like CH$_4$ flow rate, [CH$_4$], deposition temperature, T_D, and rf power. The dielectric constant, Fourier transform-infrared spectra (FT-IR), stress and water absorption for the resulting films were studied. Also in this research, the stability of k for films with varying [CH$_4$] was investigated by annealing these films from 400 - 800 °C.

EXPERIMENTAL

The method of deposition for the SiO$_2$:F:C films was plasma enhanced chemical vapor deposition, PECVD, using a gas mixture of SiH$_4$, O$_2$, CF$_4$ and CH$_4$. The following fixed conditions for all films were used: [SiH$_4$] = 1.4 sccm, [O$_2$] = 1.9 sccm, [CF$_4$] = 2.9 sccm and total pressure = 1.7 Torr. Three different series of deposition conditions were used. The first series was the variation of [CH$_4$] from 0 sccm to 8 sccm keeping the rf power and T_D constant at 20 W and 300 °C, respectively. The second and third series were the variation of either T_D, from 100 to 500 °C or rf power from 10 to 50 W while fixing constant either the rf power at 20 W or T_D at 300 °C. In the latter two cases, the maximum [CH$_4$] = 8 sccm was utilized. The films were deposited simultaneously on Silicon crystalline substrates with resistivity lower than 0.02 Ω cm for the current-voltage measurement, C-V, and those with resistivity higher than 1000 Ω cm for the measurement of Fourier transform-infrared spectroscopy. Quartz substrates were also used for the measurement of stress. The deposition times were adjusted to produce films with thickness of around 0.3 - 0.4 μ m for the first series of films and around 1 μ m for the second and third series. MOS diodes were fabricated by thermally evaporating Al electrodes with a diameter of 1 mm for films deposited on substrates with resistivity of 0.02 Ω cm for the measurement of C-V. The first series of films were annealed for 30 min in a N$_2$ atmosphere at a temperature, T_A, from 400 - 800 °C and the changes in the dielectric constant as a function of T_A were focused.

RESULTS AND DISCUSSION

Figure 1 shows the typical FT-IR spectra for as-deposited (measured within 30 minutes after deposition) SiO_2:F:C film over the range 400 - 4000 cm^{-1}. Basically, the spectra represent those of the SiO_2 film as shown by the dominant absorption bands arising from the stretching, bending and rocking modes of Si-O bonds found at 1050-1150, 800 and 450 cm^{-1}, respectively. Fluorine and carbon doping to the film results in the appearance of Si-F absorption band at around 930 cm^{-1} and carbon related bands like Si-CH$_3$ stretching absorption located at around 1250 cm^{-1}. Moisture-related absorption is found at around 3650 cm^{-1} due to Si(O-H) stretching mode.

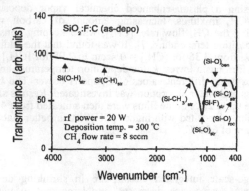

Fig. 1. FT-IR absorption spectrum of as-deposited SiO_2:F:C film. The absorption bands are shown using arrows and the deposition condition is also included.

Figure 2 shows the relative dielectric constant for as-deposited films as a function of (a) [CH$_4$], (b) T_D, or (c) rf power. As reported in a previous paper [5], the addition of carbon to SiO_2:F resulted in the decrease of k. For the present films, it was found that the addition of CH$_4$ decreases k from 3.36 to 2.95 for [CH$_4$] = 0 sccm (SiO_2:F) and 8 sccm (SiO_2:F:C), respectively. Figure 2(b) shows that k decreases slightly from 3.06 to 2.96 for T_D from 100 °C to 400 °C, respectively. The k value, however, increases abruptly to 3.50 at a temperature of 500 °C. This abrupt change is due to the disappearance of Si-F and carbon-related bonds as will be shown on later figures (Figs. 4 and 5). The third series of samples shown in Fig. 2(c) also depict a reduction in k with increasing rf power keeping T_D fixed at 300 °C and [CH$_4$] = 8 sccm. For the film with rf power = 10 W, k has a rather high value of 3.60. The lowest k value in these films was found to be 2.85 for rf power = 30 W. For rf power = 50 W, however, k increases slightly to 3.18 again.

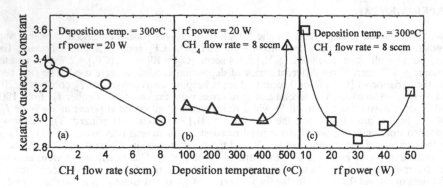

Fig. 2. Relative dielectric constant for as-deposited SiO_2:F:C films as a function of (a) CH$_4$ flow rate, [CH$_4$], (b) deposition temperature, T_D, or (c) rf power.

58

The higher values of k will be shown later to interpret in terms of lower density of Si-F or Si-CH$_3$ and other carbon-related bonds. Thus, these results suggest that the dielectric constant for the carbon-doped films were lower compared with SiO$_2$:F films. Under optimized T_D and rf power conditions, the dielectric constant was reduced further.

The concentration of Si-O bonds, N_O, was estimated from the sum of the integrated intensities, I_{SiO}, of both the 1050-1100 cm^{-1} and the 1150 cm^{-1} components of the Si-O absorption using the relationship $N_O = A_{SiO}I_{SiO}$, where $A_{SiO} = 1.5 \times 10^{19}$ cm^{-2} [6]. Figure 3 shows N_O for the films as a function of (a) [CH$_4$], (b) T_D, or (c) rf power. The N_O for films deposited with increasing [CH$_4$] or rf power, as shown in Fig. 3(a) and 3(c) decreases monotonically but the condition of increasing [CH$_4$], however, was more effective for decreasing N_O. On the other hand, increasing T_D as shown in Fig. 3(b) under fixed rf power and [CH$_4$] leads to increasing N_O.

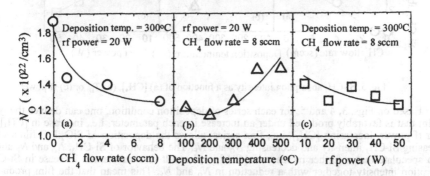

Fig. 3. Density of Si-O bonds as a function of (a) [CH$_4$], (b) T_D, or (c) rf power.

The concentration, N_F, of Si-F bonds was estimated from the intensity, I_{SiF}, of the Si-F stretching spectra around 930 cm^{-1} using the relationship $N_F = A_{SiF}I_{SiF}$, where $A_{SiF} = 3.4 \times 10^{19}$ cm^{-2} [7]. Figure 4 shows the concentration of Si-F bonds for the three different series of samples. Increases in either [CH$_4$] or T_D have the effect of a decrease in N_F as shown in Fig. 4(a) and 4(b). In the result at $T_D = 500$ °C, the N_F value becomes zero. On the other hand, for the third series of samples, the increasing rf power increases N_F (Fig. 4(c)). We can see from Figs. 3 and 4 that increases in T_D or rf power cause the opposite effects for N_O and N_F.

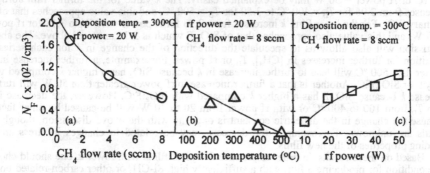

Fig. 4. Density of Si-F bonds as a function of (a) [CH$_4$], (b) T_D, (c) rf power.

Figure 5 shows the Si-CH$_3$ absorption intensity as a function of three different conditions. As expected, increasing [CH$_4$] increases the Si-CH$_3$ absorption intensity (Fig. 5(a)). However, the increase in T_D or rf power leads to the decrease in Si-CH$_3$ absorption intensity as depicted in Figs.

5(b) and 5(c). Under the highest $T_D = 500$ °C condition, the Si-CH$_3$ absorption is found to disappear.

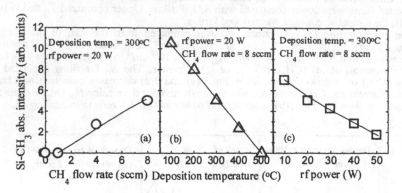

Fig. 5. Si-CH$_3$ absorption intensity as a function of (a) [CH$_4$], (b) T_D, or (c) rf power.

Based on Figs. 3, 4 and 5, for each series of deposition condition, one can deduce the type of film that is favorably produced under an increase in each parameter, i.e., increase in [CH$_4$] or T_D or rf power. Increasing [CH$_4$], i.e., first series, allows the deposition of SiO$_2$:F:C films with increasing Si-CH$_3$ intensity and decreasing N_O and N_F. The behavior of Si-CH$_3$, N_O and N_F allow us to speculate that a further increase in [CH$_4$], will result in a substantial increase in Si-CH$_3$ absorption intensity together with a reduction in N_O and N_F. This mean that the film produced becomes more like that of SiO$_2$:C as [CH$_4$] is further increased. In the second series of deposition condition, increasing T_D also produced SiO$_2$:F:C with increasing N_O and decreasing N_F and Si-CH$_3$ intensity. However, as T_D increases to 500 °C, the film deposited becomes more like that of SiO$_2$. This is confirmed from the result that at a $T_D = 500$ °C, only Si-O band can be observed and no Si-F and Si-CH$_3$ absorption bands were detected by FT-IR. For the third series of deposition condition, increasing rf power also produced SiO$_2$:F:C films with increasing N_F and decreasing N_O and Si-CH$_3$ intensity. A further increase in rf power greater than 50 W might lead to further increase in N_F together with a decrease in Si-CH$_3$ intensity and N_O. Thus, the type of film with high rf power becomes more like that of SiO$_2$:F.

The abrupt changes in the dielectric constant for films produced in extreme cases like $T_D = 500$ °C or rf power = 50 W may be explained easily. The k value for the former film abruptly increases from 2.96 for $T_D = 400$ °C to 3.50 for $T_D = 500$ °C, and this value approaches that of an ordinary SiO$_2$. For the latter film, k increases from 2.93 for rf power = 40 W to 3.18 for rf power = 50 W, and this value also approaches that of SiO$_2$:F which is 3.36 as stated above. The above discussion will also allow us to speculate the direction of the change in k for each series of condition for further increases in [CH$_4$], T_D or rf power. For example, a further increase in T_D greater than 500 °C will lead to further increase in k because SiO$_2$ has a higher k compared with SiO$_2$:F or SiO$_2$:F:C. Another is that a further increase in rf power greater than 50 W will further increase k because SiO$_2$:F has a higher k compared with SiO$_2$:F:C. However, the decrease in k with T_D from 100 to 400 °C or with rf power from 20 to 40 W will be caused by other factors because the change in the dielectric constant is opposite with the above discussion, though the details are unknown: For example, the decrease may be related to changes in stress and/or bonding properties of the new films.

Based on these results, in order to produce SiO$_2$:F:C films with lower k, one should choose the condition for producing a film with a sufficiently high Si-CH$_3$ or other carbon-related bonds together with sufficient amount of N_F or F bonded to C. If N_F is very low as in the case of rf power = 10 W (Fig. 4(c)), k appears to be very high as shown in Fig. 2(c). The decrease in k for films deposited with varying [CH$_4$] was discussed previously [8]. In that article, the well-known sum rule has been used for interpreting how the decrease in k is related to the decrease in the IR effective dynamic charge of Si-O dipoles, $e_{SiO}{}^*$, in connection with the decrease in I_{SiO}. It was suggested that an increase in A_{SiO} arising from the decrease in $e_{SiO}*$ was caused by the incorporation of F and C atoms in the films. So that, the decrease in $e_{SiO}*$ along with the direct contribution of C-F and Si-F bonds incorporated to SiO_2, can explain the decrease in k with increasing F or C.

Figure 6 shows the stress, estimated using Stoney's equation, for the three series of films. Negative values for stress denote compressive stress while positive values mean tensile stress. The increases in T_D or rf power increase the stress for the films while the increase in [CH_4] decreases the stress. As stated earlier, the decrease in k for the films deposited with T_D from 100 to 400 °C or with rf power from 20 to 40 W may be affected by the changes in stress.

Figure 7 shows effects of the water absorption in SiO_2:F:C films connected with the Si(O-H) absorption band around 3650 cm^{-1}. The FT-IR spectra were measured after exposure of the films to air for 7 days. The direction of change for the Si(O-H) absorption intensity also depends on the type of film. For the first series of condition, water absorption decrease with [CH_4] because carbon doping to SiO_2:F film increases moisture tolerance [5]. The increase in T_D also decreases Si(O-H) intensity as shown in Fig.7(b) because SiO_2 has a higher moisture resistance compared with SiO_2:F film. However, the increase in rf power increases the Si(O-H) intensity (Fig. 7(c)) because of the high water absorption property of SiO_2:F films. Thus, it is found that the increase in the Si(O-H) intensity corresponds well with the increase in N_F as observed in Fig. 4.

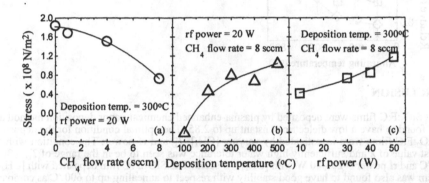

Fig. 6. Stress for the SiO_2:F:C films as a function of (a) [CH_4], (b) T_D and (c) rf power.

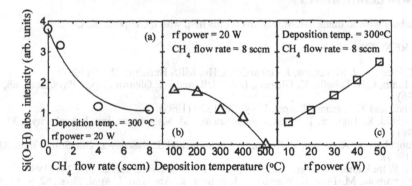

Fig. 7. Si(O-H) absorption intensity for films exposed to air for 7 days as a function of (a) [CH_4] (b) T_D, and (c) rf power.

Figure 8 which is also a result in a previous paper [8], shows the changes in k for SiO_2:F:C films with various [CH_4] values as a function of annealing temperature, T_A, from 400 - 800 °C. Note that these films were deposited with T_D = 300 °C and rf power = 20 W and were annealed just after deposition. The figure shows that the k for films with [CH_4] = 8 sccm appears to exhibit a relatively slower increase in k with increasing T_A between 300 and 500 °C, as compared with those films deposited with lower [CH_4]. This result suggests that the films deposited with higher [CH_4] were more thermally stable with respect to a change in the dielectric constant. Further annealing of the films with T_A > 500 °C results in a further increase in k. Under increasing T_A from 600 to 800 °C, k abruptly increase to 3.7 to 3.9 whose magnitude is close to the values observed for ordinary SiO_2 [9]. The abrupt increase has been discussed in a previous article [8] and was interpreted in terms of desorption of F and C.

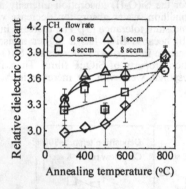

Fig. 8. Relative dielectric constant for SiO_2:F:C films deposited with different CH_4 flow rate as a function of annealing temperature. Deposition temperature was fixed at 300 °C and rf power at 20 W.

CONCLUSION

SiO_2:F:C films were deposited by plasma-enhanced chemical vapor deposition method and were found to have a low dielectric constant up to 2.85. The optimal condition for the deposition of SiO_2:F:C films were investigated by varying [CH_4], T_D and rf power. The condition with the lowest value of k and sufficiently high water tolerance was seen to be in the range of T_D = 300 - 400 °C and rf power = 20 - 40 W with [CH_4] = 8 sccm. The SiO_2:F:C film deposited with [CH_4] = 8 sccm was also found to have good stability with respect to annealing up to 600 °C as compared to films deposited with lower [CH_4] values.

ACKNOWLEDGEMENTS

The authors wish to thank Mr. Masuyama for his help with the experiments.

REFERENCES

1. E. T. Ryan, A. J. Mckerrow, J. Leu and P.S. Ho, MRS Bulletin, **22**, 49 (1997).
2. S.J. Limb, C.B. Labelle, K. Gleason, D.J. Edell, and E.F.Gleason, Appl. Phys. Lett., **68**, 2810 (1996).
3. K. Endo and T. Tatsumi, J. Appl. Phys., **78**, 1370 (1995).
4. Y. Uchida, K. Taguchi, T. Nagai S. Sugahara and M. Matsumura, Jpn. J. Appl. Phys., **37**, 6369 (1998).
5. J. Lubguban Jr., A. Saitoh, Y. Kurata, T. Inokuma, and S. Hasegawa, Thin Solid Films, **337**, 67 (1998).
6. L He, T. Inokuma, Y. Kurata and S. Hasegawa, J. Non-Cyrst. Solids, **185**, 249 (1995).
7. K. Yamamoto, M. Tsuji, K. Washio, H. Kasahara, K. Abe, Jpn. J. Appl. Phys., **52**, 925 (1983).
8. J. Lubguban Jr., Y. Kurata, T. Inokuma and S. Hasegawa (unpublished).
9. A.C. Adams, Solid State Technology, (1983), p. 135.

ANALYSIS OF A TEOS / OXYGEN PLASMA: INFLUENCE OF ENERGY AND PARTICLE FLUX ON THE DEPOSITION PARAMETERS

M. L. Pereira da Silva[*], A. N. Rodrigues da Silva[*], J. J. Santiago-Aviles [**]
*Laboratório de Sistemas Integráveis Departamento de Engenharia Eletrônica USP, SP, BRAZIL.
**University of Pennsylvania, Dept. of Electrical Engineering, 200 S, 33rd. St. Philadelphia, PA 19104, santiago@ee.upenn.edu

ABSTRACT

In this work we looked at deposition parameters such as rate and film microstructure during the plasma enhanced CVD processing of TEOS $\{Si (OCH_2 CH_3)_4\} + O_2$. This is a multi-component oxide due to the inevitable accidental inclusion of carbonatious contaminants. We decided to characterize the plasma parameters , namely the electron energy (from 50 to 600 eV), and the flux of the reactive species. The deposition chamber was modified by the introduction of a stainless steel tubular ring between the electrodes, such that when a positive bias is applied it is possible to inject electrons into the plasma. A dual role for the tubular ring is the transport of oxygen to different locations in the plasma, and to monitor the influence of the oxygen flux on deposition. The experimental results showed that by increasing the voltage bias on the ring from 0 to 600 V, the oxide deposition rate is enhanced. For low precursor concentration (TEOS), the deposited films shows micro-structural improvement as evidenced by FTIR spectroscopy. We monitored the formation of carbon compounds by their RAMAN signature, and one can see that for small ring bias, the concentration of carbon contaminants is large and it decreases with increasing electron energy. A reaction model consistent with our experimental results must consider the need of oxygen ions in the oxidation of the precursor. Since an increase in the electron flux hinders the formation of oxygen ions, a pressure decrease must be utilized to improve the chemical properties of the film such as the formation of carbonatious contaminants. The inescapable conclusion is that higher density plasmas favor the processing of organo-silicates.

INTRODUCTION

Oxide films deposition utilizing an oxygen plasma and e organosilicate such as TEOS, is a common practice in micro-electronics and micro-fabrication [1-3]. This oxide precursor is particularly utilized for its good step coverage [4-5] an essential characteristic for interconnections and multi-level technology. Even though TEOS technological importance may be high, its reaction mechanisms during plasma deposition are not yet well understood. Previous studies show that carbon compounds incorporation in the film are possible and that these compounds degrade the film electrical characteristics as well as preventing the complete oxidation of the precursor molecule [6-9]. These studies do not usually consider the electron energy effects on the chemical reactions, particularly its effect on the multiple ion / molecule interactions that occur during deposition[10-12]. This article reports our findings characterizing the main features of the films when two of the most important plasma parameters are systematically modified, namely the electrons energy and the gas flowing into the plasma chamber, in particular oxygen.

Mat. Res. Soc. Symp. Proc. Vol. 606 © 2000 Materials Research Society

EXPERIMENTAL

Set-up and Instrumentation

The experimental set-up has been previously described elsewhere [7]. It is fundamentally a home built multi-chambered plasma enhanced CVD reactor. For the plasma characterization measurements, the plasma (reaction) chamber was modified by inserting an electrically insulated copper ring as shown in figure 1. It is possible to bias this ring with potentials from –600 to + 600 volts, which allow us to modulate the concentration and kinetic energy of the plasma electrons. Our initial results suggest that reactions in the gas phase are suppressed by low plasma electrons energy, and as a consequence reactions in the substrate surface predominates. To validate this thesis we decided to explore if the predominance of these substrate surface reactions was due to oxygen species difficulties in crossing the plasma sheath and adsorbing on the substrate surface, or due to the low energy input to the plasma. To study this, we substituted the copper ring for a 1/8" SS tube with 1 mm holes an inch apart and oriented towards the electrode where the substrate lies. Using this configuration an oxygen flux of up to 500 sccm can be allowed to impinge on the substrate. So we can introduce oxygen in two different places, one close and another relatively far from the substrate, and modulate both the oxygen input and electrons energy individually and simultaneously.

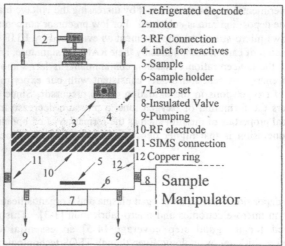

1-refrigerated electrode
2-motor
3-RF Connection
4- inlet for reactives
5-Sample
6-Sample holder
7-Lamp set
8-Insulated Valve
9-Pumping
10-RF electrode
11-SIMS connection
12 Copper ring

Sample
Manipulator

Figure 1. Schematic of plasma reaction chamber design.

Films were deposited on 3 " p-type, <100>, 10-20 Ohms Si wafers under the following deposition conditions: Electrode temperature of 350C, Chamber pressure of 1 Torr, TEOS flux of 10^{-2} NI, O_2 flux of $2.5x10^{-1}$ to $5x10^{-1}$ NI, Ar flux of 10^{-2} NI, Power from 200 to 400 W, distance between the electrodes of 2.5 cm, and distance from ring to substrate of 1.5 cm. Argon was utilized when optical characterization was desired, and was used as reference (actinometry). The deposited films were analyzed using the following characterization techniques, namely FTIR, Raman and ellipsometry..

RESULTS

Electrons Energy Analysis: in Gaseous Phase

The optical spectroscopy analysis show that for this system the addition of TEOS precursor to the plasma quench the overall plasma emission intensity. A plausible explanation is the reduction in electrons energy by their inelastic collision with TEOS molecules and the formation of precursor ions. A positive bias to the copper ring (inducing the removal of electrons from the plasma) did not significantly altered the $Ar + O_2$ spectra as shown in figure 2a, That was not the case for the mixture of $Ar + O2 + TEOS$, as shown in figure 2b. In this case as seen by normalizing the traces of O_2 and CO/CO_2 by the Ar trace one can note an increase in the CO/CO_2 and a decrease of the O_2 emission intensity, which possibly indicates that the TEOS + O_2 occurs in low electron density (concentration) conditions.

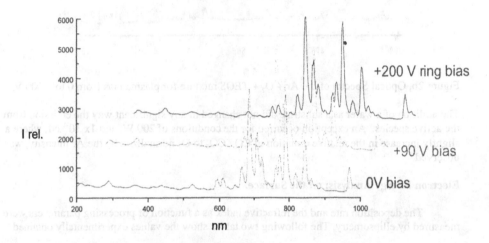

Figure 2a, Optical spectra of the $Ar + O_2$ mixture when the plasma was biased with potentials from 0 to +200 V.

When we apply a negative bias (introducing electrons into the plasma) a significant perturbation is observed in the plasma, mostly when the bias exceed the –400 V. Under these conditions the mixture behaves as if the electron input favors the emission from certain particular species, mostly CO / CO_2 . This may indicate that the energy increment was utilized almost exclusively for the TEOS + O_2 reaction.

Of particular interest is the effect on the plasma as one varies the TEOS concentration. Experiments with input flows in the range from 2.5×10^{-3} to 5×10^{-2} NI , power input of 300W and ring bias of –600V show that TEOS addition reduces the O_2 emissions and favors that of CO / CO_2 up to concentrations of 2×10^{-2} NI as the oxidation reaction tends to be enhanced under these conditions. At concentrations of 2.5×10^{-2} NI the O_2 emissions re-appear, perhaps indicating difficulties for the reaction to occur at high TEOS concentrations. A plausible explanation is the diminishing high energy electron concentration due to TEOS decomposition.

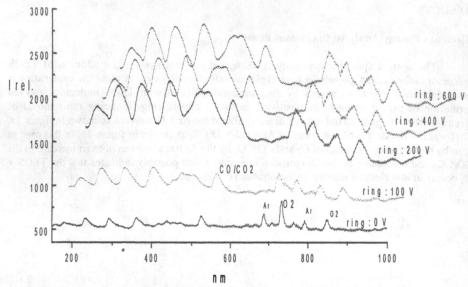

Figure 2b, Optical Spectra of the $Ar + O_2 + TEOS$ mixture for plasma bias from 0 to + 600 V

The addition of a wafer as a substrate did not changed in any significant way the emission from the active species. An exception occurred for the conditions of 200 W, and $1x\ 10^{-2}$ NI, where a slightly increase in the relative emission of CO / CO_2 to O_2 normalized by the Ar intensity, was observed.

Electron Energy Analysis: on the Surface.

The deposition rate and the refractive index as a function of processing parameters were measured by ellipsometry. The following two tables show the values experimentally obtained.

Conditions:			
For 300 W			
Ring Voltage (V)	**Refractive Index**	**Deposition rate (nm/min)**	
		center	periphery
0	1450	180	173
600	1451	241	236
For 200 W			
Ring Voltage (V)	**Refractive Index**	**Deposition rate (nm/min)**	
		center	periphery
0	1450	221	208
600	1453	310	208

Table 1.   Deposition rate and refractive index as a function of the applied bias for 0.01 NI of TEOS and 0.25 NI of O_2, 350 C, and 1 Torr.

Conditions: 0.250 Nl of O_2, 350 °C, 1 Torr, 300 W, 600 V ring bias		
For 0.005 Nl of TEOS	Refractive index	Deposition rate (nm/min) center periphery
	1.454	89.3	89.8
For 0.020 Nl of TEOS	Refractive index	Deposition rate (nm/min) center priphery
	1.465	134	135
For „025 Nl of TEOS	Refractive index	Deposition rate (nm/min) center periphery
	1.460	270.4	295.7

Table 2.   Deposition rate and refractive index as a function of the TEOS concentration for 0.24 NI of O_2 , 350 C , 1 Torr, 300W and 600 V bias to the ring.

Conditions: 0.010 Nl of TEOS, .250 Nl of O_2, 350 °C, 1 Torr	
For 300 W	
Ring bias (V)	Raman Scattering Microscope observations
0	Scattered graphite nodules were observed on the surface, Some of them fairly substantial, as large as 50 - 70 μm diameter.
600	No graphite nodules observed
For 200 W	
0	Graphite nodules in quantities larger than for 300 W were observed, although of a smaller size, with diameters in the range from 1-5 μm.
600	No nodules detected, except a few small ones far in the periphery.

Table 3.Characteristics of deposited film as observed by a Raman Scattering Microscope, these parameters were measured as function of bias for 0.010 Nl of TEOS, 0.250 Nl of O_2, 350 °C, 1 Torr.

Conditions: 0.250 Nl of O_2, 350 °C, 1 Torr, 300 W, 600 V ring bias	
For 0.005 Nl of TEOS	Raman Scattering Microscope observations
Small nodules concentrations, mostly in the periphery, they may be related to defects on the substrate, all their diameters smaller than 5 μm.	
For 0.020 Nl of TEOS	Raman Scattering Microscope observations
Some nodules start appearing in the center, the periphery concentration substantially increases. All small nodules (1-5 μm.)	
for 0.025 Nl of TEOS	Raman Scattering Microscope observations
Homogeneous distribution through the sample surface of clear and opaque nodules.	

Table 4: Raman Scattering Microscope measurements as a function of TEOS concentration da for 0.250 Nl of O_2, 350 °C, 1 Torr, 300 W, 600 V ring bias.

Note that the ring voltage variations does improve the deposition rate, although is does not significantly influence the refractive index, suggesting that the input of energetic electrons may be increasing the concentration of chemically active species positively influencing the deposition.

The Raman scattering measurements gave us a better perspective of the phenomena. The results are summarized in table 3. In this case we observe that for low ring bias, a bias voltage increase may supply electrons with high enough energy as to directly oxidize the TEOS molecule in the gaseous phase. It may be that the electrons mediate in the production of other active species that react on the solid surface; this way eliminating carbon radicals that may lead to the generation of graphite nodules on the surface.

CONCLUSIONS

On the gaseous phase, the addition of TEOS seems to severely disturb the species energy distribution in the chamber, and at the same time seems to act as a buffer, reducing the possibility of large variations (with the exception of large TEOS concentration). An increase in electrons concentrations apparently do not favor the formation of chemically active Oxygen species, as we observed in our spectroscopic measurements. As a consequence the reduction in pressure usually lead to better film chemical characteristics, primarily the incorporation of carbon. These experiments suggest to us that higher density plasma may be more adequate for the deposition of organo-silicates.

For the solid surface, the increase in ring bias during deposition lead to the formation of larger amounts of active species, with a consequence of an increase from 30 to 49% in the deposition rate. The increase in plasma power (from 200 to 300 W) was not reflected in a proportional deposition rate, although surface phenomena seems to be of importance, as graphite nodules were nucleated and grown with a severity seemingly proportional to the substrate defects density. The frequency and size of the graphite nodules was reduced by the application of an increasing ring bias, perhaps indicating that energetic electrons mediates the TEOS oxidation reaction.

REFERENCES

1. K. Petersen, "*MEMS: What lies ahead?*", Digest of Technical papers, Transducers '95, Eurosensors IX, June 25 - 29, 1995, Stockholm, Sweden, Vol. 1, 1995, 894 - 897.
2. A. Heuberger, "*Silicon Microsystems*", Microelectronic Engineering 21 445 – 458 (1993).
3. J. N. Zemel e R. Furlan, "Microfluidics", Chapter 12 of "Handbook Of Chemical and Biological Sensors", Richard F. Taylor, Jerome S. Schultz, Ed. Institute of Physics Publishing, Bristol and Philadelphia, , pp. 317 – 347, (1996).
4. L. Chin, E. P. van de Ven, *Solid State Technol.*, April, 119 (1988).
5. C. G. Magnella, T. Ingwersen, *V-MIC Conf. Proc.*, June, 366 (1988)
6. M. L. P. Silva, A. R. Cardoso, *Revista Brasileira de Aplicações de vácuo*, **16**, 2 (1997) 39
7. M. L. P. Silva, A. R. Cardoso, J. J. Santiago- Aviles, *MRS Fall Meeting*, 1997
8. A. R. Cardoso, M. L. P. Silva, J.J. Santiago- Aviles, *MRS Fall Meeting*, 1997
9. M. Silva..J.M. Riveros, , *Int. J. Mass Sprectrom. Ion Process*, **165/166** 83-95 (1997).
10. N. H. Morgon, A. B. Argenton, M. L. P. Silva, J. M. Riveros, *J. Am. Chem. Soc.*, **119**, 1708 (1997).
11. , M. Silva, J.M. Riveros, *J. Mass Spectrom.* **30(5)**, 733-740 (1995)
12. M. Silva, J.M. Riveros, *Proceedings of X SBMicro*, **1**, 567-576 (1995).

METAL ORGANIC CHEMICAL VAPOR DEPOSITION
OF Co-, Mn-, Co-Zr AND Mn-Zr OXIDE THIN FILMS

D. BARRECA [*], F. BENETOLLO [**], M. BOZZA[**], S. BOZZA[**], G. CARTA[**],
G. CAVINATO [*], G. ROSSETTO[**] and P. ZANELLA[**]
[*]DCIMA and CSSRCC-CNR, University of Padova, Via Marzolo 1, 35131 Padova, Italy
[**]ICTIMA-CNR, Corso Stati Uniti 4, Padova 35127 Italy

ABSTRACT

Deposition of thin films of Co- and Mn- oxides as well as of their mixtures with ZrO_2 have been carried out by MOCVD using $Co(C_5H_5)_2$, $Mn(C_5F_6HO_2)_2(THF)_2$ and $(C_5H_5)_2Zr(CH_3)_2$ as precursors. XRD and XPS analyses of the obtained deposits are reported. Introduction of water vapor into the reactor chamber during the flow of the precursors improved their decomposition efficiency and the quality of the films.

INTRODUCTION

Thin films of materials based on Co and Mn oxides as well as on their mixed oxides with ZrO_2 are attracting increasing scientific interest due to their wide applications in modern technology. Light batteries and advanced environmental catalysts are significant examples [1,2]. In order to synthesize high quality materials one of the most suitable methods is MOCVD. However, while this technique has been currently adopted for the deposition of ZrO_2 thin films [3], quite a few reports concerning the deposition of Co and Mn oxides and especially of the Co-Zr and Mn-Zr oxide mixtures have been discussed.

In this paper we describe the vapor phase deposition of Co and Mn single oxides by using respectively $Co(C_5H_5)_2$ and $Mn(C_5F_6HO_2)_2(THF)_2$, which to our knowledge, have been never used before [$Co(C_5H_5)_2$ had been used for Co metal and Co silicide deposition] [4,5]. The same compounds together with $(C_5H_5)_2Zr(CH_3)_2$ have been used for the deposition of the $Co-ZrO_2$ and $Mn-ZrO_2$ mixed oxides.

EXPERIMENT

$Co(C_5H_5)_2$ and $Mn(C_5F_6HO_2)_2(THF)_2$ has been obtained by following procedure well developed in our institute [6]. The MOCVD system was a normal hot walls apparatus and the growth parameters were adjusted as reported in the table I.

Table I. Growth condition for Co-, Mn-, Zr-oxides.

Material	Co oxide	Mn oxide	ZrO_2-Co oxide	ZrO_2-Mn oxide
Source temperature	60 °C	115 °C	85 °C	110 °C
Pressure	19.4 Torr	5 Torr	0.6 Torr	0.75 Torr
N_2 carrier flux	8 scc/min	8 scc/min	8 scc/min	8 scc/min
O_2-H_2O flux	25 scc/min	25 scc/min	25 scc/min	25 scc/min
Growth time	4 hr	1 hr	1.5 hr	40 min
H_2O consumption	1.83 g	1.10 g	2.12 g	0.62 g
Substrates	Polished stainless steel, Si (100), glass	Aluminum, glass	Si (100)	Si (100)

69

RESULTS

Deposition of Co-oxide thin films

Co(C$_5$H$_5$)$_2$ has been preferred as precursor to CO- based volatile compounds in view of its appreciable volatility [4,5] and absence of toxicity. It can be easily prepared and purified by sublimation. Our deposition experiments have been carried in the 450-550 °C temperature range. The best results have been obtained at 475 °C. It must be noted that only at a pressure higher than 19 Torr an oxide deposit was obtained, while at lower pressure the precursor was collected unchanged in a cooled trap after the reaction chamber. Moreover by using only dry O$_2$ as a reactant gas the deposits were inhomogeneous; their quality improved substantially by using a mixture of O$_2$-H$_2$O vapor instead of pure O$_2$. The deposits resulted nanocrystalline and were characterized unequivocally as Co$_3$O$_4$ with no evidence of CoO present as shown by XRD analysis [Fig. 1]. The substrates influenced both the crystallographic orientation and the dimension of the crystallites [23 nm on stainless steel; 50 nm on silicon (100) and 63 nm on glass]; see table II.

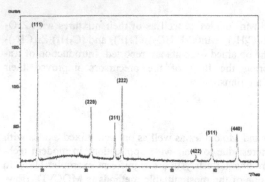

Fig. 1. XRD pattern of Co$_3$O$_4$ deposited at 475 °C on glass.

Table II. XRD for Co$_3$O$_4$ film on different substrate.

Crystall. Planes	I (%) Stainless Steel	I (%) Si (100)	I (%) Glass
(111)	30.57	17.06	100.00
(220)	47.60	32.71	38.54
(311)	100.00	100.00	27.87
(222)	15.55	10.10	53.97
(400)	12.06	14.15	0.68
(422)	8.38	7.24	4.41
(511)	21.64	21.51	18.67
(440)	23.56	25.16	17.13

The XPS spectra [Fig. 2] show the presence of only carbon, cobalt and oxygen in the film surface, but after a 5 min. sputter cleaning the C atomic percentage is reduced from about 30% to below the XPS detection limit. This proves that carbon arises only by atmospheric contamination and is not due to residuals of undecomposed precursor incorporated in the oxide matrix. Hence, this confirms that the Co(C$_5$H$_5$)$_2$ has a clean decomposition pattern in presence of oxygen and moisture. The Co 2p surface peak shows a shape and BE position similar for all the cobalt oxide films [Fig. 3]. The BE (780.1-780.3 eV) values are close to those already reported

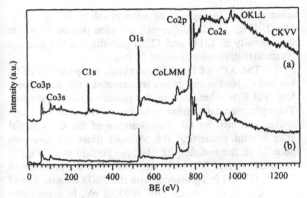

Fig. 2. XPS wide scan spectra of a Co_3O_4 thin film deposited on glass: (a) as grown; (b) after 5' sputter cleaning. Note that the C 1s peak disappears.

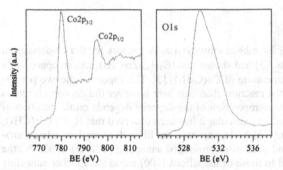

Fig. 3. XPS Co 2p and O 1s surface peak for a Co_3O_4 film deposited on glass at 475 °C in an O_2-H_2O mixture.

Deposition of Mn-oxide thin films

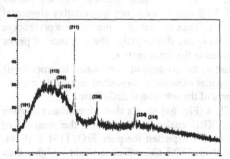

Fig. 4. XRD pattern of γ-Mn_3O_4 deposited on glass at 475 °C.

for Co_3O_4 [7]. The presence of this species which is the unique Co-O phase is confirmed by the low intensity of the shake-up satellites at BE which are greater than 9 eV with respect to the principal peaks [8,9]. Furthermore, the BE distance between the Co $2p_{3/2}$ and Co $2p_{1/2}$ components is 15.5 eV, in good agreement with the literature data for Co_3O_4 [10]. The O 1s peak [Fig. 3] is distorted towards higher BEs due to the presence of hydrated species (-OH and H_2O) adsorbed on the surface [11], which may arise from the presence of water vapor in the reaction atmosphere. The main component at 529.8 eV is due to the oxygen of oxide [12]. The O/Co surface ratio (ca. 2) is greater than expected due to the presence of hydration. Results concerning the deposition of Co oxide(s) generally agree on the composition Co_3O_4, which is also the thermodynamically stable form at room or moderately high temperatures.

Here manganese bis(hexafluoro)acetylacetonate bis tetrahydrofurane was used as precursor, it can be quantitatively sublimed without decomposition (80-110 °C temperature, 0.4 Torr pressure). In the mass spectrum the higher mass peak corresponds to the $Mn(C_5F_6HO_2)_2^+$ ion, evidently originated after dissociation of the two coordinated THF molecules. The decomposition subsequently proceeds through dissociation of organic fluorinated fragments to the metal ion Mn^{2+}. The films deposited from $Mn(C_5F_6HO_2)_2(THF)_2$ and a mixture of O_2-H_2O vapor at 475 °C on glass

71

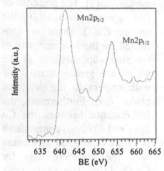

Fig. 5. *XPS Mn 2p surface peak for a Mn₃O₄ film deposited on Al at 475 °C.*

substrate were composed of nanocrystalline tetragonal γ-Mn₃O₄ [Fig. 4] with differently oriented planes, in which the intensity of (211) and (220) prevails and the average crystallites dimension is about 20 nm.

The XPS of these films reveals the presence of O, Mn and C. No fluorine peaks are detected, indicating therefore that F is eliminated in the precursor decomposition. The carbon contamination is limited to the outermost layers, as confined by the disappearance of the C 1s signal after a mild sputtering; the obtained films are therefore pure. In all the samples the Mn $2p_{3/2}$ surface peak [Fig. 5] has a BE around 641.4 eV, which confirm the presence of Mn₃O₄ [11,13] in agreement with the XRD results. The O 1s peak main component falls at 530.2 eV, in accordance with the literature [11]. In this case the presence of hydrated components at higher BEs is much more enhanced with respect to pure Co₃O₄ thin films: in fact, the O/Mn surface ratio is about 3.

Co-deposition of Co-ZrO₂

Co(C₅H₅)₂ and (C₅H₅)₂Zr(CH₃)₂ have been employed as precursors for the co-deposition of Co and Zr thin films. Previous results [3] had shown that (C₅H₅)₂Zr(CH₃)₂ has an appreciable vapor pressure in the same range of temperature of Co(C₅H₅)₂ [3]. The vapors of the two precursors do not interfere before mixing in the reaction chamber; here however the deposition rate is different for the two precursors, thus the composition of the deposits depends on the position of the substrate inside the reaction chamber. In particular it has been observed that (C₅H₅)₂Zr(CH₃)₂ decomposes in greater amount than Co(C₅H₅)₂. In any case the films deposited on silicon substrate at 450 °C resulted amorphous and even after prolonged annealing (4 hour) at 600 °C the only XRD peaks detected corresponded to those of the silicon (100) substrate. Further annealing at 700, 800 and 900 °C caused increasing crystallization of the deposits and evolution of crystalline species. At 700 °C the film was mainly composed of t-ZrO₂ [(111), (220) and (311)] but a modest peak characteristic of m-ZrO₂ (-202) is also present together with a peak which could be indifferently be ascribed to CoO (111) or Co₃O₄ (311) as the these peaks have almost the same value. Increasing the temperature has no significant effects on the plane orientation [except the appearance of new peaks related to Co₃O₄ (220) and CoO (220) orientations], but influences the t-ZrO₂ crystallite dimensions (14 nm at 700 °C, 23 nm at 800 °C and 29 nm at 900 °C). No increase in the fraction of m-ZrO₂ phase is observed. XPS and fluorescence quantitative elemental analysis indicate the presence of 63 % Zr and 37 % Co. Thus the cobalt is present at a percentage far higher than its solubility limit in ZrO₂ (< 15%) [14] and this explains the presence of peaks corresponding to Co₃O₄ and CoO species not dissolved in the ZrO₂ matrix.

Two aspects appear worthy of consideration: a) the co-deposit is obtained in amorphous phase at the same temperature at which the single oxides separately deposited appear in crystalline form; b) the Co-oxides act as efficient stabilizers of the tetragonal ZrO₂ form.

The XPS surface survey for the Co-ZrO₂ films [Fig. 6a] shows that adventitious carbon percentage is moderately low (~30%). Even if the XRD analyses have shown the presence of crystalline ZrO₂, the Zr $3d_{5/2}$ peak BE is lower than that expected for pure ZrO₂ (181.2 eV vs. 182.2 eV) [11]. The Co $2p_{3/2}$ principal line falls at about 780.3 eV, which is slightly higher than the reported value for Co₃O₄ [8,9]. Moreover, the shake-up satellites (S) are now more intense than in the case of a pure Co₃O₄ film [Fig. 7] and fall at BE higher than 6 eV with respect to the

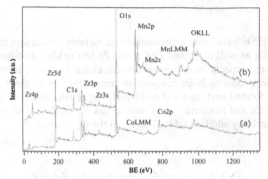

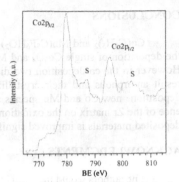

Fig. 6. XPS wide scan spectra of mixed oxides thin films: (a) CoO-Co₃O₄-ZrO₂ (as grown) deposited at 450 °C on Si (100); (b) MnO₂-ZrO₂ deposited at 475 °C on Si (100).

Fig. 7. XPS Co 2p surface peak of a mixed oxide thin film (Co-Zr-O) deposited at 450 °C on Si (100). The shape-up satellites indicate the presence of CoO.

spin-orbit components [Fig. 3] [7]. The intensity of the shake-up satellites and their position are used as a fingerprint for the recognition of the Co (II) high spin centers in CoO [15].

The BE distance between the Co $2p_{3/2}$ and Co $2p_{1/2}$ components is 15.8 eV, which is an intermediate value between that of CoO (16 eV) [7a, 7c] and that of Co₃O₄ (15.5 eV) [10].

Therefore, it can be concluded that a mixture of the two cobalt oxides is present in the thin film sample, in accordance with the XRD results. The shift of the Zr 3d peak at lower BE and the shift of the Co 2p lines at higher BE might indicate the presence of a Co-O-Zr interaction, with a high ionic character of the Co-O bond [16]. This effect is in accordance with the presence of both Co₃O₄ and CoO which was not detected in the single phase Co-O films. Cobalt monoxide may act as a stabilizing agent towards the t-ZrO₂ phase (XRD). Quantitative Co/Zr analyses indicate the presence of 63% Zr and 37% Co.

Co-deposition of Mn-ZrO₂

Co-deposition of Zr and Mn oxide has been carried out with $Cp_2Zr(CH_3)_2$ and $Mn(C_5F_6HO_2)_2(THF)_2$ precursors. As observed for Co-Zr system, also in the case of Mn-Zr the deposited layer at 475 °C result amorphous. Annealings at 600, 700, 800 and 900 °C causes progressive crystallization. XRD indicates the presence of ZrO₂, mainly tetragonal but also monoclinic, as well as of MnO₂ [peaks (101) and (210)] with traces of other Mn oxides. While the rise of temperature causes increase of t-ZrO₂ crystallite dimensions, (4 nm at 700 °C, 14 nm at 800 °C and 22 nm at 900 °C for tetragonal and 15 nm at 800 °C and 24 nm at 900 °C for the monoclinic) it does not seem to alter the related amount of the monoclinic and the tetragonal phases.

The XPS surface survey for a Mn-Zr-O mixed film on Si (100) (as grown) is reported in Fig. 6b. Carbon contamination is very low (about 15%) already on the surface. Differently from the case of the Co-Zr-O samples, the Zr $3d_{5/2}$ peak is located at 182.2 eV, which is exactly the literature value for ZrO₂ [11]. The Mn $2p_{3/2}$ photopeak is found at 642.2 eV, which is the BE of MnO₂ [11]. These results are in excellent agreement with those of the XRD analyses and seem to indicate the absence of Mn-O-Zr interaction. Therefore, MnO₂-ZrO₂ nanocomposites have been obtained. Quantitative Co/Mn analyses indicated the presence of 25% Zr and 75% Mn.

CONCLUSIONS

$Co(C_5H_5)_2$ and $Mn(C_5F_6HO_2)_2(THF)_2$ have been demonstrated as suitable precursors for the deposition of single Co_3O_4 and Mn_3O_4 as well as mixed Zr-Co and Zr-Mn oxide thin films. However in the co-deposition two different aspects have been observed: a) the as grown layers result amorphous and their crystallization requires high temperature annealing; b) in the co-deposition new Co and Mn species are formed such as CoO and MnO_2 suggesting some influence of the Zr matrix on the oxidation of Co and Mn ions. In all depositions the quality of all the deposited materials is improved significantly by addition of H_2O vapor to the O_2 reactant gas.

ACKNOWLEDGMENTS

The authors would like to acknowledge Mr. A. Aguiari for his technical assistance.

REFERENCES

1.   M. Isai, K. Yamaguchi, H. Iyoda, H. Fujiyasu, Y. Ito, J. Mater. Res. **14**, 1653 (1999).
2.a. H.C. Zeng, J. Lin, W.K. Teo, J.C. Wu and K.L. Tan, J. Mater. Res. **10**, 545 (1995).
  b. J. Ma, G.K. Chuah, S. Jaenicke, R. Gopalakrishnan and K.L. Tan, Berichte der Bunsen Gesells chaft- Physical Chemical Chemistry **100**, 585 (1996).
3.   S. Codato, G. Carta, G. Rossetto, G.A. Rizzi. P. Zanella, P. Scardi and M. Leoni, Chem. Vap. Deposition **5**, 159 (1999).
4.   G.J.M. Dormans, J. of Crystal Growth **108**, 806 (1991).
5.   G.J.M. Dormans, G.J.B.M. Meekes and E.G.J. Staring, J. of Crystal Growth **114**, 364 (1991).
6.   Our manuscript in preparation.
7.a. J. Haber and L.Ungier, J. Electron Spectrosc. Relat. Phenom. **12**, 305 (1977).
  b. C.A. Strydom and H.J. Strydom, Inorg. Chim. Acta **159**, 191(1989).
  c. N.S. McIntyre, D.D. Johnston, L.L. Coatsworth, R.D. Davidson and J.R. Brown, Surf. Interf. Anal. **15**, 265 (1990).
8.   D.C. Frost, C.A. McDowell and I.S. Woolsey, Mol. Phys. **27**, 1473 (1974).
9.   V.S. Jiménez, J.P. Espinós and A.R. González-Elipe, Surf. Interf. Anal. **26**, 62 (1998).
10.   C.V. Schenck, J.G. Dillard and J.W. Murray, J. Colloid Interf. Sci. **95**, 398 (1983).
11.   J.F. Moulder, W.F. Stickle, P.E. Sobol and K.D. Bomben, *Hanbook of X-Ray Photoelectron Spectroscopy*, J. Chastain, Perkin Elmer Corporation, Eden Prairie, MN, USA.
12.   S.J. Cochran and F.P. Larkins, J. Chem. Soc. Faraday Trans. 1 **82**, 1721 (1986).
13.   V. Di Castro and G. Polzonetti, J. Electron Spectrosc. Relat. Phenom. **48**, 117 (1989).
14.   P. Wu, R. Kershaw, K. Dwight and A. Wold, Mat. Res. Bull. **23**, 475 (1988).
15.   N.S. McIntyre, D.D. Johnston, L.L. Coatsworth, R.D. Davidson and J.R. Brown, Surf. Interf. Anal. **15**, 265 (1990).
16.   H.C. Zeng, J. Lin, W.K. Teo, J.C. Wu and K.L. Tan, J. Mater. Res. **10**, 545 (1995).

CHEMICAL VAPOR DEPOSITION OF CONFORMAL ALUMINA THIN FILMS

BRADLEY D. FAHLMAN [a] and ANDREW R. BARRON [a,b*]
[a] Department of Chemistry, Rice University, Houston, Texas 77005
[b] Department of Mechanical Engineering and Materials Science, Rice University, Houston, Texas 77005

ABSTRACT

Deposition of highly conformal alumina thin films has been carried out by hydrolysis of the liquid alane precursor, $AlH_3(NMe_2Et)$. Deposition onto Si wafers, quartz and carbon fibers were all carried out utilizing a hot-wall atmospheric pressure chemical vapor deposition (APCVD) system, while deposition onto ceramic particles was accomplished in a simple fluidized-bed APCVD reactor. Films were characterized by SEM, microprobe and electrical conductivity measurements. Growth rates were on the order of 40 - 80 Å.min^{-1} at 165 °C. The conformality of the films was illustrated using silicon wafers that were etched prior to deposition.

INTRODUCTION

Alumina thin films are widely utilized in the microelectronics industry due to their high chemical inertness, large thermal conductivity and high radiation resistance.[1,2] The efficiency of alumina in forming ion diffusion barriers allows these coatings to be used in devices such as LCDs, electroluminescent displays and solar cells. Other examples for alumina films in microelectronic applications include passivation of bipolar devices, buffer layers in silicon-on-insulator (SOI) devices.[3] For the applications of insulating materials as isolation layers, an important consideration is step coverage: whether a coating is uniform with respect to the surface. Uniform step coverage results when reactants or reactive intermediates are able to migrate rapidly along the surface before reacting. When the reactants adsorb and react without significant surface migration, deposition is dependent on the mean free path of the gas and minimal surface migration occurs. For insulators, in general low pressure chemical vapor deposition (LPCVD) has highly uniform coverage, in contrast, atmospheric pressure CVD (APCVD) and plasma enhanced CVD (PECVD) offer generally poor step coverage.[4] There is a need therefore, for a CVD process for alumina thin films with good conformal coverage and low temperatures.

Precursors such as $AlMe_3$, $Al(acac)_3$, and $AlR_x(O^iPr)_{3-x}$, have all been utilized along with various reactive gases (e.g., O_2, N_2O and air) in both APCVD and LPCVD of alumina thin films.[5,6,7,8] Deposition temperatures for APCVD techniques are typically between 300 - 900 °C.[9] For comparable techniques, the utilization of N_2O instead of O_2 decreases the deposition temperature by as much as 200 °C,[10] while LPCVD usually occurs at temperatures slightly below those of the equivalent APCVD system.

The aluminum hydride complex, $AlH_3(NMe_2Et)$, has been successfully utilized for the CVD of pure metallic aluminum films.[11,12,13] This precursor is attractive since it is a liquid at room temperature which allows for the facile control of transport properties relative to solid precursors,[14] and the lack of Al-C bonds allows for low carbon incorporation into the resulting films. Further, the liquid is non-pyrophoric (though reacts readily with moisture) and possesses high volatility (ΔH_{sub} = 28.5 kJ.mol^{-1}).[15] Herein we report a method of utilizing this precursor under controlled hydrolytic conditions for the APCVD of conformal alumina thin films.

RESULTS AND DISCUSSION

Growth of alumina films was carried out in a hot-wall APCVD system similar to that previously described, see Experimental.[16] Table 1 summarizes the results using silicon wafers as substrates. Above 170 °C, $AlH_3(NMe_2Et)$ precursor thermally decomposes yielding only

---

* To whom correspondence should be addressed (http://python.rice.edu/~arb/Barron.html)

Table 1. APCVD of alumina onto silicon wafers.

deposition temp. (°C)	precursor flow (L.min⁻¹)	water flow (L.min⁻¹)	comments
155	0.50	0.80	none
165	0.40	0.80	no white film, some birefringence
165	0.50	0.80	birefringence, no white film
165	0.63	0.86	some birefringence, mainly white
170	0.50	0.80	some birefringence, mainly white
170	0.74	0.74	mainly white, some Al metal
170	0.63	0.86	less white, some Al metal
175	0.50	0.80	mainly Al metal

aluminum metal. Also, if the precursor flow rate was too high, a white granular coating was observed due to extremely rapid deposition of alumina onto the substrate. The optimum ratio of water vapor to alane precursor was determined to have been reached when the films were conformal, uniform, featureless and displayed birefringence.

As may be seen from the high-resolution SEM image of the film shown in Figure 1, the films are granular in nature, appearing to be comprised of discrete alumina particles approximately 200 nm in diameter. The films also possess a high degree of homogeneity with only minor variations in film thickness across the entire surface of the substrate. Deposition rates on the order of 40 - 80 Å.min⁻¹ were determined by SEM analysis. The films were found to be amorphous in nature, as determined by X-ray diffraction, and possessed a resistance value of *ca.* 10^{12} Ω.cm.

The conformal nature of the as deposited films was demonstrated by film growth on Si wafers that were pre-etched (see Experimental). As may be seen from the SEM images in Figure 2, the etched features are faithfully reproduced after film growth. In addition, the cross sectional FESEM images shown in Figure 3 indicate the alumina film accurately follows the surface features down to the μm scale. Although this degree of step coverage is observed for APCVD occurring *in vacuo*, it is rather uncommon for APCVD.[4] Thus, we propose that the facile hydrolysis of the alane precursors must result in the formation of highly mobile reactant species on the substrate surface during the deposition process.

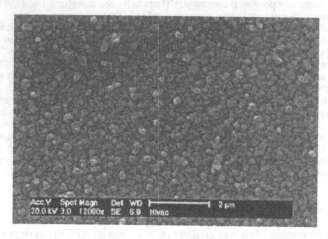

Figure 1. SEM image of an alumina film on a silicon substrate.

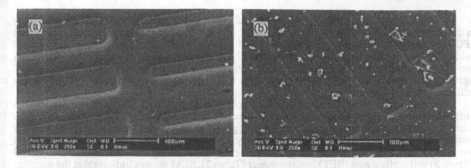

Figure 2. SEM images of etched Si wafers before (a) and after (b) APCVD of alumina.

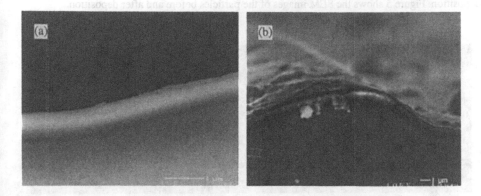

Figure 3. Cross-section FESEM images of alumina deposited on etched Si wafers.

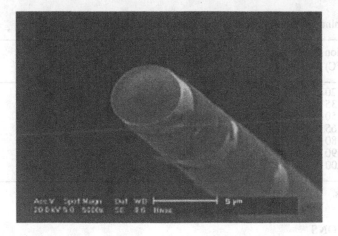

Figure 4. SEM image of alumina coating the surface of a carbon fiber.

As a further demonstration of the conformal nature of the coatings we have deposited films on carbon fibers and ceramic particles. Lengths of fiber tows were lain inside the horizontal APCVD chamber in a manner previously described.[17] In contrast the ceramic particles were coated inside a simple fluidized bed reactor. Unfortunately, under the conditions studied, the deposition rate was too rapid for the fiber experiments resulting in the majority of the fibers being "glued" together. However, as illustrated in Figure 4, some isolated fibers were evident which possessed a sufficiently-thick layer of alumina.

In contrast, APCVD of alumina on ceramic particles inside a fluidized bed results in a uniform continuous encapsulating coating. A summary of the deposition experiments is given in Table 2. A deposition temperature above 180 °C, gray aluminum metal formed through thermal decomposition of the precursor rather than hydrolysis while below 150 °C no deposition suggests that the precursor is not decomposing. The presence of a coating was determined from SEM (Figure 5) and surface area (BET) measurements (Table 2). SEM images indicate that the particles are being coated, and the increasing surface area, calculated from BET measurements, is most likely the result of a coating of a certain thickness giving rise to a larger overall particle following deposition. Figure 5 shows the SEM images of the particles before and after deposition.

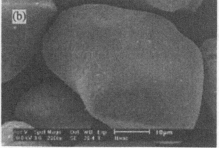

Figure 5. SEM images of ceramic particles before (a) and after (b) APCVD of alumina.

Table 2. Aluminum Oxide Deposition onto Ceramic Particles.

deposition temp. (°C)	SEM	surface area[a] (m^2g^{-1})
120	no coating	0.087
135	no coating	0.090
150	coating	0.117
165	coating	0.152
180	coating	0.185
190	no coating	0.089
200	no coating	0.091

[a] BET analysis.

CONCLUSIONS

The extremely low deposition temperature makes this co-vaporization technique very attractive relative to comparable APCVD methods. Further, homogeneous alumina films, which possessed a high degree of conformality, were deposited relatively free from atomic incorporation. The alumina films deposited on carbon fibers and ceramic particles showed a homogeneous thickness and illustrate the usefulness of this co-precursor system to yield protective coatings on substrates such as paint pigments and electroluminescent particles.

EXPERIMENTAL

Synthesis of AlH$_3$(NMe$_2$Et) was carried out by the literature method.[18] Silicon (100) wafers, were cleaned with 10% HF in ethanol, rinsed with deionized water and dried using KimWipes. The wafers were then rinsed with Microposit C-50 Primer and heated at 90 °C for 30 min. to ensure the complete removal of water residue. Microposit 1813 photoresist was applied onto the silicon samples using a spin coater set at 6500 rpm for 20 sec. The samples were then "soft baked" at 90 °C for 30 min and allowed to cool to room temperature. The patterns, on 35 mm film negatives, were exposed onto the silicon pieces using U.V. radiation for 1 min. The patterns were developed by dipping the samples in Microposit MF320 or AZ400K developer solutions for 1 min. The samples were rinsed with water and "hard baked" at 120 °C for at least an additional 30 min. After allowing the silicon pieces to return to room temperature, the patterns were wet/dry etched using a variety of techniques. For wet etching, the wafers were dipped into an acidic mixture of HF/CH$_3$COOH/HNO$_3$ (3:3:5 ratio) for 20 seconds. For plasma etching, a Technics PE II-A Plasma System and a Technics PD II-A Deposition System were utilized. Two gas mixtures were used, one consisting of CF$_4$ with 6% N$_2$O, the other consisting of CF$_4$ with 10% O$_2$. In accord with studies on silicon plasma etching, the latter gas mixture resulted in faster etching rates with sharper features.

Scanning electron micrographs were obtained on a Philips Electroscan XL30 ESEM-FEG and a Jeol 6320F SEM. EDAX and microprobe analyses were obtained on a Cameca SX-50 relative to calibration standards. BET measurements were performed on a Coulter SA3100 surface area and pore size analyzer.

Chemical Vapor Deposition

Depositions onto quartz, carbon fiber and silicon wafers were carried out in an atmospheric pressure laminar-flow hot-wall glass reactor (Figure 6). Silicon (100) substrates were cleaned with 10% HF in ethanol and rinsed with deionized water, while fused quartz substrates were rinsed with acetone prior to use. The entire system was evacuated to a pressure of 1 x 10^{-3} Torr overnight. The CVD chamber, loaded with substrates, was heated to the desired deposition temperature and purged with argon for about 1 hour prior to deposition. Argon gas was used as the carrier for the precursor and water vapors which reacted only in the region around the substrate. Deposition times varied between 1 - 3 hours depending on the thickness of the film desired.

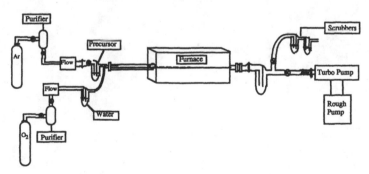

Figure 6. Schematic for horizontal hot-wall APCVD.

ACKNOWLEDGMENTS

Financial support for this work is provided by the Office of Naval Research and the National Science Foundation. Christopher D. Jones and Milton Pierson are gratefully acknowledged for assistance with surface area and microprobe analyses, respectively.

REFERENCES

1   N. C. Tombs, H. A. Wegener, R. Newman, B. T. Kenny, and A. J. Coppola, *Proc. IEEE.* **55**, 1168 (1967).
2   K. H. Zaininger and A. S. Waxman, *IEEE Trans. Electron. Devic,* **16**, 333 (1963).
3   S. Hashimoto, J. L. Peng, and W. M Gibson, *Appl. Phys. Lett.* **47**, 1071 (1985).
4   A. R. Barron in *CVD of Non-Metals*, Ed. by W.S. Rees, Jr., VCH, New York, 1996, pp. 262-313.
5   T. H. Huas and M. Armgarth, *J. Electron. Mater.* **16**, 27 (1987).
6   T. Maruyama and S. Arai, *Appl. Phys. Lett.* **60**, 322 (1992).
7   J. Saraie, J. Kwon, and Y. Yodogawa, *J. Electrochem. Soc.* **132**, 890 (1985).
8   R. G. Gordon, K. Kramer, and X. Liu, *Mat. Res. Soc. Symp. Proc.* **446**, 383 (1997).
9   M. G. Simmonds, W. L. Gladfelter, N. Rao, W. W. Szymanski, K. H. Ahn, and P. H. McMurry, *J. Vac. Sci. Technol.* **A9**, 2782 (1991).
10   K. M. Gustin and R. G. Gordon, *J. Electronic Mater.* **17**, 509 (1988).
11   M. G. Simmonds, E. C. Phillips, J. W. Hwang, and W. L. Gladfelter, *Chemtronics* **5**, 155 (1991).
12   D. M. Frigo, G. J. M. van Eijden, P. J. Reuvers, and C. J. Smit, *Chem. Mater.,* **6**, 190 (1994).
13   T. W. Jang, W. Moon, J. T. Back, and B. T. Ahn, *Thin Solid Films,* **333**, 137 (1998).
14   A. R. Barron and W. S. Rees, Jr., *Adv. Mater. Opt. Electr.* **2**, 271 (1993).
15   B. D. Fahlman and A. R. Barron, unpublished results.
16   E. G. Gillan and A. R. Barron, *Chem. Mater.,* **9**, 3037 (1997).
17   C.C. Landry and A. R. Barron, *Carbon,* **33**, 381 (1995).
18   Y. Senzaki, D. Uhrhammer, E. C. Phillips and W. L. Gladfelter in *Inorg. Synth.,* **31**, 74 (1997).

Chemical Vapor Deposition of
Non-Oxide Ceramics

MONOMERIC CHELATED AMIDES OF ALUMINUM AND GALLIUM: VOLATILE, MISCIBLE LIQUID PRECURSORS FOR CVD

SEÁN T. BARRY, ROY G. GORDON and VALERIE A. WAGNER
Harvard University Chemical Laboratories
Cambridge, MA 02138

ABSTRACT

New precursors were developed for the chemical vapor deposition (CVD) of aluminum nitride (AlN) and gallium nitride (GaN) at low temperatures. Synthetic methods for the new materials will be reported, along with their analyses and spectral characterization. The precursors are volatile (180 mTorr at 45-55 °C), low-viscosity (10 centipoise) liquids, so they are more convenient as vapor sources than previously available solid sources. They are thermally stable to temperatures well above their vaporization temperatures, so their vaporization is highly reproducible and leaves no residue. Unlike previously available liquid precursors, they are not pyrophoric, so they are safer to handle. AlN films formed by reaction with ammonia at around 200 °C are amorphous, transparent insulators that are good barriers to diffusion of water, oxygen, and other materials.

INTRODUCTION

The nitrides of aluminum and gallium have been deposited from a variety of metal-containing sources. Vapors of the metal alkyls react with ammonia at elevated temperatures (typically over 1000 °C) to form films of the metal nitrides.[1] These metal alkyls are spontaneously flammable in air, and the handling and shipping of these pyrophoric liquids presents serious safety hazards. The dialkylamides of aluminum[2] and gallium[3] allow CVD of the metal nitrides at much lower temperatures, 200 to 400 °C, and are much safer to handle because they are not pyrophoric. However, aluminum and gallium dimethylamides are solids, so they are less convenient to handle and vaporize than liquid precursors. The purpose of this paper is to introduce non-pyrophoric liquid precursors for CVD of AlN and GaN. These materials have the general formula shown in Figure 1. The chelating amide ligand is designed to keep them monomeric. This increases their volatility compared to the dimeric dimethylamides, and helps to keep them liquid, rather than solid.

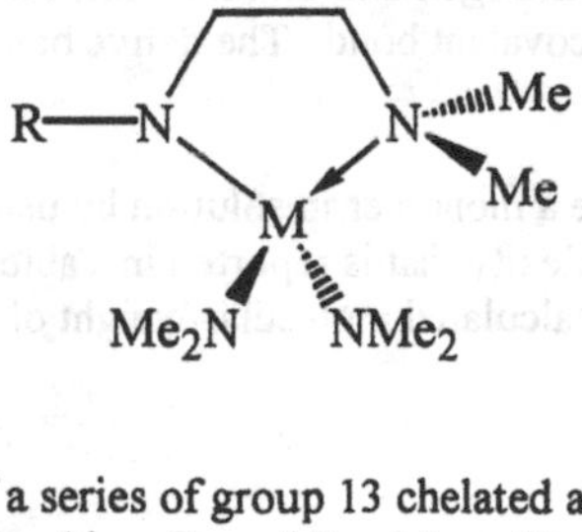

Figure 1. The general formula of a series of group 13 chelated amides designed to be low viscosity, monomeric liquids. M = Al or Ga and R = Me or Et.

Recently, the compound $(Me_2NCH_2CH_2NMe)Al(NMe_2)_2$ was synthsized.[4] Unfortunately, there was no physical characterization nor reactivity reported for this material, and its extent of oligomerization was not addressed. This work will expand what is known about this compound, as well as exploring other diamine ligands and examining similar reactivity for gallium.

EXPERIMENTAL AND RESULTS

All syntheses were performed using standard Schlenk techniques. All NMR spectra were taken on either a Bruker AM 500 MHz or a Bruker AM 300 MHz NMR spectrometer. Diethylether was dried over Na/K alloy and used freshly distilled. Both diamines were purchased from Aldrich Chemical Company and distilled before use. The hexakis(dimethylamido)digallium was made according to the seminal literature preparation[5]; the hexakis(dimethylamido)dialuminum was a gift from Aldrich Chemical Company.

Transamination from the starting material $[M(NMe_2)_3]_2$ (where M = Al, Ga) proved to be facile. The reaction occurred at room temperature under stirring for both $[Al(NMe_2)_3]_2$ and $[Ga(NMe_2)_3]_2$.

$$[M(NMe_2)_3]_2 + (Me)_2NCH_2CH_2NH(R) \rightarrow [(Me)_2NCH_2CH_2NR]M(NMe_2)_2 + HNMe_2 \qquad (1)$$

The syntheses were performed neat, with the diamine initially behaving as the solvent and the liquid product acting as the solvent towards the end of the reaction.

Synthesis of $Al(N(CH_3)CH_2CH_2N(CH_3)_2)(N(CH_3)_2)_2$ (1).

In a 100-mL flask, $[Al(N(CH_3)_2)_3]_2$ (2.0 g, 12.6 mmol) and $(CH_3)_2NCH_2CH_2NH(CH_3)$ (1.63 mL, 12.6 mmol) were mixed, and the clear, slightly orange solution was stirred overnight. Volatile materials were removed *in vacuo*, and clear, colorless 1 (2.54 g, 88 % yield) was distilled out at 65-70 °C at 300 mTorr with the distillation condenser just above ambient temperature. C: 51.76% found, 52.14% calc., H: 11.82% found, 11.92% calc., N: 24.32% found, 23.92% calc. ^{1}H NMR: (2.85 ppm, singlet, 12H), (2.70 ppm, double doublet, 2H), (2.15 ppm, double doublet, 2H), (2.03 ppm, singlet, 3H), (1.75 ppm, singlet, 6H)

When N,N,N '-trimethylethylenediamine (H3meda) was added to solid $[Al(NMe_2)_3]_2$ at room temperature, a slightly orange, clear liquid formed immediately with effervescence. After stirring over night at room temperature, 1 was distilled off in high yield between 65-70 °C. It was a clear, colorless liquid that froze around 26 °C (Table I). Compound 1 crystallized from the melt, and a single crystal X-ray diffraction study revealed the material to be a monomer with two molecules in the unit cell. Of the nitrogen-aluminum bonds, three were within the range 1.805-1.810 Å, which is expected for a covalent bond. The dative bond from the tertiary amine in the diamine was longer (2.036 Å).

Compound 1 was found to be a monomer in solution by using the melting point depression of *p*-xylene. The molecular complexity that is reported in Table I is the ratio of the experimental molecular weight divided by the calculated molecular weight of the monomer.

Table I. Physical data for liquid aluminum and gallium compounds. Note compound **1** (*) is a solid at room temperature and melts around 26°C.

Compound	Observed T_{bp} T (°C)/P (mTorr)	Corrected T_{bp} (°C)	Molecular Complexity	Viscosity (centipoise)
Al(3meda)(NMe₂)₂ **(1)**	65-70/300	250-255	1.04	10.22*
Al(dmeeda)(NMe₂)₂ **(2)**	65-70 /300	250-255	1.10	12.92
Ga(3meda)(NMe₂)₂ **(3)**	48-55/180	247-254	1.33	16.60
Ga(dmeeda)(NMe₂)₂ **(4)**	65-70/180	260-265	1.04	10.08

Synthesis of Al(N(CH₂CH₃)CH₂CH₂N(CH₃)₂)(N(CH₃)₂)₂ (2).

In a 100-mL flask, [Al(N(CH₃)₂)₃]₂ (2.0 g, 12.6 mmol) and (CH₃)₂NCH₂CH₂NH(CH₂CH₃) (2.0 mL, 13.2 mmol) were mixed, and the clear, slightly orange solution was stirred overnight. Volatile materials were removed *in vacuo*, and clear, colorless **2** (2.69 g, 93 % yield) was distilled out at 65-70 °C at 300 mTorr. C: 50.17% found, 49.98% calc., H: 11.62% found, 11.65% calc., N: 24.64% found, 25.90% calc. ^{1}H NMR: (3.04 ppm, quartet, 2H), (2.82 ppm, singlet, 12H), (2.75 ppm, double doublet, 2H), (2.15 ppm, double doublet, 2H), (1.78 ppm, singlet, 6 ppm), (1.32 ppm, triplet, 3H)

When N,N-dimethyl-N '-ethylethylenediamine (Hdmeeda) was added to solid [Al(NMe₂)₃]₂, a clear orange liquid immediately formed. Effervescence was not obvious, but after stirring overnight Al(dmeeda)(NMe₂)₂ **(2)** distilled off at 65-70 °C at 300 mTorr as a clear, colorless liquid (Table 1). Compound **2** was also a monomer in *p*-xylene.

Synthesis of Ga(N(CH₃)CH₂CH₂N(CH₃)₂)(N(CH₃)₂)₂ (3).

In a 100-mL flask, GaCl₃ (5.0 g, 28.4 mmol) was cooled to 0 °C, and 50 mL of diethylether was slowly added. In a 200-mL flask, LiN(CH₃)₂ (4.34 g, 85.1 mmol) was suspended in 50 mL of diethylether and cooled to 0 °C. The GaCl₃ solution was added by cannula to the LiN(CH₃)₂ suspension with immediate precipitation. The mixture was refluxed at 45 °C over night. The solids were filtered off, and (CH₃)₂NCH₂CH₂NH(CH₃) (3.70 mL, 28.4 mmol) was added by syringe. The solution was refluxed at 45 °C for 3 hours and the volatile materials were removed *in vacuo*. Clear, colorless **3** (5.08 g, 81% yield) was distilled off at 65-70 °C at 180 mTorr. C: 41.73% found, 41.35% calc., H: 9.72% found, 9.93% calc., N: 21.63% found, 20.24% calc. ^{1}H NMR: (3.03 ppm, double doublet, 2H), (2.84 ppm, singlet, 12H), (2.50 ppm, double doublet, 2H), (2.21 ppm, singlet, 3H), (2.17 ppm, singlet, 6H)

When H3meda was added to [Ga(NMe₂)₃]₂, effervescence was immediate, and a slightly yellow, clear liquid formed. After stirring overnight, clear colorless **3** distilled off at 48-55 °C at 180 mTorr. A melting point depression experiment in *p*-xylene showed **3** to have a molecular complexity of 1.33, suggesting a monomer/dimer equilibrium. This equilibrium does not seem to affect the distillation temperature, which was similar to the rest of the reported compounds. However, the higher viscosity of **3** is possibly due to this monomer/dimer exchange (Table I).

<u>Synthesis of Ga(N(CH$_2$CH$_3$)CH$_2$CH$_2$N(CH$_3$)$_2$)(N(CH$_3$)$_2$)$_2$ (4).</u>

In a 100-mL flask, [Ga(NCH$_3$)$_2$]$_2$ (2.29 g, 11.3 mmol) and (CH$_3$)$_2$NCH$_2$CH$_2$NH(CH$_2$CH$_3$) (1.8 mL, 11.4 mmol) were mixed. The clear, yellow solution was allowed to stir overnight. Volatile materials were removed *in vacuo*, and clear, colorless **4** (2.39 g, 77% yield) distilled off at 48-55 °C at 180 mTorr. C: 44.27% found, 43.98% calc., H: 10.49% found, 9.97% calc., N: 20.89% found, 20.52% calc. ^{1}H NMR: (3.05 ppm, multiplet, 2H), (2.94 ppm, triplet, 2H), (2.84 ppm, singlet, 12H), (2.51 ppm, double doublet, 2H), (2.19 ppm, singlet, 6H), (1.25 ppm, triplet, 3H)

This monomer/dimer exchange does not seem to occur in Ga(dmeeda)(NMe$_2$)$_2$ (**4**). After Hdmeeda and [Ga(NMe$_2$)$_3$]$_2$ were subjected to the aforementioned reaction conditions, clear, colorless **4** distilled off at 65-70 °C. It was a monomer in *p*-xylene, unlike compound **3**.

Thermal decomposition studies.

Thermolyses of these compounds were performed in sealed heavy-walled NMR tubes. The solvent used was d$_{12}$-mesitylene with some mesitylene as an internal standard. The decomposition was followed by the loss of an isolated signal in the ^{1}H NMR spectrum (a 1.94 ppm singlet for compounds **1** and **3**; a 1.34 ppm multiplet in the case of **2** and **4**).

The decomposition of compound **1** occurred at 275 °C with a half-life of 29 minutes. The loss of the peak followed a linear path over the time frame measured. In contrast, compound **2** showed no decomposition at this temperature and only started to decompose at 300 °C with a half-life of 10.5 minutes. Both of these decompositions produced orange, soluble byproducts. Peaks for these byproducts were evident in the NMR, but were too numerous to offer any information about their structure.

Decomposition for **3** occurred at 212 °C with a half-life of 50 minutes. During the course of the themolysis, a gray insoluble material formed, and the solution turned orange. Again, the soluble byproducts were uncharacterizable. Compound **4** decomposed at the same temperature with a half-life of 37 minutes. A gray precipitate formed in the course of this thermolysis as well.

Reactivity with ammonia

All four compounds showed reactivity with ammonia at room temperature. In a typical reaction, 0.5g of material was taken up in 5 mL of dry ether in an inert atmosphere. Ammonia gas was bubbled into the solution and an intractable white solid precipitated immediately. Characterization of the white solid by IR showed a strong absorption in the N-H stretching region in each case.

CVD of aluminum nitride from 2 and ammonia.

Thin films were grown by atmospheric pressure chemical vapor deposition (APCVD). Compound **2** was diluted with mesitylene (2.08 g of **2** in 3.47 g mesitylene) and vaporized by ultrasonic nebulization. The high-frequency (1.4 MHz) ultrasonic system created a cloud of tiny droplets (approximately 20-30 microns in diameter) in a reproducible and well-controlled manner.[6] The resulting fog was carried by a low flow of nitrogen (0.5 L/min, filtered by an oxygen scavenger) into a glass reaction tube surrounded by a tube furnace at 200 °C. Ammonia diluted by nitrogen in a 1:1 molar ratio was also introduced into the reaction tube. This mixture was used to purge the CVD system for 3 h prior to deposition. The precursor concentration in the deposition gas stream was 0.41 mol%, the ammonia concentration was 14 mol%, and the total flow rate was 0.70 L/min. Films were deposited on silicon substrates resting on the bottom of the glass tube. The deposition occurred over 6 minutes, and then the furnace was allowed to cool to room temperature under a flow of N_2/NH_3. Rutherford Backscattering (RBS) was used to determine the elemental composition of the films.

DISCUSSION

In this system, transamination occurs at room temperature to produce volatile liquid mixed amides of aluminum and gallium. All of the compounds have viscosities below 20 centipoise at room temperature, which makes them very easy to handle in a CVD reactor. Also, their low boiling points allow for fast vaporization and easy transport to the substrate.

The relatively low boiling points and viscosities of these materials suggest that intermolecular interaction is low. With the exception of compound **3**, these compounds are monomers in a non-coordinating solvent. This is due to bidentate coordination of the diamine ligand. This allows for the coordinative saturation of the metal center. Indeed, the X-ray diffraction study of **1** demonstrates this (Figure 2).

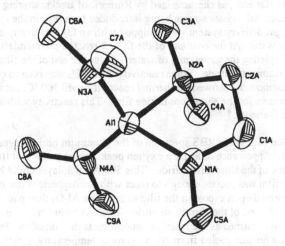

Figure 2. The molecular structure of **1** determined by X-ray diffraction from a single crystal. The thermal ellipsoids represent 50% probability.

Both dimethylamido moieties are planar, with the sum of the angles with nitrogen at the vertex totaling 360.0° and 359.6°. Planarity (359.0°) was also found in the amido end of the bidentate ligand. This is likely an average over several different orientations of the more usual umbrella structure.

The amino end of the bidentate ligand was found to be a distorted tetrahedron, with the Al-N-C (ring) angle being narrowed at the expense of the other two Al-N-C vertices. The dihedral angle in the N-C-C-N moiety was 40.23°, which approaches a typical dihedral angle in an alkane ring (55°). The N-Al-N ring angle is 87°, which is surprisingly small for the distorted tetrahedron around aluminum. Finally, the angles between the amido nitrogen of the bidentate ligand and the nitrogen of the dimethylamido moeities were all between 115-116°. This suggests that the ring is strained by the dihedral twist of the ethylene bridge, and this is pulling the amido nitrogen closer to the amino nitrogen of the bidentate ligand and forcing a distortion of the aluminum tetrahedron.

In the case of compound **3**, the molecular weight data and NMR spectrum suggest a monomer/dimer equilibrium. This is likely a steric effect, with the extra bulk of the ethyl group in **4** hindering the metal center just enough to prevent dimerization. Barron *et. al.* have done extensive work on bidentate ligands with different heteroatoms and have found that the strength of the metal-nitrogen bond and tetrahedral geometry of the coordinated amine promote dissociation from a 5-coordinate metal center to a monomer as steric bulk increases.[7]

All the compounds synthesized were found to be miscible with each other, as well as with nonpolar, aprotic organic solvents such as hydrocarbons.

Deposition of a thin film of aluminum nitride from compound **2** was possible at 200 °C in a flow of ammonia carried by nitrogen. The film was entirely aluminum oxide close to the introduction of precursor (near end) and became predominantly aluminum nitride at the other end of the substrate (far end), as characterized by Rutherford Backscattering (RBS). This was likely due to an incidental oxygen source being introduced by a leak in the system or by the gas stream. Since the gas delivery system was equipped with an O_2 scavenger, it is likely that the oxygen source was water. At the near end of the film, **2** reacted preferentially with water to form aluminum oxide, depleting the gas stream of water. At the far end of the film, **2** reacted with ammonia to form aluminum nitride. Since reactivity with NH_3 was seen to occur easily at room temperature and compound **2** showed no thermal reactivity until 300 °C, ammonia was the obvious nitrogen source for the aluminum nitride film. This reactivity with ammonia is well-documented in the literature.[8]

A more detailed look at the RBS spectrum of the aluminum nitride reveals some interesting features (Figure 3). Appearance of a sharp oxygen peak demonstrates that there was a veneer of Al_2O_3 on the surface of the aluminum nitride. This 500 Å thick layer could have been caused by two factors. If the film was porous enough to react with atmospheric water or dioxygen, this layer could have formed upon exposing the film to air. The Al_2O_3 then passivated the surface to prevent reaction of the rest of the aluminum nitride layer. Another possible explanation is that the Al_2O_3 surface layer formed as the reaction conditions at the end of the deposition changed. That is to say, as the furnace cooled from 200°C to room temperature, reaction with incidental water slowed at the near end of the film, and subsequently water became available for reaction at the far end of the film.

The RBS analysis also showed an aluminum nitride film with a thickness of 7000 Å. Interestingly, the simulation gave an Al:N ratio of 1:2.2. This overabundance of nitrogen could be due to the presence of hydrogen (undetectable by RBS) in the film from incomplete reaction with the ammonia.

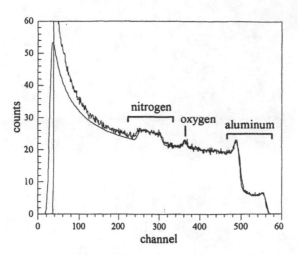

Figure 3: RBS spectrum for an Aluminum nitride film deposited using Al(dmeeda)(NMe₂)₂ (2) and NH₃ at 200°C. The thin, smooth line is a calculated fit to the spectrum.

CONCLUSIONS

Volatile liquid compounds of aluminum and gallium were synthesized, and their physical properties measured. These materials are low viscosity liquids that distill easily under vacuum. They are much safer to handle than the metal alkyls and react at much lower temperatures. These materials are thermally stable up to 200 °C and react readily with ammonia even at room temperature. They are highly suitable for CVD of aluminum nitride, gallium nitride, or their alloys, which has been demonstrated by synthesizing an aluminum nitride film.

ACKNOWLEDGMENT

This work was supported in part by the National Science Foundation under Grant No. CHE 95-10245.

REFERENCES

1. See, for example, C.-M. Zetterling, M. Ostling, K. Wongchotigul, M. G. Spencer, X. Tang, C. I. Harris, N. Nordell and S. S. Wong, J. Appl. Phys. **82**, 2990 (1997).
2. R. G. Gordon, U. Riaz and D. M. Hoffman, J. Mater. Res. **7**, 1679 (1992).
3. R. G. Gordon, D. M. Hoffman and U. Riaz, (Mat. Res. Soc. Symp. Proc. **204**, Pittsburgh, PA, 1991) p. 95.
4. B. A Vaartstra, US Patent 5,908,947 (1999).
5. H. Nöth,; P. Konrad, , Z. Noturforsch., **30b**, 681 (1975).
6. R. G. Gordon, F. Chen, N. J DiCeglie, Jr., A. Kenigsberg, X. Liu, D. J. Teff and J. Thornton, Mat. Re. Soc. Symp. Proc. **495**, pp. 63 (1998).
7 C. N. McMahon, J. A. Francis, S. G. Bott, A. R. Barron, J Chem. Soc., Dalton Trans., 67 (1999).
8. for an example, see: J. F. Janik, and R. L.Wells, Chem. Mater., **10**, 1613-1622 (1998).

LOW TEMPERATURE CHEMICAL VAPOR DEPOSITION OF TITANIUM NITRIDE THIN FILMS WITH HYDRAZINE AND TETRAKIS-(DIMETHYLAMIDE)TITANIUM

CARMELA AMATO-WIERDA*, EDWARD T. NORTON, JR.*, and DERK A. WIERDA**
*Materials Science Program, University of New Hampshire, Durham, NH 03824, ccaw@cisunix.unh.edu
**Department of Chemistry, Saint Anselm College, Manchester, NH 03102

ABSTRACT

Hydrazine and tetrakis-(dimethylamido)titanium have been used as precursors for the low temperature chemical vapor deposition of TiN thin films between 50°C and 200°C at growth rates between 5 to 35 nm/min. At hydrazine to TDMAT ratios of 50:1 and 100:1 the resulting films show an increase in the Ti:N ratio with increasing deposition temperature. They contain 2% carbon, and varying amounts of oxygen up to 36% as a result of diffusion after air exposure. The low temperature growth is improved when hydrazine-ammonia mixtures containing as little as 1.9% hydrazine are used. Their Ti:N ratio is almost 1:1 and they contain no carbon or oxygen according to RBS. The TiN films grown from pure hydrazine or the hydrazine-ammonia mixture have some crystallinity according to x-ray diffraction and their resistivity is on the order of 10^4 $\mu\Omega$ cm. The low temperature growth is attributed to the weak N–N bond in hydrazine and its strong reducing ability. In these films, the Ti:N ratio is approximately 1:1.

INTRODUCTION

TiN thin films can be deposited both by physical and chemical vapor deposition. Physical deposition methods, mostly reactive sputtering and pulsed laser deposition, have poor conformality in sub-micron features on integrated circuits, and they do not offer the large area uniformity of coverage required in some hard coatings applications.[1-8] Sputtered TiN is currently used in ULSI devices to prevent the interaction between the aluminum interconnect metal and the silicide contact layers on silicon. However, this TiN technology does not meet all the resistivity and conformality requirements of future devices with copper metallization.[3-5] This has lead to investigations of CVD TiN barriers with improved barrier properties, including conformality and large coating areas . Primarily two precursor systems have been used: TiCl₄ and ammonia, or tetrakis(dimethylamido)titanium and ammonia.[9-27] The first system requires temperatures too high (500-700 °C) for most device applications. In the case of the organometallic precursors, it is possible to obtain films with a combination of low impurity levels, low resistivity, and sufficient conformality for 0.25 μm technology.

We present here the low temperature chemical vapor deposition of TiN between 50-200 °C by using hydrazine (N_2H_4) as the nitrogen source precursor, along with tetrakis-(dimethylamido)titanium (TMT). This is significantly lower than the 300-450 °C temperatures typically reported for this process. In addition, as little as 1.9% (molar percent) hydrazine in a hydrazine-ammonia mixture is required to obtain TiN films at these temperatures. These lower deposition temperatures may provide TiN films with improved step coverage for sub-0.25 μm metallization schemes, since conformality is expected to improve with decreasing deposition temperature according to computer simulations of CVD.[23] This is a result of the decreased sticking coefficient of the precursors at lower deposition temperatures. Additional advantages of hydrazine compared to ammonia are its lower hydrogen to nitrogen ratio and its lower decomposition temperature (as low as 140 °C).[28]

EXPERIMENT

The liquid precursors, tetrakis(dimethylamido)titanium (99%) and hydrazine (98%) were purchased from Strem Chemicals and Aldrich, respectively. Hydrazine was treated with molecular sieves (3 Å) prior to use. These chemicals were handled in a dry nitrogen glovebox. The safety precautions for hydrazine are similar to those for toxic, air-sensitive, and pyrophoric compounds.[29] In particular, the explosive range of hydrazine vapors *in air* is 4.7-100 volume%; however, this hazard can be minimized in air free, high vacuum CVD processes. Silicon wafers were subject to the following cleaning steps prior to deposition: 10 minute soak in NH_4OH/H_2O_2 (1:4) at 50°C, 2 minute rinse in distilled water, 2 minute soak in HF/H_2O_2 (1:10), and a final two minute distilled water rinse.

The films were deposited using a custom-made low pressure hot wall CVD reactor with a resistively heated quartz tube. A gas inlet system with mass flow controllers, bubblers, and pressure control valves allows accurate delivery of known molar quantities of precursors and gases. The substrates are loaded into the reactor and a series of three pump/purge cycles removes air and water from the reactor. A pump/purge cycle consists of pumping the reactor to less than 0.001 torr and then filling it to atmospheric pressure with argon. After these pump/purge cycles, the reactor is purged with argon for 1 hour. Simultaneously, the bubbler and the precursor inlet tubing is heated to 55 °C. Following the purge cycle, the desired flows of argon carrier, argon diluent, and reactant gases are started. The reactor pressure was throttled to the desired pressure. Following deposition, reactant gases were turned off. The substrates were cooled under an argon flow and transferred from the reactor to the glovebox when they reached ambient temperature.

RESULTS

TiN films with thicknesses between tens of nanometers and two microns were deposited using hydrazine between 50 °C and 200 °C with deposition rates ranging from 5 to 35 nm/min. In a control experiment, no deposition was obtained at these temperatures if the hydrazine was replaced with ammonia, or if the precursor was used alone. Films less than 50 nm thick were gold, while thicker films were metallic and mirror-like or showed interference patterns. The films were smooth, scratch resistant, and adhesive by the Scotch tape test. The resistivity of the films was on the order of 10^4 $\mu\Omega$ cm, which in the range previously reported by TiN films made by CVD with the TMT precursor.

Figure 1 shows a representative RBS spectrum (using a 2.0 MeV He^{2+} beam) of a TiN film deposited on Si at 100°C using TMT with pure hydrazine (Figure 1a) and TMT with a 1.9% hydrazine-ammonia mixture (Figure 1b). The atomic composition of the film made with pure hydrazine is $Ti_{33}N_{29}C_{36}O_2$, which results in an atomic ratio of Ti:N equal to 1.14:1. The oxygen is incorporated into these films during air exposure after deposition. The diffusion of oxygen into TiN films prepared from TMT and ammonia has previously been demonstrated.[19,21-22,24-25] It is attributed to low density, porous films containing coordinatively unsaturated titanium which readily bonds to oxygen. These films show an increasing resistivity with time after they are exposed to air. The increased resistivity results from the increasing oxygen content of the films. The RBS results from TiN films made from TMT and a 1.9% hydrazine-ammonia mixture (Figure 1b) contain no oxygen or carbon signals, which means that less than 2-3 atom % of these elements is present. The composition of these films is $Ti_{52}N_{48}$, resulting in an atomic ratio of Ti:N equal to 0.92:1.

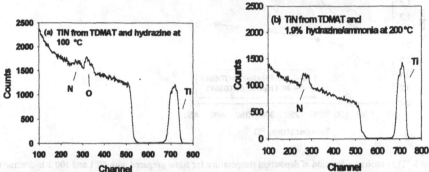

Figure 1. RBS spectrum of TiN deposited on silicon with TMT and (a) pure hydrazine at 100°C and (b) 1.9% hydrazine-ammonia mixture at 200°C.

Figure 2 compares the deposition rates of TiN films made using hydrazine and ammonia. Using TMT with ammonia results in no appreciable deposition at 200°C; whereas, the processes with hydrazine and the 2.7% hydrazine-ammonia mixture have deposition rates of 10 nm/min and 25 nm/min, respectively. These growth rates have been normalized for 0.1 sccm TMT flow. CVD with hydrazine results in growth rates at 200°C comparable to those obtained at 350°C with ammonia. More interestingly, the hydrazine-ammonia mixture deposits approximately twice as fast compared to using hydrazine alone with TMT.

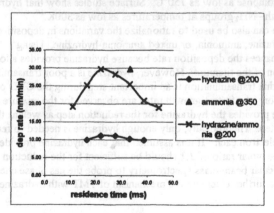

Figure 2. Deposition rate profiles for TiN films made with ammonia, hydrazine, and mixed ammonia-hydrazine

Analysis of the film composition ratio as a function of deposition temperature shows (Figure 3)an increase in Ti:N ratio with increasing deposition temperature. Presumably increased reaction temperatures cause increased TDMAT reaction and therefore more incorporation into the films.

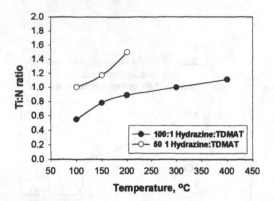

Figure 3. Ti:N ratio as a function of deposition temperature for films prepared with 50:1 and 100:1 hydrazine to TDMAT ratios.

The low temperature growth of TiN films using TMT and hydrazine can be rationalized on the basis of the following mechanistic speculation. Reaction of TMT with hydrazine provides a lower energy pathway (compared to the reaction with ammonia) for the reduction of the titanium from Ti(IV) in the precursor to Ti(III) in titanium nitride. This can be accomplished by heterolytic cleavage of the N–N bond in hydrazine, which produces NH_2 radicals that can donate a lone electron to the metal. This makes hydrazine is very strong reducing agent. Additionally, the N–N bond in hydrazine is very weak with a bond energy equal to 167 kJ/mol compared to 386 kJ/mol for the N–H bond in ammonia.[31] Previous mass spectrometry data indicate that hydrazine begins to thermally decompose as low as 150°C. Surface studies show that hydrazine can be used to cover a surface with –NH_2 groups at temperatures as low as 300K.[30]

This reduction mechanism can also be used to rationalize the variations in deposition rates among films made with hydrazine, ammonia, or mixed ammonia-hydrazine. Using hydrazine instead of ammonia enhances the deposition rate because hydrazine provides a lower temperature pathway for reduction of the titanium. However, hydrazine is a poor transamination reagent and it is well established that transamination is an important step during the CVD of TiN from TMT and ammonia. Therefore, the fastest growth rates are observed for the mixture of hydrazine-ammonia. The mixture provides the hydrazine for the reduction step as well as the ammonia for the transamination step. Theoretically, only enough hydrazine is needed to reduce all the TMT molecules with one electron each. If it is assumed that each hydrazine provides two NH_2 radicals, then TMT:hydrazine molar ratio of 2:1 should be sufficient for the reduction process. Experiments using molecular beam mass spectrometry to probe the gas phase chemistry during CVD of TiN are planned to further elucidate the mechanism of TMT with hydrazine.

CONCLUSION

In conclusion, TiN films have been produced by CVD between 50°C and 200°C with TMT and hydrazine. More significantly, the highest deposition rates have been observed with a

1.9% hydrazine-ammonia mixture. Further studies are underway in our laboratory to systematically characterize the TiN films made with TMT and hydrazine as a function of process conditions, including molecular beam mass spectrometry studies to elucidate the gas phase chemistry of TMT with hydrazine.

ACKNOWLEDGEMENTS

The authors gratefully acknowledge financial support provided by the National Science Foundation (DMR-9631794 and DMR-9875062) and the University of New Hampshire.

REFERENCES
1. Tu, K.N., Mayer, J.W., Poate, J.M., and Chen, L.J. (eds.) *Advanced Metallization for Future ULSI* (Materials Research Society Press, Pittsburgh, PA, 1996), Vol. 427.
2. Kattelus, H.P. and Nicolet, M. in *Diffusion Phenomena in Thin Films and Microelectronic Materials*, edited by Gupta, D. and Ho, P.S. (Noyes Publications, Park Ridge, NJ, 1988), p. 432.
3. Smith, P. M., Custer, J. S., Jones, R. V., Maverick, A. W., Roberts, D. A., Norman, J. A. T., Hochberg, A. K., Bai, G., Reid, J. S., Nicolet, M. A. Conference Proceedings ULSI XI (Materials Research Society, Pittsbrugh, PA, 1996), p. 249.
4. Reid, J. S. *Amorphous ternary diffusion barriers for silicon metallizations*. Ph.D. Thesis, California Institute of Technology, May, 1995.
5. Smith, P. M. and Custer, J. S. Appl. Phys. Lett. **70**, 3116 (1997).
6. Baliga, J. Semiconductor International **3**, 76 (1997).
7. Wang, S. MRS Bulletin **19**, No. 8, 30 (1994).
8. Eizenberg, M. MRS Bulletin **20**, No. 11, 38 (1995).
9. Sugiyama, K., Pac, Sangryul, P., Takahashi, Y., and Motojima, S. J. Electrochem. Soc. **122**, 1545 (1975).
10. Fix, R.M., Gordon, R.G., and Hoffman, D. M. Chem. Mater. **2**, 235 (1990).
11. Fix, R.M., Gordon, R.G., and Hoffman, D. M. Chem. Mater. **3**, 1138 (1991).
12. Musher, J.M. and Gordon, R.G. J. Mater. Res. **11**, 989 (1996).
13. Musher, J.M. and Gordon, R.G. J. Electrochem. Soc. **143**, 736 (1996).
14. Hoffman, D.M. Polyhedron, **13**, 1169 (1994).
15. Dubois, L.H., Zegarski, B.R., and Girolami, G.S. J. Electrochem. Soc. **139**, 3603 (1994).
16. Prybyla, J.A., Chiang, C.-M., and Dubois, L.H. J. Electrochem. Soc. **140**, 2695 (1993).
17. Dubois, L.H. Polyhedron **13**, 1329 (1994).
18. Intemann, A., Koerner, H., and Koch, F. J. Electrochem. Soc. **140**, 3215 (1993).
19. Eizenberg, M., Littau, K., Ghanayem, S., Mak, A., Maeda, Y., Chang, M., and Sinha, A.K. Appl. Phys. Lett. **65**, 2416 (1994).
20. Weber, A., Nikulski, R., Klages, C.P., Gross, M.E., Brown, W.L., Dons, E., and Charatan, R.M. J. Electrochem. Soc. **141**, 849 (1994).
21. Raaijmakers, I. J. *Thin Solid Films* **247**, 85 (1994).
22. Raaijmakers, I. J. and Yang, J. *Applied Surface Science* **73**, 31 (1993).
23. Sun, S.C. and Tsai, M.H. *Thin Solid Films* **253**, 440 (1994).
24. Katz, A., Feingold, A., Pearton, S.J., Nakahara, S., Ellington, M., Chakrabarti, U.K., Geva, M., and Lane, E. J. Appl. Phys. **70**, 3666 (1991).
25. Katz, A., Feingold, A., Nakahara, S., Pearton, S.J., Lane, E., Geva, M., Stevie, F.A., and Jones, K. J. Appl. Phys. **15**, 993 (1992).

26. Paranjpe, A. and IslamRaja, M. J. Vac. Sci. Technol. B. **13**, 2105 (1995).
27. Faltermeir, C., Goldberg, C., Jones, M., Upham, A., Manger, D., Peterson, G., Lau, J., Kaloyeros, A.E., Arkles, B., and Paranjpe, A. J. Electrochem. Soc. **144**, 1002 **(1997)**.
28. Fujieda, S., Mizuta, M. and Matsumoto, Y. Advanced Materials for Optics and Electronics, **6**, 127 (1996).
29. Schiessl, H.W. Aldrichimica Acta, **13**, 33 (1980).
30. Truong, C.M., Chen, P.J., Corneille, J.S., Oh, W.S., and Goodman, D.W. J. Phys. Chem. **99**, 8831 **(1995)**.
31. Huheey, J.E. Inorganic Chemistry, 3rd ed. (Harper and Row, New York, 1983), p. A30-A31.

ASPECTS OF GAS PHASE CHEMISTRY DURING CHEMICAL VAPOR DEPOSITION OF Ti-Si-N THIN FILMS WITH Ti(NMe2)4 (TDMAT), NH3, and SiH4

CARMELA AMATO-WIERDA*, EDWARD T. NORTON, JR.*, and DERK A. WIERDA**
*Materials Science Program, University of New Hampshire, Durham, NH 03824,
ccaw@cisunix.unh.edu
**Visiting scientist, Department of Chemistry, Saint Anselm College, Manchester, NH 03102

ABSTRACT

Silane activation, predominantly in the gas phase, has been observed during the chemical vapor deposition of Ti-Si-N thin films using Ti(NMe2)4, tetrakis(dimethylamido)titanium, silane, and ammonia at 450°C, using molecular beam mass spectrometry. The extent of silane reactivity was dependent upon the relative amounts of Ti(NMe2)4 and NH3. Additionally, each TDMAT molecule activates multiple silane molecules. Ti-Si-N thin films were deposited using similar process conditions as the molecular beam experiments, and RBS and XPS were used to determine their atomic composition. The variations of the Ti:Si ratio in the films as a function of Ti(NMe2)4 and NH3 flows were consistent with the changes in silane reactivity under similar conditions.

INTRODUCTION

Ti-Si-N is a refractory amorphous ternary nitride which is a promising candidate for diffusion barrier applications in future metallization schemes of integrated circuits.[1-4] One chemical vapor deposition (CVD) process for Ti-Si-N films uses the Ti(NR2)4 (R = Me or Et), NH3, and SiH4 precursor system and enables deposition at temperatures below 450°C, which is low enough for microelectronics applications.[1,4] The Ti(NR2)4 (R = Me or Et) + NH3 process (without the silane) has been extensively studied because it is used to deposit TiN in the semiconductor industry.[5-12]

It is interesting that silane reacts at these low temperatures given that silane and ammonia typically require temperatures between 600-700°C to form Si3N4.[13] This suggests that silane reactivity, and therefore silicon incorporation into the films, is facilitated by the presence of Ti(NR2)4. Both the electrical and mechanical properties of Ti-Si-N films depend strongly on their silicon content. Therefore, it is important to determine if and how the gas phase chemistry controls silicon incorporation into these films.

Using molecular beam mass spectrometry, we have demonstrated that silane is activated, predominantly in the gas phase, by Ti(NMe2)4, tetrakis-dimethylamidotitanium (TDMAT) and ammonia at 450°C during the CVD of Ti-Si-N films. More importantly, we have correlated the extent of this silane reaction under several conditions to variations in atomic composition in Ti-Si-N deposited by thermal CVD.

EXPERIMENT

The experimental apparatus, shown schematically in Figure 1, is a high temperature flow reactor coupled to a four-stage differentially pumped molecular beam mass spectrometer. The details of the apparatus are described in a previous publication. In short, an aperture at the end of the reactor samples the gas flow during CVD and forms it into a molecular beam, which is then skimmed and collimated before entering the ion source of a quadrupole mass spectrometer. In the experiments below, the silane (mass=32) is monitored at m/e=30 (parent ion minus two hydrogen atoms) because this is the most intense peak in the silane mass spectrum. The silane signals are normalized to argon as an internal standard to account for drifting instrument sensitivity. This normalized silane signal is proportional to the concentration of silane in the gas phase. The instrument's detection limit for silane is 200 ppm.

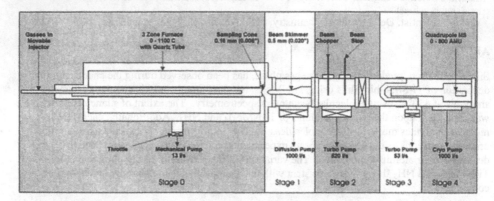

Figure 1. Schematic of molecular beam mass spectrometer coupled to the high temperature flow/CVD reactor.

A gas inlet system meters gas flows to the reactor and consists of mass flow controllers, a TDMAT bubbler, pressure control valves and capacitance manometers. Gases include TDMAT, argon carrier and diluent gas, ammonia, and silane (1.5% in Ar). A feedback-controlled throttle valve on the exhaust port regulates the reactor pressure. For all experiments, the reactor temperature is 450°C, reactor pressure is 6.5 torr, and the total gas flow rate is 785 sccm, which corresponds to a residence time of 35 ms under the laminar flow conditions used in this work.

RESULTS

Figure 2 shows the percent silane reacted as a function of varying $Ti(NMe_2)_4$ precursor flows (0.0, 0.1, and 0.3 sccm). The two data sets were taken at different surface-to-volume ratios, 1.33 cm^{-1} (circles) and 2.04 cm^{-1} (squares). The ammonia flow was constant at 10 sccm. Upon addition of $Ti(NMe_2)_4$, the percent silane reacted increases. For the low surface area case, 9.3% of the silane has reacted upon addition of 0.1 sccm of $Ti(NMe_2)_4$. This increases to 21.3% for 0.3 sccm of $Ti(NMe_2)_4$. The extent of silane reaction per unit $Ti(NMe_2)_4$ decreases with increasing $Ti(NMe_2)_4$.

Table 1 expresses the moles of silane reacted per mole of $Ti(NMe_2)_4$ used in the flow ($silane_R$/TDMAT). This ratio decreased from 9.1 to 7.1 as the $Ti(NMe_2)_4$ flow was tripled from 0.10 to 0.30 sccm, indicating that less silane reacts per unit $Ti(NMe_2)_4$ as more $Ti(NMe_2)_4$ is used in the process. Another interesting aspect of the ratios in Table 1 is that each $Ti(NMe_2)_4$ molecule is capable of activating or reacting with more than one silane molecule.

In order to estimate the contribution of a heterogenous component toward silane reactivity, the surface-to-volume (S/V) ratio of the reactor was increased by using a quartz insert. Figure 2 shows that the extent of silane reaction decreases significantly at the higher S/V ratio, implying that silane reactivity occurs predominantly in the gas phase and in fact, is inhibited by large surface areas. The larger surface area adsorbs more Ti(NMe$_2$)$_4$, thus making it unavailable for gas phase reaction with silane.

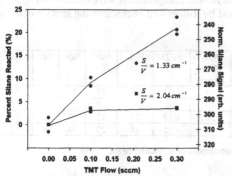

Figure 2. Silane mass spectrometer signal (at m/e = 30) as a function of Ti(NMe$_2$)$_4$ flow during the CVD of Ti-Si-N from Ti(NMe$_2$)$_4$ + NH$_3$ + SiH$_4$. The reactor conditions are: temperature = 450°C, pressure 6.5 torr, NH$_3$ flow = SiH$_4$ flow = 10 sccm, residence time = 35 ms. The circles are for S/V = 1.33 cm^{-1} and the squares are for S/V = 2.04 cm^{-1}. The lines serve only to guide the eyes.

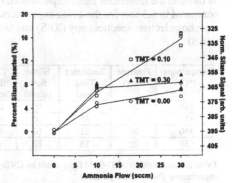

Figure 3. Silane mass spectrometer signal (@ m/e = 30) as a function of NH$_3$ flow during the CVD of Ti-Si-N from Ti(NMe$_2$)$_4$ + NH$_3$ + SiH$_4$. The reactor conditions are: temperature = 450°C, pressure 6.5 torr, SiH$_4$ flow = 10 sccm. Residence time = 35 ms. The circles are with TDMAT = 0.0 sccm, the squares with TDMAT = 0.1 sccm, and the triangles with TDMAT = 0.3 sccm. The lines serve only to guide the eyes

TMT Flow		S / V = 1.33 cm^{-1}		S / V = 2.04 cm^{-1}	
scc/ min	mmol/ min	silane$_R$ TMT	% Silane Reacted	silane$_R$ TMT	% Silane Reacted
0.00	0.0000	0	0.0%	0	0.0%
0.10	0.0045	9.3	9.3%	3.2	3.2%
0.30	0.0134	7.1	21.3%	1.2	3.7%

Table 1. Percent (%) silane reacted as a function of Ti(NMe$_2$)$_4$ flow and S/V for the reaction of Ti(NMe$_2$)$_4$ + SiH$_4$ + NH$_3$. The reactor conditions are: temperature = 450°C, pressure 6.5 torr, NH$_3$ flow = SiH$_4$ flow = 10 sccm, residence time = 35 ms.

Figure 3 shows the percent silane reacted as a function of NH$_3$ flow (0, 10, and 30 sccm) for various Ti(NMe$_2$)$_4$ flows, 0.0 (squares), 0.1 sccm (circles), and 0.3 sccm (triangles). Silane reacts to a small extent in the presence of NH$_3$ with no Ti(NMe$_2$)$_4$. Addition of 0.1 sccm of Ti(NMe$_2$)$_4$ increases the extent of silane reaction compared to the no Ti(NMe$_2$)$_4$ situation for all ammonia flows. At 0.1 sccm of Ti(NMe$_2$)$_4$ and 10 sccm of ammonia, 6.9% of the silane reacts and this increases to 16% at 30 sccm of ammonia. Surprisingly, an increase in Ti(NMe$_2$)$_4$ to 0.3 sccm does not further increase the percent silane reacted. Rather, the percent silane reacted at 0.3 sccm of Ti(NMe$_2$)$_4$ drops back down to 8.2% at 30 sccm of ammonia. This is the approximately the same extent of reaction that occurred at 0.1 sccm Ti(NMe$_2$)$_4$ and 10 sccm ammonia. It is interesting to note that the ratio of Ti(NMe$_2$)$_4$ to ammonia is the same for these two cases, i.e., 0.3 sccm Ti(NMe$_2$)$_4$:30 sccm ammonia equals 0.1 sccm Ti(NMe$_2$)$_4$:10 sccm ammonia. This suggests that the extent of silane reaction is dependent on the relative amounts of Ti(NMe$_2$)$_4$ and ammonia.

Ti-Si-N films were deposited in a separate hot-wall thermal CVD reactor to determine how well the silane reactivity studies correlate with silicon incorporation into thin films. Table 2 shows the atomic composition of these films. The deposition conditions in Table 2 are the same as those in the molecular beam experiments, except for reactor pressure which was 20 torr compared to 6.5 torr for the beam experiments. The atomic compositions were determined by x-ray photoelectron spectroscopy (XPS) and confirmed by Rutherford backscattering spectrometry (RBS).

Temperature (°C)	Pressure (torr)	Residence Time (ms)	Silane flow (sccm)	TMT flow (sccm)	NH₃ flow (sccm)	Si/Ti	%Si	%Ti	%N	%C	%O
450	20	35	10	0.1	10	1.03	27	17	35	8	14
450	20	35	10	0.3	10	0.6	17	28	31	9	15
450	20	35	10	0.1	30	1.38	23	17	33	3	25
450	20	35	10	0.3	30	0.55	16	29	34	8	12

Table 2. Composition of Ti-Si-N films made by CVD using conditions of molecular beam experiments (% = atomic percent)

The Si:Ti ratio in the films decreases from 1.03 to 0.6 as $Ti(NMe_2)_4$ is increased in the presence of 10 sccm ammonia. This behavior is repeated at 30 sccm of ammonia, in which the Si:Ti ratio decreases from 1.38 to 0.55 as $Ti(NMe_2)_4$ is increased. These results are consistent with the molecular beam mass spectrometry studies in which the silane reacted per unit $Ti(NMe_2)_4$ decreased when $Ti(NMe_2)_4$ was increased from 0.1 to 0.3 sccm. For constant $Ti(NMe_2)_4$ at 0.1 sccm, the Si:Ti ratio increases from 1.03 to 1.38 as ammonia is increased. In contrast, at $Ti(NMe_2)_4$ equal to 0.3 sccm, the change in the Si:Ti ratio is negligible (0.6 to 0.55) as ammonia is increased. These changes in Si:Ti ratio are also in agreement with how the extent of silane reaction varied with ammonia flow in the molecular beam experiments.

DISCUSSION

An interpretation of this combination of silane reactivity and thin film deposition studies is that $Ti(NMe_2)_4$ activates silane in the presence of NH_3, allowing silicon to become incorporated into the growing films. A possible mechanism for silane activation by $Ti(NMe_2)_4$ and ammonia involves formation of some type of titanium imido complex. Previously Cundari et. al. have performed *ab initio* studies of silane (and methane) activation by group IVB imido complexes, including $H_2Ti=NH$.[14] The existence of titanium imido complexes is supported by experimental evidence in the literature.[11a,c,12,15-21] The gas phase kinetic studies of $Ti(NMe_2)_4$ and NH_3 by Weiler are particularly relevant because they support the presence of titanium imido complexes, such as $Ti(=NH)(NMe_2)_2$, resulting from ammonia transamination and successive dimethylamine eliminations (1-2).[12]

Based on these previous observations as well as the results presented here, it is reasonable to postulate that the first step of silane activation occurs according to the following mechanism (3):

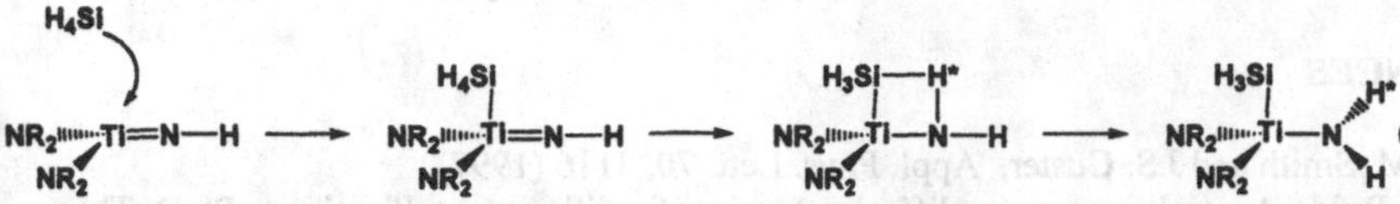

The imido species is generated according to mechanisms (1) and (2) above. The silane activation step is an addition of the Si–H bond across the Ti=N bond through the four-center transition state described by Cundari.[14]

    Although not conclusive, this mechanism is consistent with our silane reactivity studies. Increasing Ti(NMe₂)₄ causes more silane to react by increasing the amount of the imdio complex available. If incomplete transamination of Ti(NMe₂)₄ is assumed under the conditions of these flow experiments, the mechanism also explains why increasing ammonia did not increase silane reactivity for the case of 0.3 sccm Ti(NMe₂)₄. There was insufficient ammonia *relative* to the increased Ti(NMe₂)₄. If complete transamination is assumed the role of ammonia in silane activation becomes less clear.

    Although a titanium imido species may activate more than one silane molecule (up to four) according to the proposed mechanism, the mechanism does not account for how each TDMAT molecule can activate seven or more silane molecules, as observed in our experiments. The alternative mechanism (4) shows a pathway for multiple silane activation by one TDMAT molecule. Of course, some of the above observations could be a result of surface processes.

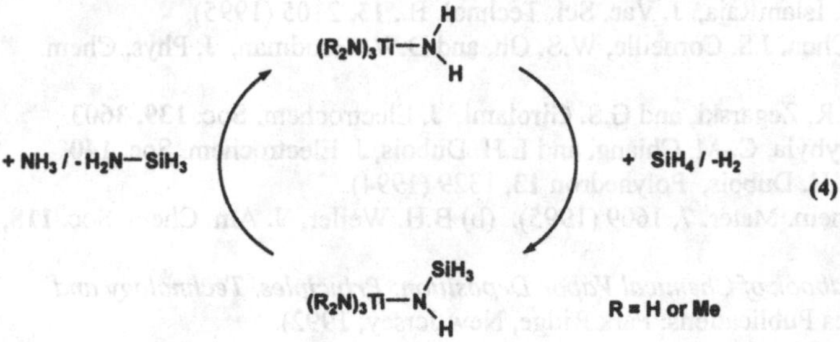

CONCLUSIONS

    The most significant result from this work is the observation that Ti(NMe₂)₄ activates silane in the presence of ammonia at 450 °C. The extent of silane reaction is dependent upon the relative amounts of Ti(NMe₂)₄ and ammonia. Additionally, each Ti(NMe₂)₄ molecule activates multiple silane molecules. Ti-Si-N thin films have been deposited in which the variations of the Ti:Si ratio with Ti(NMe₂)₄ and ammonia flows are consistent with observations of silane reactivity under similar conditions. This demonstrates that knowledge of the gas phase mechanisms that occur during CVD of Ti-Si-N ultimately can be useful as a means of controlling film composition. Further experiments are underway to further determine how the gas phase chemistry of this process influences film properties.

ACKNOWLDEGEMENTS
The authors gratefully acknowledge financial support provided by the National Science Foundation (DMR-9631794 and DMR-9875062) and the University of New Hampshire.

REFERENCES

1. P. M. Smith and J.S. Custer, Appl. Phys. Lett. **70**, 3116 (1997).
2. J.S. Reid, Amorphous ternary diffusion barriers for silicon metallizations. Ph.D. Thesis, California Institute of Technology, May, 1995.
3. (a) X. Sun, J.S. Reid, E. Kolawa, and M.A. Nicolet, J. Appl. Phys. **81**, 656 (1997).
    (b) X. Sun, J.S. Reid, E. Kolawa, and M.A. Nicolet, *J. Appl. Phys.* **81**, 664 (1997).
4. I.J. Raaijmakers, Thin Solid Films **247**, 85 (1994).
5. K. Sugiyama, Pac, P. Sangryul, Y. Takahashi, and S. Motojima, J. Electrochem. Soc., **122**, 1545 (1975).
6. (a) R.M. Fix, R.G. Gordon, and D.M. Hoffman, Chem. Mater. **2**, 235 (1990).
    (b) R.M. Fix, R.G. Gordon, and D.M. Hoffman, Chem. Mater. **3**, 1138 (1991).
    (c) J.M. Musher and R.G. Gordon, J. Mater. Res. **11**, 989 (1996).
    (d) J.M. Musher and R.G. Gordon, J. Electrochem. Soc., **143**, 736 (1996).
    (e) D.M. Hoffman, Polyhedron, **13**, 1169 (1994).
7. (a) A. Katz, A. Feingold, S.J. Pearton, S. Nakahara, M. Ellington, U.K. Chakrabarti, M. Geva, and E. Lane, J. Appl. Phys. **70**, 3666 (1991). (b) A. Katz, A. Feingold, S. Nakahara, S.J. Pearton, E. Lane, M. Geva, F.A. Stevie, and K. Jones, J. Appl. Phys. **15**, 993 (1992).
8. S.C. Sun and M.H. Tsai, Thin Solid Films, **253**, 440 (1994).
9. A. Paranjpe. and M. IslamRaja, J. Vac. Sci. Technol. B., **13**, 2105 (1995).
10. C.M. Truong, P.J. Chen, J.S. Corneille, W.S. Oh, and D.W. Goodman, J. Phys. Chem. **99**, 8831 (1995).
11. (a) L.H. Dubois, B.R. Zegarski, and G.S. Girolami, J. Electrochem. Soc. **139**, 3603 (1992). (b) J.A. Prybyla, C.-M. Chiang, and L.H. Dubois, J. Electrochem. Soc. **140**, 2695 (1993). (c) L.H. Dubois, Polyhedron **13**, 1329 (1994).
12. (a) B.H. Weiler, Chem. Mater. **7**, 1609 (1995). (b) B.H. Weiler, J. Am. Chem. Soc. **118**, 4975 (1996).
13. H.O. Pierson, *Handbook of Chemical Vapor Deposition: Principles, Technology and Applications*, (Noyes Publications: Park Ridge, New Jersey, 1992).
14. (a) T.R. Cundari, J. Am. Chem. Soc. **114**, 10557 (1992).
    (b) T.R. Cundari and M.S. Gordon, J. Am. Chem. Soc. **115**, 4210 (1993).
    (c) T.R. Cundari, and J.M. Morse, Chem. Mater. **8**, 189 (1996).
15. (a) C.H. Winter, P.H. Sheridan, T.S. Lewkebandara, M.J. Heeg, and J. W. Proscia, J. Am. Chem. Soc., **114**, 1095 (1992). (b) T.S. Lewkebandara, P.H. Sheridan, M.J. Heeg, A.L. Rheingold, and C.H. Winter, Inorg. Chem. **33**, 5879 (1994).
16. J.E. Hill, R.D. Profilet, P.E. Fanwick, I.P. Rothwell, Angew. Chem. Int. Ed. Engl., **29**, 664 (1990).
17. H.W. Roesky, H. Voelker, M. Witt, and M. Noltemeyer, Angew. Chem. Int. Ed. Engl., **29**, 669 (1990).
18. C.C. Cummins, C.P. Schaller, G.D. Van Duyne, P.T. Wolczanski, A.W.E. Chan, and R. Hoffman, J. Am. Chem. Soc. **113**, 2985 (1991).
19. W.A. Nugent, and B.L. Haymore, Coordination Chem. Rev. **31**, 123 (1980).
20. W.A. Nugent, and R.L. Harlow, J.C.S. Chem. Comm., 579 (1978).
21. C.C. Cummins, S.M. Baxter, and P.T. Wolczanski J. Am. Chem. Soc. **110**, 8731 (1988).

MOLECULAR BEAM MASS SPECTROMETRY STUDIES OF THE THERMAL DECOMPOSITION OF TETRAKIS(DIMETHYLAMINO)TITANIUM

Carmela C. Amato-Wierda[†], Edward T. Norton Jr., Derk A. Wierda[*]
University of New Hampshire, Materials Science Program, Durham, NH
[*]Saint Anselm College, Department of Chemistry, Manchester, NH
[†]Author to whom correspondence should be addressed.

ABSTRACT

Tetrakis(dimethylamino)titanium (TDMAT) is an important precursor for TiN, TiCN, and TiSiN thin films in chemical vapor deposition. In order to better understand how the gas phase chemistry influences the formation of these films, the decomposition of TDMAT has been studied in a high-temperature flow reactor (HTFR) by molecular beam mass spectrometry (MBMS). Two kinetic regimes have been observed as a function of temperature. Rate expressions and mechanistic implications will be presented. Further studies are in progress to identify the gas phase species relevant to the decomposition mechanism of TDMAT.

INTRODUCTION

Tetrakis(dimethylamino)titanium, TDMAT, is an important precursor for the chemical vapor deposition (CVD) of TiN, TiSiN, and TiCN thin films. TiN and TiSiN thin films are used for diffusion barriers in microelectronic circuits[1-4] and TiCN thin films are used as hard coatings for cutting tools and wear resistant parts. Gas phase chemistry can play an important role in the CVD process under typical reaction conditions[5].

Understanding the kinetics and mechanism of the gas phase reactions in these processes will lead to a better understanding of the CVD process, and an improvement in film properties. As a basis for understanding more complex reactions such as those for the deposition of TiN and TiSiN, the focus of this research is upon the thermal decomposition of TDMAT.

EXPERIMENT

A schematic of the molecular beam mass spectrometer used in this experiment is shown in Figure 1. It consists of a low pressure, hot-wall CVD reactor, coupled to a molecular beam

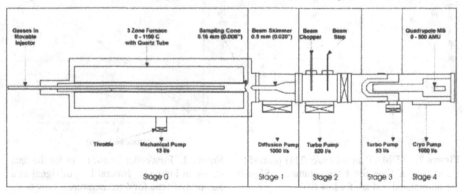

Figure 1. Schematic of the 4-stage molecular beam mass spectrometer.

Mat. Res. Soc. Symp. Proc. Vol. 606 © 2000 Materials Research Society

sampling system for the quadrupole mass spectrometer. The gasses are injected through a temperature controlled, moveable injector into the reactor. The reactor is a quartz tube 60 cm long with an ID of 3.4 cm. It is heated by a furnace in three 20 cm zones. At the end of the reactor is a sampling cone with a 0.16 mm diameter aperture. Pressure control is by a throttle valve with feedback from a capacitance manometer mounted on the reactor chamber. Most of the gasses are pumped away by a chemical-plasma grade mechanical pump vented to an exhaust system.

The gasses passing through the sampling cone enter the low pressure region of the next stage and form a molecular beam. It undergoes a reversible, adiabatic expansion which lowers the temperature to a few degrees Kelvin. This stops any chemical reactions in the sample, taking a "snap-shot" of the conditions in the reactor at the residence time determined by the volume flow rate and injector position relative to the sampling aperture. The core of the beam is sampled by the skimmer and the beam passes along the axis of the system, through several stages of differential pumping, to the quadrupole mass spectrometer where is it analyzed.

The TDMAT (99.99%, Aldrich) is a liquid at room temperature and pressure, with a vapor pressure of about 0.10 torr at 25°C. It is delivered by a bubbler held at 69°C at 200 torr, maintained by a constant temperature bath and a pressure control valve/capacitance manometer system. UHP argon (99.999%, Airgas) is delivered by a mass flow controller and is bubbled through the liquid precursor. These conditions result in a 1% gas/vapor mixture being delivered to the reactor. The gas lines after the bubbler are resistively heated to above the bubbler temperature to prevent condensation of the TDMAT vapor. The movable injector is jacketed and is kept at constant temperature by water from a constant temperature bath.

For all experiments, the reactor pressure was 2.75 torr and the injector tip is held at 40 cm from the sampling aperture. The argon flow rate through the bubbler was varied from 5.0 to 122 sccm, yielding flow rates of TMT from 0.05 to 1.28 sccm, and residence times from 10.000 to 0.200 seconds. The temperatures range from 333 to 593K. TDMAT signals were monitored at m/e = 224, the parent ion signal. For the low surface to volume (low S/V) ratio experiments, the S/V = 0.118 mm^{-1}. For the high S/V experiments the reactor was packed with smaller tubes, increasing the S/V by a factor of six times (S/V = 0.708 mm^{-1}).

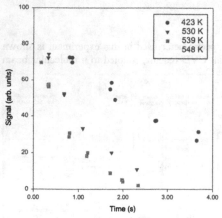 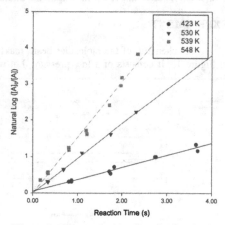

Figure 2. TDMAT signal (m/e=224) plotted as a function of time for 4 temperatures. Signals are normalized and are for low S/V.

Figure 3. First order kinetics plot for the data shown in Figure 2. Natural Log of signal as a function of time for 4 temperatures. Low S/V.

RESULTS

The first experiments reported are for determining the kinetics of the decomposition of TDMAT. Figure 2 shows a sample of the data taken for TDMAT as a function of reaction time and reactor temperature. The signals are for the parent ion of TDMAT (m/e=224) and the signal decreases as a function of time, more rapidly at higher temperatures. Only a sample is shown in this plot for clarity. Figure 3 shows the first order kinetics workup for the same sampling of data. The data presented this way is clearly linear, thus a very good first order fit. The slopes of the lines represent the rate constants (k, with units of s^{-1}) for the reaction conditions. This same analysis was performed for all of the decomposition data taken for TDMAT, at both high and low S/V.

All of the calculated rate constants are shown in Figure 4, a plot of rate constant as a function of temperature for both low S/V and high S/V. For the low S/V data (triangles), the rate constants are very low in the temperature range of 333K to 525K. At just above this temperature at approximately 550K, the trend in rate constant shows a sharp increase. The high S/V data (circles) follows a very similar trend, except that the sharp increase occurs at approximately 500K. Figure 5 is a typical Arrhenius plot. A plot of natural logarithm of rate constant (ln k) versus 1/T, and it clearly shows two regimes. Often, a plot such as this is linear or nearly linear with the slope proportional to the activation energy of the process and the intercept representing the natural log of the Arrhenius pre-exponential factor. Figure 6 shows the high temperature data fitted to the Arrhenius expression for both high and low S/V. The 99% confidence intervals are show along with the line of best fit to demonstrate how well the data fits the expression in this region. The calculated Arrhenius parameters are summarized in Table 1 for both high and low S/V and for both high and low temperature.

The second set of experiments reported are for determining species important to the decomposition of TDMAT during chemical vapor deposition. In these experiments, the mass range from 5 to 250 m/e was monitored instead of the 224 m/e peak (TDMAT parent). Data

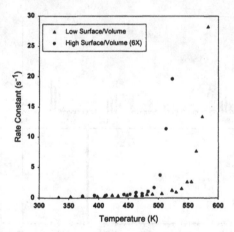

Figure 4. Plot of first order kinetic data for the experiments performed at low S/V (triangles) and at high S/V (circles) over the temperature range from 333K to 593K. Clearly shows two regimes for each data set.

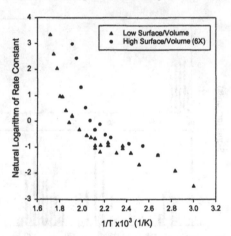

Figure 5. Arrhenius plot of natural log of rate constant versus 1/T for kinetic data summarized in Figure 4.

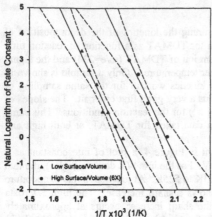

Figure 6. Arrhenius plot for high temperature data for both high and low S/V. Slope and intercept show is used to determine activation energy and pre-exponential factor. Dashed lines represent 99% confidence intervals, showing quality of fit.

Table 1. Summary of the Arrhenius parameters calculated for both high and low S/V. High temperature data is calculated using rate constants above the break point for each data set, and low temperature data is for below the break point.

High Temperature	E_a (kJ/mol)	A
S/V = 0.118 mm⁻¹	165.5	1.60E+16
S/V = 0.708 mm⁻¹	163.5	4.23E+17
Low Temperature	E_a (kJ/mol)	A
S/V = 0.118 mm⁻¹	16.25	32.28
S/V = 0.708 mm⁻¹	15.00	34.83

from six different temperatures was taken: 343K, 393K, 453K, 503K, 543K, and 553K. Using the kinetic data from the first set of experiments, four residence times were determined for each temperature by calculating the times at which 40, 50, 60, and 70 percent of the TDMAT would be decomposed. The different spectra could then be directly compared. A sample of the data from the experimental run at 503K is shown in Figure 7. The residence time of 750 ms for this temperature represents 40 percent of the TDMAT being decomposed already. It is very difficult to determine which peaks are from ionizer fragments of undecomposed TMT, parent ions of reaction products and fragments of reaction products, or both.

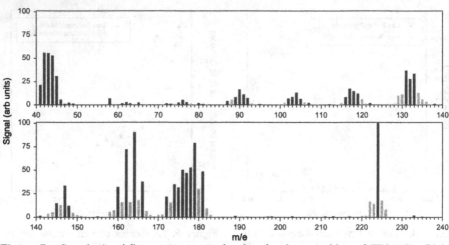

Figure 7. Sample (partial) mass spectrum showing the decomposition of TDMAT. Light colored bars represent peaks which can be eliminated as isotope peaks, simplifying data analysis. Reactor conditions: T = 503K, P = 2.73 torr, reaction time = 750 ms, S/V = 0.118 mm⁻¹.

Conclusions

The results from the decomposition kinetics are both interesting and intriguing. Interpretation of the high temperature data results in activation energies of 165.5 kJ/mol (low S/V) and 163.5 kJ/mol (high S/V) (39.6 and 39.1 kcal/mol, respectively). The difference in the pre-exponential factor is the only major difference between the two. The same trend follows for the low temperature data, but the calculated pre-exponential factors for the low temperature data do not seem to yield reasonable results. This would seem that the Arrhenius expression is not a good model for the kinetics under these conditions.

Knowledge of the mechanism involved in this process would help in understanding the kinetic data. Unfortunately, the speciation results are as of yet inconclusive. Due to the complexity of the mass spectrum in these experiments, further efforts are under way to get useful information from the data.

The effect of S/V in this experiments is apparent, but as in the cases of speciation and kinetics, more work is required to determine the significance of these effects in these CVD processes.

Acknowledgements

The authors are grateful and would like to thank the National Science Foundation (DMR-9631794 and DMR-9875062) and the University of New Hampshire for funding and support of this research.

References

1. P. M. Smith and J.S. Custer, Appl. Phys. Lett. **70**, 3116 (1997).

2. J.S. Reid, Amorphous ternary diffusion barriers for silicon metallizations. Ph.D. Thesis, California Institute of Technology, May, 1995.

3. (a) X. Sun, J.S. Reid, E. Kolawa, and M.A. Nicolet, J. Appl. Phys. **81**, 656 (1997).
    (b) X. Sun, J.S. Reid, E. Kolawa, and M.A. Nicolet, *J. Appl. Phys.* **81**, 664 (1997).

4. I.J. Raaijmakers, Thin Solid Films **247**, 85 (1994).

5. Weiller, B. H. *J. Am. Chem. Soc.* **1996**, *118*, 4975-4983

LOW TEMPERATURE THERMAL CHEMICAL VAPOR DEPOSITION OF SILICON NITRIDE THIN FILMS FOR MICROELECTRONICS APPLICATIONS

Spyridon Skordas, George Sirinakis, Wen Yu, Di Wu, Haralabos Efstathiadis and Alain E. Kaloyeros
New York State Center for Advanced Thin Film Technology and Department of Physics, The University at Albany – SUNY, Albany, New York, 12222.

ABSTRACT

Silicon nitride technology has been incorporated in ultra-large scale integration (ULSI) microchip fabrication, thin film transistors (TFT), solar cells, and many other applications in a rapidly expanding market. Nevertheless, silicon nitride technologies currently in use face considerable limitations. Low pressure chemical vapor deposition (LPCVD) occurs at relatively high temperature (>700 °C) and plasma enhanced chemical vapor deposition (PECVD), although occurring at temperatures below 300 °C, produces hydrogen-rich films and could be self-limiting in terms of conformality and damage to the devices due to ion bombardment. In the present work, successful low temperature thermal chemical vapor deposition (LTCVD) of silicon nitride is reported on 8" silicon wafers. The use of a halide-based silicon precursor, tetraiodosilane (SiI_4) has led to the deposition of high quality silicon nitride thin films at temperatures as low as 300 °C.

Characterization of resulting film properties has been performed to determine their dependence on deposition parameters by Auger Electron Spectroscopy (AES), Rutherford Backscattering Spectroscopy (RBS), Fourier Transform Infrared (FTIR), Nuclear Reaction Analysis (NRA), Ellipsometry, Capacitance-Voltage (C-V), and Current-Voltage (I-V) measurements.

INTRODUCTION

Silicon nitride is widely used in semiconductor devices and integrated circuit technologies. Due to its high dielectric constant, high breakdown strength, high resistance against diffusion, large energy bandgap and high resistance against radiation, it has applications in advanced ULSI, TFT [1,2,3], as a storage medium in nonvolatile memories [4], solar cells, and thin film electroluminescence (TFEL) displays. However, the current high thermal budget requirement in thermal CVD SiN_x has set serious limitations for most of its applications. Plasma based processes, such as PECVD and electron cyclotron resonance CVD (ECR – CVD) can grow SiN_x films at, respectively, 100 °C [5] 60 °C [6,7]. However, they can cause ion bombardment damage of the underlying device structures, and are subject to hydrogen inclusion and poor conformality. Therefore, it is desirable to develop a process to grow SiN_x thin films at low deposition temperature by conventional thermal CVD technology [8].

In this paper, the successful deposition of SiN_x ($x \geq 1.38$) films is reported by low pressure low temperature thermal chemical vapor deposition (LTCVD). Tetraiodosilane (SiI_4) was selected as the Si precursor because of its chemical simplicity and associated bonding energy considerations. Its chemical structure is conducive to bond dissociation at low temperature, with recombination being interrupted in the presence of nitrogen to yield SiN_x. Also, the Si-I bond strength is ~ 56 Kcal/mol, as compared to 69 Kcal/mol and 76 Kcal/mol for, respectively, Si-Br and Si-Cl, implying that SiI_4 could decompose at significantly lower temperature that $SiBr_4$ and $SiCl_4$ [9]. A systematic Design of Experiments (DOE) approach was implemented in a detailed investigation of the effects of key process parameters on film structural, chemical, mechanical, and electrical properties.

Mat. Res. Soc. Symp. Proc. Vol. 606 © 2000 Materials Research Society

EXPERIMENTAL APPROACH AND CHARACTERIZATION TECHNIQUES

The LTCVD SiN_x process was developed in an 8" wafer, stainless-steel, warm wall module of a four-chamber cluster tool. The reactants used were SiI_4 and ammonia (NH_3). An MKS 1153 solid source delivery system was employed to precisely and reproducibly control the delivery of the Si precursor.

In order to systematically approach the development and optimization of low temperature CVD SiN process, a DOE study was implemented. In the DOE study, the Minitab Statistical software was used to plan and execute the deposition experiments and develop a process window. The DOE approach used a response surface model (RSM) to map the entire process parameter space, thus yielding complete and highly useful results [10].

The DOE approach explored the effects on the three key process parameters, namely, ammonia gas flow rate, substrate temperature, and process pressure on the film properties. The SiI_4 precursor flow rate was kept constant at 12 sccm during processing. Table I summarizes the range of deposition parameters investigated and optimized experimental process parameters, that in terms of highest deposition rate at the optimum substrate temperature of 370 °C.

Table 1. DOE process window investigated for low temperature CVD SiN films.

Process Parameters	Range Investigated	Best parameters (In terms of highest growth rate of ~ 5.4 nm/min)
Substrate temperature (°C)	300 - 440	370
Pressure (Torr)	1 - 3	3
NH_3 flow rate (sccm)	300 - 900	600
SiI_4 flow rate (sccm)	12	12
Si source temperature (°C)	160	160

The resulting DOE-produced films were analyzed as follows:

- Film thickness and refractive index (n) were measured by ellipsometry using a Tencor UV 1080 spectroreflectometer at $\lambda = 600$ nm.
- Compositional analysis and density measurements were performed by Rutherford Backscattering (RBS) at a primary He^+ energy of 2.0 MeV.
- The hydrogen profile in the films was also investigated by Nuclear Reaction Analysis (NRA) using the nuclear reaction $^1H(^{15}N, \alpha\gamma)^{12}C$.
- The film breakdown electric field (E_{bd}) and dielectric constant were measured through, respectively, leakage current-voltage (I-V) and high frequency (1 MHz) capacitance – voltage (C-V) in a series of Metal-Nitride-Semiconductor (MNS) diodes ($Al/SiN_x/n$-Si) for SiN_x thicknesses ranging from 1080 – 2000 Å. Aluminum dots of 760 μm diameter and of 5000 Å thickness were evaporated by e-beam through a mask onto the SiN films. As ohmic contact of the Si substrates, an aluminum layer of 5000 Å thickness was also evaporated on the substrate side of the SiN_x samples.
- Stress of SiN_x films was measured on 4" Si wafers at room temperature by determining the bending of the wafer before and after film deposition. The equipment used was the Tencor FLX 2320 system.

- Chemical bonding configurations in the films were investigated by FTIR in the range of 400 – 4000 cm⁻¹.

RESULTS AND DISCUSSION

Figure 1 shows a typical RBS spectrum of 2000 Å – thick SiN$_x$ film deposited on Si substrate at temperature of 370 °C. The N/Si ratio was ~ 1.55, with an iodine incorporation of 0.74 at. %, and film density of 2.9 gr/cm³. For illustration purposes, Figure 2 shows a plot of the current density (J) as a function of the applied electric field for two Al/SiN/Si MNS diodes. Figure 2 indicates that for both samples (Samples A and B) the breakdown electric filed was higher than 5 MV/cm. The breakdown electric field of sample A was 5.6 MV/cm. Due to the limitation of the I-V measurement system (the maximum applied voltage was 100 Volts), no breakdown was observed for sample B within the limit of the experimental set-up used.

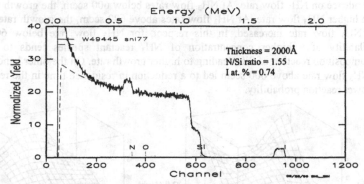

Figure 1. RBS spectrum of a 2000 Å - thick SiN$_x$ film on Si substrate, deposited at: T = 370 °C, P = 2 Torr and NH₃ flow rate = 600 sccm. The dashed curve represents the simulated spectrum.

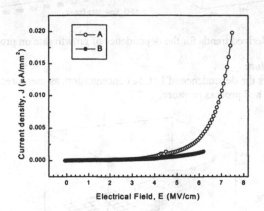

Figure 2. The leakage current density (J) vs. applied electric field (E) for two samples (Sample A: T = 370 °C, P = 3 Torr, NH₃ flow = 300 sccm and SiI₄ flow = 12 sccm and Sample B: T = 300 °C, P = 3 Torr, NH₃ flow = 600 sccm and SiI₄ flow = 12 sccm)

The detail effects of key parameters on film properties will be discussed below.

1) Growth rate:

Figure 3 displays the dependence of the SiN_x growth rate on key parameters. The growth rate varied between 14.8 to 54.4 Å/min for various growth conditions. The main behavioral trends observed were:

- Growth rate exhibited direct dependence on process pressure. This behavior could be attributed to an increase in precursor partial pressure in the reaction zone with higher process pressure. This might indicate that the growth mode is diffusion-limited as documented in the next paragraph.
- Growth rate was mostly independent of variations in substrate temperature in the range from 300 to 440 °C.
- Within the parameter space investigated, growth rate exhibited two distinct modes of dependence on NH_3 flow rate. At NH_3 flow rates below 600 sccm, the growth rate increased with higher NH_3 flow rate. At NH_3 flow rates above 600 sccm, the growth rate decreased as the NH_3 flow rate increased. In this respect, for NH_3 flow rate below 600 sccm, the availability of a higher concentration of NH_3 reactant species tends to enhance the decomposition reaction of SiI_4, leading to higher growth rate. On the other hand, the increase in NH_3 flow rate above 600 sccm led to a reduction in residence time in the reactor, leading to lower reaction probability.

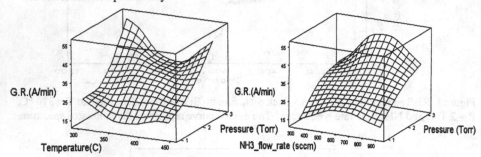

Figure 3. DOE-derived trends for the dependence of growth rate on process parameters

2) Iodine concentration:

Figure 4 shows the dependence of I at. % concentration, as measured by RBS, on substrate temperature and process pressure.

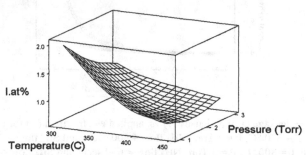

Figure 4. DOE-derived trends for the dependence of I at. % on substrate temperature and process pressure.

The main behavioral trends observed were:

- Iodine concentration exhibited a strong dependence on substrate temperature. A decrease in I at. % concentration, from 2.0 to 0.66 at. %, with an increase of substrate temperature from 300 °C to 440 °C was observed. This trend might be attributed to a more efficient precursor decomposition behavior with higher temperature, leading to a reduced I inclusion in the resulting films.

3) Stoichiometry (N/Si ratio):

The films exhibited N/Si ratio between 1.38 to 1.58 in the process regime under investigation. The main behavioral trends observed were:

- N/Si ratio increased with an increase in substrate temperature. This might be attributed to a higher probability for reaction with active nitrogen species, leading to the incorporation of more N in the films.

4) Breakdown electric field and dielectric constant:

- Within the process regime under investigation, the films exhibited almost constant breakdown electric field around 5 to 6 MV/cm.
- Dielectric constants of the films are around 5 to 8.

5) Stress

SiN_x thin films usually exhibited a high tensile stress of a few GPa [11]. Minimizing stress of a SiN_x film can improve its adhesion to the substrate and resulting device reliability. In this respect, Figure 5 shows the DOE-derived trends for the dependence of stress of SiN_x films grown in process parameters. The main behavioral trends observed were:

- Stress exhibited a marked decrease with a decrease of pressure. The decrease might be due to the fact that the lower pressure leads to lower growth rate, thus providing a more efficient pathway for stress relief in the films with the achievement of a more energetically favorable film texture.

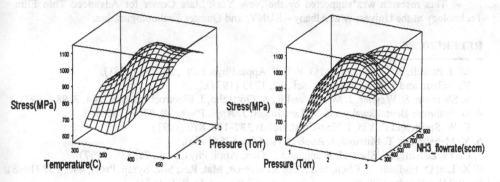

Figure 5. DOE derived trends for the dependence of film stress on process parameters.

- High substrate temperature and high NH_3 flow rate led to higher stress levels. This is consistent with the results of increased tensile stress with increasing N/Si ratio in SiN_x [12, 13].

All the FTIR spectra of the SiN$_x$ films on Si substrate showed a strong absorption band centered at around 850 – 870 cm^{-1} that was identified as the stretching vibration mode of the Si-N bond. The FTIR spectra also exhibited a strong N-H stretching mode near 3350 cm^{-1} and bending mode near 1200 cm^{-1}. No absorption band corresponding to Si-H bonds was found near 2100 cm^{-1}. Given that hydrogen was mainly bonded to nitrogen, integration of the N-H(s) related peak provided a good estimate of hydrogen concentrations [14]. It was found that the film contained 20 ± 2 at. % of hydrogen, a value in agreement with findings from hydrogen profiling measurements done by NRA. A decrease in H content of ~ 2.0 at. % was observed upon increasing the substrate temperature from 400 to 440 °C.

The film index of refraction was found in the range of 1.72 to 1.80. The increase in the index of refraction was correlated to the decrease of N/Si ratio from 1.58 to 1.38 [15].

CONCLUSIONS

It was demonstrated that silicon nitride films can be grown by thermal CVD process from SiI$_4$ and NH$_3$ at temperatures as low as 300 °C. The process was developed and optimized on an 8" wafer cluster tool platform.

A design of experiment (DOE) approach was implemented to explore the effects of the three key parameters, namely substrate temperature, process pressure and NH$_3$ flow rate, on film properties. Structural, chemical, optical, and electrical characterization of the resulting SiN$_x$ films deposited on Si wafers was completed.

The films deposited were found to be N-rich with N/Si ~ 1.38 – 1.58. The breakdown E-field of CVD SiN$_x$ films was determined to be in the range of 5 – 6 MV/cm, while the dielectric constant was found to be in the range of 5 – 8. Process conditions were identified through the DOE approach to yield films with the lowest stress level of 600 MPa.

ACKLOWLEDGMETS

This research was supported by the New York State Center for Advanced Thin Film Technology at the University at Albany – SUNY, and Quester Technologies, Inc.

REFERENCES

1. M. J. Powell, B. C. Easton, and O. F. Hill, Appl. Phys. Lett. , 38, 794 (1981).
2. K. Niihara and T. Hirai, J. Mater. Sci., 12, 1233 (1977).
3. S. Sherman, S. Wagner, J. Mucha, and R. A. Gottscho, J. Electrochem. Soc. 144, 3198 (1997).
4. D. Frohman-Bentchkwsky and M. Lenzlinger, J. Appl. Phys., 40, 3307 (1969).
5. F. W. Smith and Z. Yin, J. Non-Cryst. Solids 137-138, 879 (1991).
6. Y. Manabe and T. Mitsuyu, J. Appl. Phys., 66, 2475 (1989).
7. C. Ye, Z. Ning, M. Shen, H. Wang, and Z. Gan, Appl. Phys. Lett., 71, 336 (1997).
8. X. Lin, D. Endisch, X. Chen, and Alain Kaloyeros, Mat. Res. Soc. Symp. Proc., 495, 107 (1998).
9 CRC Handbook of Chemistry and Physics, 75th Ed., ed. D. R. Lide Jr.
10. X. Chen, G. G. Peterson, C. Goldberg, G. Nuesca, H. L. Frisch, A. E. Kaloyeros, and
    B. Arkles, J. Mater. Res., 14, No.5, 2043 (1999).
11. E. EerNisse, J. Appl. Phys., 48, 3337 (1977).
12. S. Hasegawa, Y. Amano, T. Inokuma and Y. Kurata, J. Appl. Phys., 72, 5676 (1992).
13 S. Habermehl, J. Appl. Phys., 83, 4672 (1998).
14 W. Lanford and M. Rand, J. Appl. Phys., 49, 2473 (1978).
15 T. Cotler and J. Chapple-Sokol, J. Electrochem. Soc. 140, 2071 (1993).

METHYLAMINE GROWTH OF SiCN FILMS USING ECR-CVD

C.-Y. Wen[a], J.-J. Wu[b], H.J. Lo[c], L.C. Chen[a], K.H. Chen[b], S.T. Lin[c], Y.-C. Yu[d], C.-W. Wang[d], and E.-K. Lin[d]

[a]Center for Condensed Matter Sciences, National Taiwan Unversity, Taipei, Taiwan;
[b]Institute of Atomic and Molecular Sciences, Academia Sinica, Taipei, Taiwan;
[c]Department of Mechanical Engineering, National Taiwan University of Science and Technology, Taipei, Taiwan;
[d]Institute of Physics, Academia Sinica, Taipei, Taiwan

ABSTRACT

Continuous polycrystalline SiCN films with high nucleation density have been successfully deposited by using CH_3NH_2 as carbon source gas in an ECR-CVD reactor. Fom the kinetic point of view, using CH_3NH_2 as carbon source could provide more abundant active carbon species in the gas phase to enhance the carbon incorporation in the SiCN films. The compositions of the SiCN films analyzed from Rutherford Backscattering Spectroscopy showed that higher $[CH_3NH_2]/[SiH_4]$ ratio led to higher carbon content in the films. Moreover, a lower carbon content was measured when the film was deposited at higher substrate temperature. The direct band gap of the aforementioned SiCN films determined using PzR is around 4.4 eV, indicating a wide band gap material for blue-UV optoelectronics.

INTRODUCTION

Ternary Si-C-N compound has been drawing much attention in recent years due to its potential applications in high-speed electronics and blue-ultraviolet optoelectronic devices.[1] To date, several types of SiCN compounds have been reported, such as amorphous SiCN films by pulsed laser ablation deposition and ECR-CVD [2,3], crystalline SiCN films by microwave plasma CVD [4-6], amorphous SiCN powder by laser and RF discharges [7], and nanocrystalline Si_2CN_4 by sol-gel method [8]. Among them, the ECR-CVD method has been demonstrated to achieve high nucleation density and deposited continuous polycrystalline SiCN with structure similar to that of α-Si_3N_4 [9].

In our previous studies of SiCN deposition by ECR-CVD, the addition of the H_2 into the system enhanced the carbon incorporation into the films due to the abstraction reaction by H atoms, which further enhanced active carbon species. Besides, when comparing different carbon source gas, such as CH_4, C_2H_2, and CH_3NH_2, CH_3NH_2 was most efficient for the growth of SiCN films, which is attributed to the 8 orders of magnitude higher dissociation for CH_3NH_2 than that for CH_4. Therefore, H_2 addition to CH_3NH_2 enables a source gas to effectively tailor the carbon content in SiCN films.

In this paper, a study of the relationship between the contents of the SiCN films and its corresponding properties have been performed for films deposited using CH_3NH_2 in an ECR-CVD reactor. Among the properties measured, the crystal structures of SiCN films were studied by high resolution transmission electron microscope (HRTEM). The direct band gaps of SiCN films were measured by piezoreflectance spectroscopy (PzR), which were described in our earlier work [1].

EXPERIMENTAL

The silicon carbon nitride films were deposited using an ECR-CVD reactor. Detailed description of the ECR-CVD used in this work has been reported in our recent work [9]. Substrate cleaning process includes standard HF acid deoxidization, acetone and methanol degrease, and distilled water cleaning followed by H_2 plasma etching of surface impurities. A mixture of semiconductor grade gases, H_2, N_2, CH_3NH_2, and 10% SiH_4 diluted in N_2, were used as reaction gas sources. In order to characterize the role of methylamine for the growth of SiCN films, various CH_3NH_2/SiH_4 ratios and substrate temperatures at a constant pressure of 3.2 mTorr and a microwave power of 1200 W were used for the film growth. The quantitative composition analysis was performed by Rutherford backscattering spectrometry (RBS) using 3.5 MeV 4He ion. The non-RBS cross-section of 4He ion from C and N [12,13] were linked to RUMP2 program for quantitative analysis. A Perkin Elmer Phi 1600 ESCA system was used to study the chemical bonding state of the films, in which Mg $K\alpha$ radiation of 1253.6 eV was used as the x-ray source with a linewidth of 0.7 eV. The analysis area for XPS measurement was 800 μm in diameter and the pass energy for the chemical state analysis was 11.75eV. Scanning electron microscope (SEM) was employed to get the surface morphology of the SiCN films. HRTEM investigation of the films was performed on a JEOL-4000EX microscope at an operating voltage of 400 kV. The band edge of the direct band to band transition was measured by PzR method, in which the basic principle is detecting the strain-induced changes of the interband transitions [14,15].

RESULTS AND DISCUSSION

A summary of the film compositions determined by RBS measurement for various CH_3NH_2 gas flow and substrate temperature is listed in Table I. To examine the process-composition inter-relationship, the adjusted parameters were divided into two groups. The first group is for different CH_3NH_2 flow while keeping constant substrate temperature, and the second group is different substrate temperatures under a constant reaction gas flow ratio. The C/Si ratio in the SiCN films increases with increasing $[CH_3NH_2]/[SiH_4]$ ratio and decreases at higher substrate temperature.

The amount of the carbon content incorporated in the films depends on the various active carbon species in the gas phase and gas-surface interactions at different substrate temperature. The possible reactions of CH_3NH_2 and SiH_4 in the gas phase are listed in Table II. Assuming that the reactions take place not far way from the substrates and the reaction temperatures are close to

Table I   The process parameters in the ECR CVD reactors and the results of compositions and growth rates of the SiCN films.

No.	Process parameters					Experimental results				
	Gas flow rate (sccm)				Substrate Temperature	Composition				Growth rate
	N_2	H_2	CH_3NH_2	SiH_4	(°C)	C	Si	N	C/Si	(nm/h)
1	2.5	2.5	1.0	0.5	700	12.8	32.1	55.1	0.4	30
2	2.5	2.5	1.25	0.5	700	10.8	26.9	62.3	0.4	36
3	2.5	2.5	1.5	0.5	700	15.0	27.9	57.1	0.54	23
4	2.5	2.5	1.0	0.5	650	15.1	24.7	60.2	0.61	30
5	2.5	2.5	1.0	0.5	750	10.8	28.0	61.2	0.39	26
6	2.5	2.5	1.0	0.5	800	0	40.8	59.2	0	22

Table II   The rate constant expressions of various gas reactions [16].

No.	Reactions	K	k at 1000 K
r1	$CH_3NH_2 \rightarrow CH_3 + NH_2$	$6.92 \times 10^{10} \exp(-24230/T)$ s^{-1}	2.08 s^{-1}
r2	$SiH_4 \rightarrow SiH_2 + H_2$	$3.28 \times 10^{16} \exp(-31832/T)$ s^{-1}	4.91×10^2 s^{-1}
r3	$CH_3NH_2 + H \rightarrow H_2 + CH_2NH_2$	$1.8 \times 10^{13} \exp(-2646/T)$ cm^{3}mol^{-1}s^{-1}	1.27×10^{12} cm^{3}mol^{-1}s^{-1}
r4	$CH_3NH_2 + H \rightarrow CH_3 + NH_3$	$3.9 \times 10^{14} \exp(-5773/T)$ cm^{3}mol^{-1}s^{-1}	1.21×10^{12} cm^{3}mol^{-1}s^{-1}
r5	$SiH_4 + H \rightarrow SiH_3 + H_2$	$1.39 \times 10^{13} \exp(-1400/T)$ cm^{3}mol^{-1}s^{-1}	3.43×10^{12} cm^{3}mol^{-1}s^{-1}

the substrate temperature, the dissociation rate constants of CH_3NH_2 and SiH_4, with and without H_2, are calculated at 1000K (Table II). Without H, the dissociation rate constant of CH_3NH_2 (r1) is somewhat smaller than SiH_4 (r2). When H_2 was added, the rate constants of CH_3NH_2 (r3 and r4) are comparable to that of SiH_4 (r5). Therefore, as expected, the higher $[CH_3NH_2]/[SiH_4]$ ratio would lead to the higher carbon content in the film. However, the detailed mechanisms of the gas phase and surface reactions are not yet clear, and an explanation for the temperature dependence of the carbon content is still lacking.

Typical XPS spectra of the films are shown in Fig. 1, wherein the spectra of C(1s) and N(1s) reveal that the peaks comprise more than one Gaussian peaks. The C(1s) peak can be resolved into three components centered at 284.4, 285.8, and 288.1 eV, and the N(1s) peak consists of two components centered at 398.4 and 399.8 eV. These peaks correspond to C-C, C(1s)=N, and C(1s)-O; and N(1s)-Si and N(1s)=C, respectively.   The C-C peak at 284.0 eV disappeared after sputtering with Ar$^+$ ions, which implies that C-C bonding was originated from the surface contamination. The C(1s) peak at 285.8 eV and the corresponding N(1s) peak at 399.8 eV were tentatively assigned to the C=N bonding according to Marton et al [17]. The single component spectrum at 102.7 eV was assigned to Si(2p)-N bonding [18]. It should be noticed that there is no detectable C(1s)-Si bonding signal at 282.8 eV, and the Si(2p)-C signal around 100.3 eV doenn't exist, indicating negligible Si-C bonding in the film.

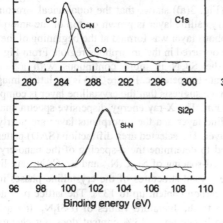

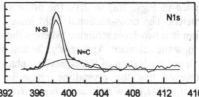

Figure 1 XPS spectra of C(1s), N(1s), and Si(2p) of the SiCN film which was deposited under the condition of No. 1.

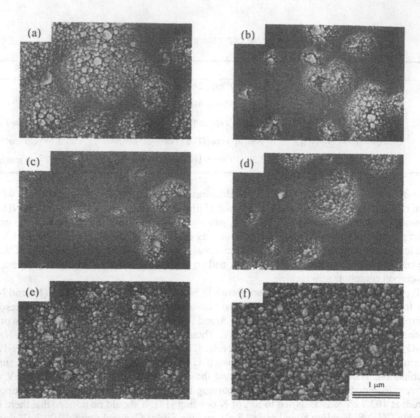

Figure 2   The SEM images of the SiCN films deposited under the conditions $[CH_3NH_2]/[SiH_4]=$ (a) 2.0, (b) 2.5, (c) 3, at 700 °C, and $[CH_3NH_2]/[SiH_4]=2$ at (d) 650 °C, (e) 750 °C, (f) 800 °C.

To investigate the morphology and the crystal structure of the deposited films, SEM and HRTEM were employed. The SiCN films contain both crystalline and amorphous phases. As presented in Figs. 2(a) to 2(f), the proportion of the two phases varies with varying process parameters. The cross-sectional TEM image (Fig. 3(a)) shows that the topological structure of the film is a two-layer structure wherein the crystalline layer is grown on top of the amorphous layer near the substrate. Apparently, the amorphous layer was formed at the beginning of the film growth and nucleation of the crystalline phase occurred in the amorphous region. From Fig. 2, it is observed that the crystalline phase in the SiCN is more continuous with a smaller $[CH_3NH_2]/[SiH_4]$ flow ratio and a higher substrate temperature. Figure 3(b) is the lattice image of the crystalline layer in Fig. 3(a). The lattice image suggests that the crystalline layer is composed of small crystallites with the size of about 20 nm. The X-ray energy dispersive spectra analysis indicates that the compositions in the crystalline layer and the amorphous layer are nearly the same for both the amorphous and crystalline layer. The selected area diffraction (SAD) pattern of the crystalline layer of each SiCN film is used to determine the d-spacing of the nano-crystals observed in the TEM images. In Table III, the d-spacing of $Si_{38}C_5N_{57}$ and $Si_{32}C_{13}N_{55}$ are listed. Comparing with the d-spacing of $\alpha\text{-}Si_3N_4$, the crystal structures of the crystalline phases in the SiCN films are close to $\alpha\text{-}Si_3N_4$, even with a carbon content up to 13%. The lattice parameters derived from the SAD patterns are also close to the literature values of $\alpha\text{-}Si_3N_4$. It's quite a surprise that the appearance of carbons in the lattice up to 13% content does not change the

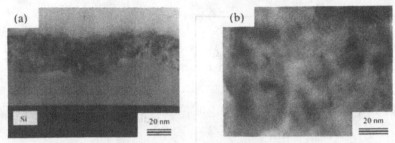

Figure 3   (a) Typical cross-sectional view of TEM image of the SiCN films. (b) The lattice image of the crystalline layer of the SiCN films.

existing framework of α-Si_3N_4. This can be explained by the fact that no Si-C bonding was observed in the XPS spectra, carbon incorporated resides solely on silicon site. Since that the size of carbon atom is smaller than that of silicon atom, a 13% substitution of silicon by carbon atoms may not affect the Si_3N_4 framework at all.

The direct band gap of the $Si_{28}C_{15}N_{57}$ film determined by PzR is shown in Fig.4, wherein the values of the band gap and the broadening parameter of the direct band-to-band transition can be obtained by detailed fitting of an PzR spectra with a form of the Cohen's expression using a first-derivative Lorentzian line shape. The band edge determined by such fitting is 4.4 eV as indicated by an arrow in the figure, which is in between the band edge of nano-crystalline $Si_{38}C_5N_{57}$ film [19] and the $Si_{35}C_{26}N_{39}$ crystals [1] with values of 4.66 eV and 3.81 eV, respectively. More data are needed to map out the whole band gap as a function of the compositions.

CONCLUSIONS

Continuous SiCN films have been successfully deposited using methylamine (CH_3NH_2) as carbon source in an ECR-CVD reactor. The carbon content in the SiCN films can be tailored by adjusting the $[CH_3NH_2]/[SiH_4]$ flow ratio and the substrate temperature. The XPS and TEM diffraction patterns suggest that the incorporated carbon resides in the Si site of the α-Si_3N_4 structure, and the lattice parameters remain up to 13% carbon content. The band edge measured by PzR indicates a direct band gap of around 4.4 eV.

Table III   The measured d-spacing from SAD patterns of $Si_{38}C_5N_{57}$ and $Si_{32}C_{13}N_{55}$, and the calculated values of α-Si_3N_4.

$Si_{38}C_5N_{57}$		$Si_{32}C_{13}N_{55}$		α-Si_3N_4			$Si_{38}C_5N_{57}$		$Si_{32}C_{13}N_{55}$		α-Si_3N_4		
d(Å)	I	d(Å)	I	d(Å)	hkl	I	d(Å)	I	d(Å)	I	d(Å)	hkl	I
6.7	s	6.77	s	6.69	100	8	2.2	m			2.244	300	6
				4.32	101	50					2.158	202	30
3.88	s	3.84	s	3.88	110	30					2.083	301	55
3.25	s	3.33	s	3.37	200	30	1.92	w	1.93	w	1.937	220	2
				2.893	201	85					1.884	212	8
				2.823	002	6	1.84	w			1.864	310	8
2.68	s	2.64	s	2.599	102	75					1.806	103	12
2.53	w	2.51	m	2.547	210	100					1.771	311	25
2.33	m	2.32	w	2.320	211	60	1.75	m			1.751	302	2
				2.283	112	8	1.66	w	1.65	w	1.637	203	8
							a=7.74 Å		a=7.82 Å		a=7.77 Å		
							c=5.85 Å		c=5.73 Å		c=5.62 Å		

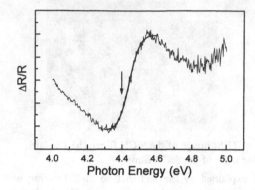

Figure 4   The PzR spectrum of $Si_{28}C_{15}N_{57}$ film, in which the fitting line is also plotted. The band gap value is 4.4 eV as indicated by the arrow.

REFERENCES

1. L.C. Chen, C.K. Chen, S.L. Wei, D.M. Bhusari, K.H. Chen, Y.F. Chen, Y.C. Jong and Y.S Huang Appl. Phys. Lett. **72**, p. 2463 (1998).
2. G. Sato, E.C. Samano, R. Machorro and L. Cota J. Vac. Sci. Technol. A **16**, p. 1311 (1998).
3. F.J. Gomez, P. Prieto, E. Elizalde and J. Piqueras Appl. Phys. Lett. **69**, p. 773 (1996).
4. L.C. Chen, C.Y. Yang, D.M. Bhusari, K.H. Chen, M.C. Lin, J.C. Lin, and T.J. Chuang Diamond and Related Mater. **5**, p. 514 (1996).
5. L.C. Chen, D.M. Bhusari, C.Y. Yang, K.H. Chen, T.J. Chuang, M.C. Lin, C.K. Chen, and Y.F. Huang Thin Solid Films **303**, p. 66 (1997).
6. G. Viera, J.L. Andujar, S.N. Sharma and E. Bertran Diamond and Related Mater. **7**, p. 407 (1998).
7. A. Badzian, T. Badzian, W.D. Drawl, and R. Roy Diamond and Related Mater. **7**, p. 1519 (1998).
8. R. Riedel, A. Greiner, G. Miehe, W. Dressler, H. Fuess, J. Bill and F. Aldinger, Angew Chem. Int. Ed. Engl. **36**, p. 603 (1997).
9. K.H. Chen, J.-J. Wu, C.-Y. Wen, L.C. Chen, C.-W. Fan, P.-F. Kuo, Y.-F. Chen, Y.-S. Huang, Thin Solid Films **355-356**, p. 203 (1999).
10. J.-J. Wu, K.H. Chen, C.-Y. Wen, L.C. Chen, J.K. Wang, Y.-C. Yu, C.-W. Wang, and E.-K. Lin, submitted to J. Chem. Mater.
11. J.-J. Wu, K.H. Chen, C.-Y. Wen, L.C. Chen, X.-J. Guo, H.J. Lo, S.T. Lin, Y.-C. Yu, C.-W. Wang, and E.-K. Lin, accepted by Diamond and Related Mater.
12. Y. Feng, Z. Zhou, Y. Zhou, G. Zhao Nul. Instr. and Meth. B **86**, p. 225 (1994).
13. Y. Feng, Z. Zhou, G. Zhao, F. Yang Nul. Instr. and Meth. B **94**, p. 11 (1994).
14. P.Y. Yu, M. Cardona, Fundamentals of Semiconductors, Springer, Berlin, 1996, pp. 307.
15. D.Y. Lin, C.F. Li, Y.S. Huang, Y.C. Jong, Y.F. Chen, L.C. Chen, C.K. Chen, K.H. Chen, D.M. Bhusari, Phys. Rev. B **56**, p. 6498 (1997).
16. F. Westly, D.H. Frizzell, J.T. Herron, R.F. Hampson, W.G. Mallard NIST Chemical Kinetics Database, Version 6.01, U.S. Department of Commerce, Technology Administration, National Institute of Standards and Technology, Standard Reference Data Program: Gaithersburg, MD.
17. D. Marton, K.J. Boyd, A.H. Al-Bayati, S.S. Todorov, J.W. Rabalais Phys. Rev. Lett. **73**, p. 118 (1994).
18. L. Kubler, J.L. Bischoff, and D. Bolmont Phys. Rev. B **38**, p. 13113 (1988).
19. C.H. Hsieh, Y.S. Huang, K.K. Tiong, C.W. Fan, Y.F. Chen, L.C. Chen, J.J. Wu, and K.H. Chen, accepted by J. Appl. Phys. (1999).

PREPARATION OF HIGH QUALITY ULTRA-THIN GATE DIELECTRICS BY CAT-CVD AND CATALYTIC ANNEAL

Hidekazu Sato[†,††], Akira Izumi[†] and Hideki Matsumura[†]

[†] Japan Advanced Institute of Science and Technology (JAIST), 1-1 Asahidai, Tatsunokuchi, Ishikawa 923-1292, Japan. hsato@jaist.ac.jp

[††] FUJITSU Limited, 1500 Mizono, Tadocho, Kuwana-gun, Mie, 511-0192, Japan.

ABSTRACT

This paper reports a feasibility of Cat-CVD system for improvement in characteristics of ultra thin gate dielectrics. Particularly, the effects of post deposition catalytic anneal (Cat-anneal) by using hydrogen (H_2)-decomposed species or NH_3-decomposed species produced by catalytic cracking of H_2 or NH_3, are investigated. The C-V characteristics are measured by MIS diode for the 4.5nm-thick Cat-CVD SiN_x and 8nm-thick sputtered SiO_2 for comparison. The small hysteresis loop is seen in the C-V curve of both SiN_x and SiO_2 films as deposition. However, it is improved by the Cat-anneal using H_2 or NH_3, and the hysteresis loop completely disappears from the C-V curves for both films. This result demonstrates that the Cat-anneal is a powerful technique to improve quality of insulating films, such as Cat-CVD SiN_x and even sputtered SiO_2 films. In addition, the leakage current of SiN_x films with 2.8nm equivalent oxide thickness is decreased by several orders of magnitude than that of the conventional thermal SiO_2 of similar EOT and the breakdown field is increased several MV/cm by Cat-anneal at 300℃.

INTRODUCTION

As semiconductor devices are scaled down to deep sub-micron dimensions, the conventional processing temperatures around 900℃ will be incompatible with the desired device structure. For example, the conventional high temperature process for formation of gate insulators changes the impurity profile in the substrate. Thus the gate insulator also must be prepared at low temperatures below 550℃[1]. Therefore the lowering growth temperature of insulator films is a key for the fabrication of the future ULSI.

The Cat-CVD method is a novel technique, in which deposition gases are decomposed by catalytic cracking reactions with a heated catalyzer placed near substrates so that SiN_x films are deposited at substrate temperatures around 300℃ without help from plasma nor photochemical excitation [2,3]. Thus, the surface of substrates and the films are not sustained plasma damage. Actually, we have already succeeded to deposit high quality SiN_x films as passivation film as thick as 300nm by this method using a gas mixture of SiH_4 and NH_3 [4]. When the flow rate of NH_3 exceeds over 50-100 times of that of SiH_4, nearly

121

stoichiometric (Si_3N_4) films are formed in which hydrogen content is as low as a few at.%. Additionally, it is known that the Cat-CVD is useful not only deposition of films but also surface modification of semiconductors, such as direct nitridation of Si [5] and GaAs [6].

In the present paper, the feasibility of Cat-CVD system for improvement in characteristics of ultra-thin gate dielectrics is studied. Particularly, the effects of Cat-anneal by using the H_2 decomposed species or NH_3 decomposed species formed by catalytic cracking of H_2 or NH_3 is investigated for SiN_x films and also for sputtered SiO_2 films.

FUNDAMENTALS FOR EXPERIMENTS

Film formation technique of Cat-CVD

The Cat-CVD apparatus is schematically illustrated in Fig.1. A tungsten wire (diameter 0.5mm ϕ, and total length 1300mm) is used as the catalyzer and placed beneath the substrate with a distance of 40mm. A catalyzer is coiled, pinned by molybdenum wires and spread widely parallel to the substrate which has an area of 65mm $\times$ 70mm. The deposition chamber (diameter 200mm, height 200mm) is made of stainless steel. The sample substrates are attached to a substrate holder which is heated by a heater. A thermocouple is mounted just beside the substrate on the substrate holder to determine the holder temperature (T_h). This includes the effect of thermal radiation from the heated catalyzer. T_h is varied from 200℃ to 400℃, but mainly kept at around 300℃ during the deposition and Cat-anneal. The catalyzer is heated electrically. The temperature of the catalyzer (T_{cat}) is estimated by both an electronic infrared thermometer placed outside of a quartz window (emissivity is 0.4) and from the temperature dependence of the electric resistivity of the catalyzer. The source gas is introduced into the chamber from many stainless steel nozzles placed below the catalyzer. The gas was decomposed at the heated catalyzer. The gas pressure (P_g) is measured by an electronic capacitance mano-meter and it is kept at about several mTorr. The chamber is pumped down to about 2.0×10^{-7} Torr by a rotary and a diffusion pump before process. The chamber pressure during deposition or Cat-anneal is controlled by main valve between the diffusion pump and the chamber.

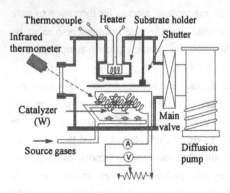

Thermocouple   Heater   Substrate holder
Infrared   Shutter
thermometer

Catalyzer
(W)   Main
                          valve
Source gases   Diffusion
                               pump

Fig.1 Schematic diagram of a Cat-CVD system.

Conditions of film formations and Cat-anneal

An n-type CZ-Si(100) wafer with a resistivity of $0.85-1.50\,\Omega\cdot cm$ is degreased and cleaned by RCA method. Then, it is dipped in 0.5% diluted HF for 0.5 min. After the cleaning, the SiN_x or SiO_2 films are formed on the Si substrate. The conditions of the SiN_x and SiO_2 are summarized in Table I and II. After the formation of insulators the Cat-anneal is performed. The conditions of the Cat-anneal are summarized in Table III.

Table I Deposition conditions of SiN_x films.

Substrate	Cz n-type Si(100) 0.85-1.50 Ω·cm
SiH_4 flow rate	1.1 sccm
NH_3 flow rate	60 sccm
Gas pressure (P_g)	10 mTorr
Power supply to catalyzer (PW_{cat})	680W
Catalyzer temperature (T_{cat})	1800-1900 ℃
Substrate holder temperature (T_{sh})	300 ℃

Table II Sputter conditions of SiO_2 films.

Substrate	Cz n-Type Si(100) 0.85 - 1.5 Ω·cm
Target	SiO_2
Suptter gas	Ar
Gas pressure (P_g)	40 mTorr
RF power	$2W/cm^2$
Substrate holder temp.	100 ℃

Table III Conditions of Cat-anneal.

PDA gas	Pressure (mTorr)	Catalyzer temp. (℃)	Substrate temp. (℃)	PDA time (min)
H_2	10	1800-1900	300	10-60
NH_3	10	1800-1900	300	10-60
He	10	1800-1900	300	60

The electrical properties of the deposition films are evaluated making an $Al/SiN_x/Si$ or $Al/SiO_2/Si$ (MIS) diode structures with the electrode area of $3 \times 10^{-2}\,mm^2$. In these samples, no post metal annealing (PMA) treatments are performed. The capacitance-voltage (C-V) characteristics are measured at the frequency of 1MHz and with the sweep rate of 0.1 V/s using a SANWA model MI-319A. The leakage current density - voltage (J-V) characteristics are measured by using a Hewlett-Packard semiconductor parameter analyzer model 4156A.

EXPERIMENTAL RESULTS

C-V characterization (Cat-CVD SiN_x)

The C-V characteristics measured on MIS diode with the Cat-CVD SiN_x films are shown in Fig.2. It is found that a small hysteresis loop is seen in the C-V curve of SiN_x films as deposition. However, it is improved by the Cat-anneal, the hysteresis loop disappears from the C-V curve. This result shows that reduction of electron charge trapped states can be achieved for SiN_x films by the Cat-anneal at 300℃.

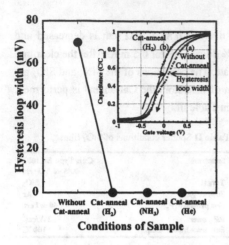

Fig. 2 The hysteresis loop width in the C-V characteristics with Cat-CVD SiN$_x$ with or without Cat-anneal. The Cat-anneal treatment time at 60min. The inset shows C-V curve. (a) Without Cat-anneal and (b) with Cat-anneal using H$_2$.

Figure 3 shows the change in dielectric constant for the Cat-CVD SiN$_x$ films by Cat-anneal as a function of the treatment time. The dielectric constant of the film is likely to increase by these Cat-anneals. We expect that the film density is increased by these Cat-anneals.

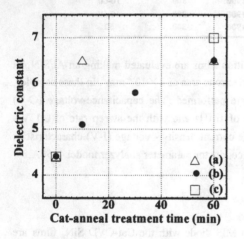

Fig. 3 Change in dielectric constant at the Cat-CVD SiN$_x$ films. (a) With Cat-anneal using H$_2$, (b) with Cat-anneal using NH$_3$ and (c) with Cat-anneal using He.

Figure 4 shows the change in interface-state density (D$_{it}$) for the Cat-CVD SiN$_x$ films by Cat-anneal as a function of the treatment time. The D$_{it}$ is reduced by Cat-anneal using H$_2$ or NH$_3$. However, in the case of Cat-anneal using He, the D$_{it}$ is not changed at all. This result implies hydrogen atoms or nitrogen atoms provided from H$_2$ or NH$_3$ would also terminate the interface dangling bonds.

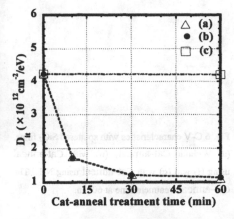

Fig. 4 Change in interface-state density at the Cat-CVD SiN$_x$ films. (a) With Cat-anneal using H$_2$, (b) with Cat-anneal using NH$_3$ and (c) with Cat-anneal using He.

J-V characterization (Cat-CVD SiN$_x$)

Figure 5 shows the leakage currents measured in Cat-CVD SiN$_x$ films and the conventional thermal SiO$_2$ films [7] whose equivalent oxide thickness (EOT) is about 2.9nm. It is found that the leakage currents are drastically decreased by the Cat-anneal using H$_2$ or NH$_3$, although nothing has changes for the case of using He. This result means H or N radicals is necessary to decrease the leakage currents. The Cat-anneal using H$_2$ or NH$_3$ for Cat-CVD SiN$_x$ films reduce the leakage currents by several orders of magnitude than the thermal SiO$_2$ of similar EOT.

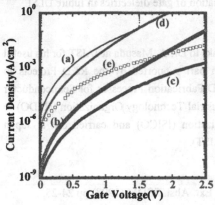

Fig. 5 J-V characteristics with Cat-CVD SiN$_x$. (a) Without Cat-anneal (EOT=2.97nm), (b) with Cat-anneal using H$_2$ (EOT=2.91nm), (c) with Cat-anneal using NH$_3$ (EOT=2.78nm), (d) with Cat-anneal using He (EOT=2.62nm) and (e) thermal SiO$_2$ (EOT=2.8nm). The Cat-anneal treatment time at 60min.

C-V characterization (Sputtered SiO$_2$)

The C-V characteristics measured on MIS diodes with the sputtered SiO$_2$ films is shown in Fig.6. A hysteresis loop is seen in the C-V curve of SiO$_2$ films as grown. However, it is improved by the Cat-anneal using H$_2$ or NH$_3$, the hysteresis loop disappears from the C-V curve. On the other hand, the flat band voltage is shifted to the negative by the Cat-anneal using NH$_3$. We consider that the SiO$_2$ film is nitrided by the Cat-anneal using NH$_3$.

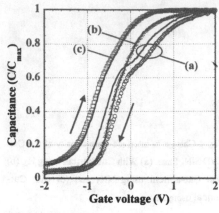

Fig. 6 C-V characteristics with sputtered SiO₂ films. (a) Without Cat-anneal, (b) with Cat-anneal using H_2 and (c) with Cat-anneal using NH_3. The Cat-anneal treatment time at 60min.

CONCLUSIONS

The Cat-anneal with Cat-CVD system is investigated for improvement of MIS characteristics for the substrate temperature at 300℃. Particularly, the Cat-anneal using H_2 or NH_3 plays remarkable role to reduce the D_{it} and the leakage currents of Cat-CVD SiN_x films by several orders of magnitude than that of the conventional thermal SiO_2 of similar EOT and to increase the breakdown field to several MV/cm. Moreover, even the quality of sputtered SiO_2 film is improved by Cat-anneal. It is concluded that the Cat-anneal technology in Cat-CVD system is the promising candidates for fabrication of gate dielectrics in future ULSI.

ACKNOWLEDGEMENTS

The authors would like to express his thanks to Dr. A. Masuda at JAIST for his useful discussions and encouragement. This work is in part supported by the R&D Projects in Cooperation with Academic Institutions "Cat-CVD Fabrication Processes for Semiconductor Devices" entrusted from the New Energy and Industrial Technology Organization (NEDO) to the Ishikawa Sunrise Industries Creation Organization (ISICO) and carried out at Japan Advanced Institute of Science and Technology (JAIST).

REFERENCES

1. Y. Saito, K. Sekine, M. Hirayama and T. Ohmi, Ext. Abst. SSDM (1998) pp.24-25.

2. H. Matsumura and H. Tachibana, Appl. Phys. Lett. **47**, 833 (1985).

3. H. Matsumura, Jpn. J. Appl. Phys. **37**, 3175 (1998).

4. S. Okada and H. Matsumura, Jpn. J. Appl. Phys. **36**, 7035 (1997).

5. A. Izumi and H. Matsumura, Appl. Phys. Lett. **71**, 1371 (1997).

6. A. Izumi, A. Masuda and H. Matsumura, Thin Solid Films, **343-344**, 528 (1999).

7. T. P. Ma, IEEE Trans. Electron Devices, **45**, 680 (1998).

GROUP III METAL SULFIDE THIN FILMS FROM SINGLE-SOURCE PRECURSORS BY CHEMICAL VAPOR DEPOSITION (CVD) TECHNIQUES

MIKE R. LAZELL,[a] PAUL O'BRIEN,[a*] DAVID J. OTWAY,[b] JIN-HO PARK[b]
a. Manchester Materials Sciences Centre and Department of Chemistry, University of Manchester, Oxford Road, Manchester, M13 9PL, UK.
b. Department of Chemistry, Imperial College of Science, Technology and Medicine, Exhibition Road, London, SW7 2AY, UK. *Email: j.h.park@ic.ac.uk; paul.obrien@man.ac.uk*

ABSTRACT

Several single-source precursors including $In(SOCN^iBu_2)_3$, $In(S_2CNMeHex)_3$ and $Ga(S_2CNMeR)_3$, (R = Et, Bu, Hex) have been prepared and used for the deposition of Group 13 metal sulfide thin films. The α- and β-In_2S_3 thin films on borosilicate glass and α-Ga_2S_3 thin films on GaAs(111) single crystal substrates were prepared from the precursors by various chemical vapour deposition (CVD) techniques. These semicondcuting materials have been characterized by XRD, SEM, XPS and EDAX.

INTRODUCTION

There has been recent interest in the preparation of group 13 chalcogenide materials. In particular thin films of the type ME or M_2E_3 [M = Al, Ga, In, Tl; E = S, Se, Te] have been shown to possess attractive properties for a variety of optical and electronic applications [1,2]. Indium sulfide is a mid band-gap semiconducting material and depending on stoichiometric ratios between indium and sulfur, the band-gap varies between 2 and 2.44 eV. The plethora of different techniques has been utilized to deposit these materials. The most widely studied growth technique is metal-organic chemical vapour deposition (MOCVD), either at low or atmospheric pressure (LP- or AP-MOCVD), or related methods such as aerosol-assisted (AACVD), or plasma-enhanced (PECVD) chemical vapour deposition.

Single-source precursors have often been used for the deposition of group 13 chalcogenide thin films (*i.e.* both the metal and chalcogen atoms are contained within the precursor complex); a number of different chalcogen containing ligands have been employed for the preparation of such precursor complexes, including thiolates, [3,4], selenolates, [5], thiocarboxylates, [6], thiocarbonates, [7], dithiocarbamates, [8,9] and diselenocarbamates [10].

Recently, we have reported the deposition of indium sulfide (β-In_2S_3) and gallium sulfide (GaS) thin films by LP-MOCVD and AACVD, grown from novel indium and gallium monothiocarbamato complexes, $In(SOCNEt_2)_3$, [11], $In(SOCN^iPr_2)_3$, [12] and $Ga(SOCNEt_2)_3$ [13].

Herein we report the deposition of gallium sulfide thin films (α-Ga_2S_3) deposited on GaAs(111) single crystal substrates using the metal dithiocarbamato complexes, $Ga(S_2CNMeEt)_3$, $Ga(S_2CNMeBu)_3$, $Ga(S_2CNMeHex)_3$ by LP-MOCVD and films of indium sulfide (β-In_2S_3) deposited on glass using $In(S_2CNMeHex)_3$ by AACVD. We have also continued our investigations into monothiocarbamato systems and report the deposition of cubic-In_2S_3 thin film from the di-*iso*-butylmonothiocarbamato complex, $In(SOCN^iBu_2)_3$, on borosilicate glass substrates by LP-MOCVD.

EXPERIMENTAL

Chemicals

Carbon disulfide, carbonyl sulfide, sodium hydroxide and secondary amines were purchased from Aldrich chemical Company Ltd and methanol and THF were from BDH. These were used with no further purification. Gallium (III) sulfide and indium (III) sulfide were gifts from Epichem Ltd.

Instrumentation

^{1}H NMR spectroscopy was carried out on a Bruker AM270 Fourier-transform instrument. Elemental analysis was performed by the Imperial College Chemistry Departmental Service. All manipulations and reactions were carried out in an inert atmosphere using Schlenk techniques and a vacuum line. X-Ray diffraction (XRD) patterns were measured using a Siemens D500 series automated powder diffractometer using Cu-Kα (radiation at 40 kV/40 mA) with a secondary graphite crystal monochromator. Samples were mounted on a glass slide (5 x 5cm) and scanned from 10 –60 ° in steps of 0.04° with a count time of 2 s. Energy Dispersive X-ray microanalysis (EDAX) was performed on a LINK QX2000 energy dispersive X-ray microanalysis unit. For scanning electron microscopy (SEM) the samples were gold-coated using a Balzers SCD 030 sputter coating unit before study with a JEOL JEM-1200 EX-II microscope.

Preparation of single-source precursors

The single-source precursors, In(S$_2$CNMeHex)$_3$, In(SOCN^{i}Bu$_2$)$_3$, Ga(S$_2$CNMeHex)$_3$, Ga(S$_2$CNMeBu)$_3$ and Ga(S$_2$CNMeEt)$_3$ have been prepared by the literature method [9,11] and were characterized by elemental analysis, IR and NMR.

Preparation of thin films by AACVD

Typically, In(S$_2$CNMeHex)$_3$ (0.3 g) was dissolved in THF (30 ml) at room temperature, and injected into the growth apparatus, with the substrate at various temperatures. The system was allowed to run for 1 hr., with a constant nitrogen flow rate of 200 sccm (carrying the precursor from flask to substrate), at which point the substrates were collected.

Preparation of thin films by LP-MOCVD

Low-pressure growth experiments ($\approx 10^{-2}$ Torr) were carried out using an Edwards model E2M8 vacuum pump system and the films were deposited on borosilicate glass and GaAs(111) substrates by LP-MOCVD. Growth experiments have been described elsewhere [14].

RESULTS AND DISCUSSION

Growth experiments were carried using LP-MOCVD for the deposition of indium sulfide from In(SOCN^{i}Bu$_2$)$_3$ and of gallium sulfide from Ga(S$_2$CNMeR)$_3$ (R = Et, Bu, Hex). AACVD was used for growing indium sulfide from In(S$_2$CNMeHex)$_3$. In(S$_2$CNMeHex)$_3$ has been reported to grow cubic α-In$_2$S$_3$ at 500 °C by LP-MOCVD.[9] However, employing AACVD, the deposition temperature can be reduced to as low as 350 °C resulting in tetragonal β-In$_2$S$_3$. Films grown at 350 to 450 °C indicated only single-phase β-In$_2$S$_3$ by XRD after one hour growth. Higher temperatures (450 - 475 °C) yielded films that were orange-red in color and well adhered to the glass substrate surface. All films strongly adhered to the substrates.

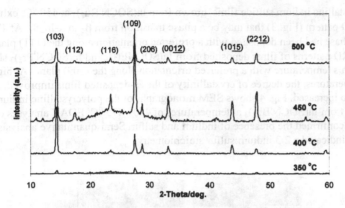

Fig. 1 XRD pattern and SEM image of β-In$_2$S$_3$ grown on glass from In(S$_2$CNMeHex)$_3$ by AACVD (temperatures indicate deposition temperatures).

Fig. 2 SEM images of β–In$_2$S$_3$ films deposited using In(S$_2$CNMeHex)$_3$ at 400 °C on glass substrates by AACVD.

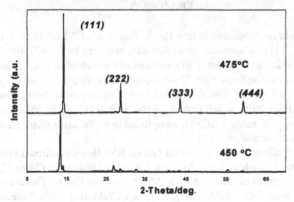

Fig. 3 XRD patterns of α-In$_2$S$_3$ grown on glass at 450 - 475 °C from In(SOCN^{t}Bu$_2$)$_3$ by LP- MOCVD.

Polycrystalline indium sulfide films, grown from In(SOCN^{i}Bu$_2$)$_3$ at 450 °C, exhibited an interesting XRD pattern (Fig. 3) that may be a phase transition from β- to α-In$_2$S$_3$. At 475 °C, cubic α-In$_2$S$_3$ phase had been deposited with a preferred orientation along the (111) plane. In contrast, the XRD patterns of films deposited from In(SOCNEt$_2$)$_3$ and In(SOCN^{i}Pr$_2$)$_3$ showed β-In$_2$S$_3$ phase at this temperature with a preferred orientation along the (109) plane. At higher deposition temperatures, the degree of crystallinity of the as-deposited films improved, and the particle size also increased. Fig. 4 shows SEM monographs of the polycrystalline indium sulfide films deposited from In(SOCN^{i}Bu$_2$)$_3$ at temperatures of 450 - 475 °C. EDAX analyses of the as-deposited films confirmed the presence of indium and sulfur. Semi-quantitative analysis of the EDAX profiles indicated a 2:3 indium:sulfur stoichiometry.

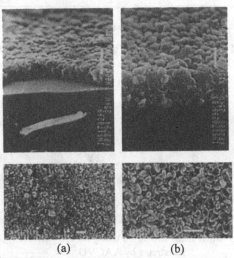

(a)   (b)

Fig. 4 SEM images of α-In$_2$S$_3$ grown on glass at 450 (a) and 475 (b) °C from In(SOCN^{i}Bu$_2$)$_3$ by LP-MOCVD.

There was little or no deposition below 450 °C from Ga(S$_2$CNMeR)$_3$, (R = Et, Bu, Hex). Films grown on GaAs(111) single crystal substrates were analyzed by EDAX showing a gallium to sulfur ratio of close to 2:3 on all cases. XRD analyses of the thin films (Fig. 5) grown from the compounds indicated that α-Ga$_2$S$_3$ (JCPDS #30-577) were deposited on GaAs(111) with a strong reflection along the (002) direction. The XRD results are much improved over films grown from precursors such as Et$_2$GaS$_2$CNEt$_2$, which gave amorphous films [8]. Films obtained from a gallium thiocarboxylate, [15], using AACVD, were found to be the same phase (α-Ga$_2$S$_3$) but with a different preferred orientation (312).

XPS analysis of gallium sulfide film from Ga(S$_2$CNMeHex)$_3$ confirmed that only gallium (2p$_1$, 2p$_3$ and Auger lines L$_2$M$_{45}$M$_{45}$, L$_3$M$_{23}$M$_{45}$ and L$_3$M$_{45}$M$_{45}$) and sulfur (2s and 2p$_3$) are present at the surface. Analyses of the films by SEM are shown in Fig. 6. The average particle size is *ca.* 1 µm grown from Ga(S$_2$CNMeHex)$_3$, and Ga(S$_2$CNMeBu)$_3$ but 1.2 µm from Ga(S$_2$CNMeEt)$_3$. The growth rate of the films is found to be *ca.* 3 µm/hr grown from Ga(S$_2$CNMeHex)$_3$, 2.4 µm/hr from Ga(S$_2$CNMeBu)$_3$ and 2 µm/hr from Ga(S$_2$CNMeEt)$_3$. Films grown from Et₂GaS₂CNEt₂, [8] or [('Bu)₂Ga(S'Bu)], [16], were dominated by spherical particles, a different morphology from those observed in this study.

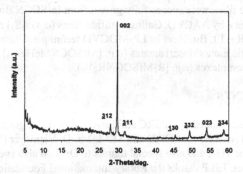

Fig. 5 XRD pattern of α-Ga₂S₃ film grown on GaAs(111) at 500 °C from Ga(S₂CNMeHex)₃ by LP- MOCVD.

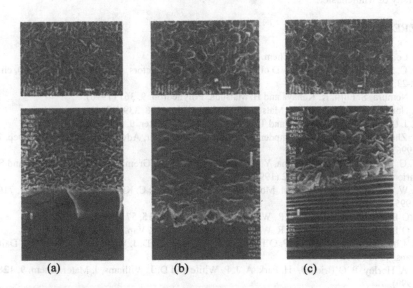

Fig. 6 SEM images of gallium sulfide thin films grown from (a) Ga(S₂CNMeEt)₃, (b) Ga(S₂CNMeBu)₃ and (c) Ga(S₂CNMeHex)₃ at 500 °C by LP-MOCVD.

It is interesting to note that the precursor [Ga(SOCNEt₂)₃], under similar conditions, gives films of the cubic phase of GaS with a **1:1** stoichiometry. The present work leads to the more sulfur rich phase Ga₂S₃. It is tempting to conclude that the different phases depend upon the supply of sulfur from the precursor, which may suggest that the amount of an element delivered by a single-molecule precursor, is one factor important in controlling the phase deposited.

CONCLUSIONS

Group III sulfide films been prepared from $[M(S_2CNRR')_3]_n$ by LP-MOCVD and AACVD. Indium sulfide films were successfully grown from $In(SOCN^iBu_2)_3$ by LP-MOCVD and from $In(S_2CNMeHex)_3$ by AACVD. Gallium sulfide films (α-Ga_2S_3) have been prepared from $Ga(S_2CNMeR)_3$, (R = Et, Bu, Hex) by LP-MOCVD technique. Future directions include the preparation of asymmetric monothiocarbamates (e.g. $[M(SOCNMeHex)_3]_n$) and growth from polymeric mixed alkyl complexes (e.g. $[R_2M(SOCNR_2)]_n$).

ACKNOWLEDGEMENTS

This work has been funded by EPSRC and Leverhulme Trust grants to P.O.B. We would like to thank Mr. K. Pell (QMW) for EDAX and SEM, Mr. D. Niemeyer (UCL) for XPS, and Dr. M. Odlyha (Birkbeck) for TGA and also special thanks to Dr. J. Walsh and Mr. G. A. Horley for initial precursor syntheses. J.H.P thanks the Rotary International Foundation for a studentship. P.O.B is the visiting Sumitomo/STS Professor of Materials at IC and Professor of Inorganic Materials at The Manchester Materials Science Centre and The Chemistry Department, University of Manchester.

REFERENCES

1. D. Coucouvanis, Prog. Inorg. Chem. **26**, 301 (1979).
2. A. C. Jones and P. O'Brien, CVD of Compounds Semiconductors, (VCH, Weinhem, 1997), ch 1, pp. 22-23; ch. 7, pp. 307-309.
3. R. Nomura, S. Fujii, K. Kanaya and H. Matsuda, Polyhedron **9**, 361 (1990).
4. R. Nomura, K. Konishi and H. Matsuda, Thin Solid Films **198**, 339 (1991).
5. H. J. Gysling, A. A. Wernberg and T. N. Blanton, Chem. Mater. **9**, 900 (1992).
6. G. Zheng, K. Kunze, M. J. Hampden-Smith and E. N. Duesler, Adv. Mater., Chem. Vap. Dep. **2**, 242 (1996).
7. V. G. Bessergenv, E. N. Ivanona, Y. A. Kovalevskaya, S. A. Gromilov, V. N. Kirichenko and S. V. Larionov, Inorg. Mater. **32**, 592 (1996).
8. S. W. Haggata, M. A. Malik, M. Motevalli, P. O'Brien and J. C. Knowles, Chem. Mater. **7**, 716 (1995).
9. P. O'Brien, D. J. Otway and J. R. Walsh, Thin Solid Films **315**, 57 (1998).
10. P. O'Brien, D. J. Otway and J. R. Walsh, Adv. Mater., Chem. Vap. Dep. **3**, 227 (1997).
11. M. Chunggaze, G. A. Horley, P. O'Brien, A. J. P. White and D. J. Williams, J. Chem. Soc. Dalton Trans. 4205 (1998).
12. G. A. Horley, P. O'Brien, J. -H. Park, A. J. P. White and D. J. Williams, J. Mater. Chem. **9**, 1289 (1999).
13. G. A. Horley, M. R. Lazell and P. O'Brien, Adv. Mater., Chem. Vap. Dep. **5**, 203 (1999).
14. M. A. Malik and P. O'Brien, Adv. Mater. Opt. Electron. **3**, 71 (1994).
15. G. Shang, M. J. Hampden-Smith and E. N. Duesler, J. Chem. Soc. Chem. Commun. 1733 (1996).
16. A. N. MacInnes, M. B. Power, A. R. Barron, Chem. Mater. **5**, 1344 (1993).

IRON SULFIDE (FeS2) THIN FILMS FROM SINGLE-SOURCE PRECURSORS BY AEROSOL-ASSISTED CHEMICAL VAPOR DEPOSITION (AACVD)

PAUL O'BRIEN,[a*] DAVID J. OTWAY,[b] JIN-HO PARK [a,b]
a. Manchester Materials Sciences Centre and Department of Chemistry, University of Manchester, Manchester, M13 9PL, UK.
b. Department of Chemistry, Imperial College of Science, Technology and Medicine, Exhibition Road, London, SW7 2AY, UK. *Email: j.h.park@ic.ac.uk; paul.obrien@man.ac.uk*

ABSTRACT

Dialkyl (or mixed alkyl)-dithiocarbamato iron(III) complexes have been used for the deposition of iron sulfide thin films using chemical vapor deposition techniques. The single-source precursors used in this work have been prepared by the reaction of $FeCl_3$ with dialkyldithiocarbamate sodium salts and characterized by a number of analytical techniques. Good quality thin films of FeS_2 have been prepared from the single-source metal organic precursor, $[Fe(S_2CNMe^iPr)_3]$, by AACVD. XRD patterns of the films indicated crystalline iron sulfide (FeS_2) grown at between 375 – 450 °C. SEM images show the films to have reasonable morphology and to be crystalline.

INTRODUCTION

In recent years there has been a significant increase in interest in metal sulfide materials that have potential for use in the electronics industry or photovoltaic applications. Pyrite (FeS_2) has an indirect band gap of 0.95 eV and an absorption coefficient of 6×10^5 cm^{-1} in the visible range.[1] FeS_2 has been prepared as a polycrystalline thin film by several techniques including atmospheric and low-pressure metal-organic chemical vapor deposition,[2,3,4,5,6] the sulfurization of iron oxides,[7] flash evaporation,[8] ion beam and reactive sputtering,[9] plasma-assisted sulfurization of thin iron films,[10] vapor transport,[11] chemical spray pyrolysis,[12] and vacuum thermal evaporation.[13] Among these techniques MOCVD is one of the most promising methods for the preparation of crystalline iron disulfide films of good quality and a well-defined stoichiometry.

Schleigh *et al.* reported the preparation of iron disulfide using iron pentacarbonyl [$Fe(CO)_5$], hydrogen sulfide (H_2S) and tert-butyl sulfide as precursors by LP-MOCVD. However, relatively high deposition temperatures (480 – 525 °C) were needed; lower temperatures resulted in poor morphology of the films and impurities were observed. Meester *et al.* also studied thin films of iron disulfide prepared from iron(III)acetylacetonate, [$Fe(acac)_3$], tert-butyl disulfide and hydrogen. In their studies, films were deposited on glass, titanium dioxide and silicon at temperatures from 300 to 340 °C. However, there is no literature available regarding the use of true single-source precursors for the deposition of iron sulfide thin films. Metal dithiocarbamate complexes of formula, $M(S_2CNRR')_n$, [M = Cd(II), Zn(II), In(III)], have been shown to be promising single-source precursors for the deposition of metal sulfide thin films by CVD techniques. CdS, ZnS and In_2S_3 have been prepared by CVD methods from such single-source precursors.

In this paper we report the syntheses of single source dialkyl and mixed alkyl-dithiocarbamato iron(III) complexes and the deposition of iron disulfide thin films using the

Mat. Res. Soc. Symp. Proc. Vol. 606 © 2000 Materials Research Society

above precursors by LP-MOCVD and AACVD techniques and as analyzed by XRD, EDAX and SEM.

EXPERIMENTAL

Chemicals

Carbon disulfide, sodium hydroxide, methylisopropyl amine (Me^iPrNH), dibutyl amine ($^nBu_2NH$) and iron (III) chloride were purchased from Aldrich chemical Company Ltd. Methanol and THF were from BDH and used without further purification.

Instrumentation

X-Ray diffraction (XRD): XRD patterns were measured using a Siemens D500 series automated powder diffractometer using Cu-Kα (radiation at 40 kV/40 mA) with a secondary graphite crystal monochromator. Samples were mounted on glass slides (5 × 5cm) and scanned from 10 –60 °C in steps of 0.04° with a count time of 2 sec. Energy Dispersive X-ray microanalysis (EDAX) was performed on a LINK QX2000 energy dispersive X-ray microanalysis unit. Scanning Electron microscopy (SEM): the samples were gold-coated using a Balzers SCD 030 sputter coating unit before electron microscopy was carried out on a JEOL JEM-1200 EX-II microscope.

Preparation of single-source precursors

The precursors were prepared by literature methods[14] and analyzed by CHN and IR. Solutions of sodium salts of the ligands were prepared by adding 0.05 mol of CS_2 to a solution of 0.05 mol of the amine in ethanol (50 cm³). 6 M NaOH (10 cm³) was added with stirring. The complexes were prepared by mixing 0.017 mol of 60% w/v $FeCl_3$ aqueous solution with the solution from the ligand preparation. A black-brown precipitate immediately formed and was recovered by vacuum filtration, washed with ethanol and air dried. The complex was recrystallised by dissolution in hot $CHCl_3$ (30 cm³) (in a fume hood), vacuum filtration, and addition of ethanol (30 cm³) to the filtrate. Black or dark brown solids formed on cooling; the solids were recovered by vacuum filtration and were washed with ethanol and air dried.

Preparation of FeS₂ thin films by AACVD

Typically the precursor (0.3 g) was dissolved in THF (30 ml) at room temperature and injected into the growth apparatus with the substrate at various temperatures. The system was allowed to run for 2 h., with a constant nitrogen flow rate of 200 sccm (carrying the precursor from the flask to the substrate), at which point the substrates were removed.

Preparation of FeS₂ thin films by LP-MOCVD

The reactor was a cold wall, horizontal reactor. Glass, GaAs(111) and InP(111) substrates were heated by a tungsten halogen lamp and precursor by a Carbolite tube furnace with the system under a dynamic vacuum at approx. 10^{-2} Torr. The iron complexes as above (~ 20 mg) were used in each experiment.

RESULTS AND DISCUSSION

The precursors used in this work have two advantages; they are air stable and soluble in most organic solvents. The preparation of the sodium salt *in situ* and the subsequent precipitation of the iron salt is a clean and efficient method of producing high yields of crystalline compounds. Two different types of CVD techniques were employed; low-pressure chemical vapor deposition (LP-MOCVD) and aerosol-assisted chemical vapor deposition (AACVD).

LP-MOCVD did not prove for the growth of iron sulfide films using $[Fe(S_2CN^nBu_2)_3]$ and $[Fe(S_2CNMe^iPr)_3]$ and very poor morphology was obtained on glass, GaAs(111) or InP(111). The temperature range for growth of the film by LP-MOCVD was between 350 to 500 °C and the compounds tended to sublime onto the cool sides of the reactor tube. However, the preparation of iron sulfide films by AACVD was more promising.

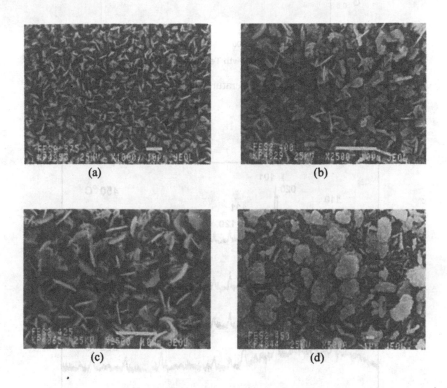

Fig. 1 SEM images of FeS$_2$ films deposited by AACVD from $[Fe(S_2CNMe^iPr)_3]$ at (a) 375 °C, (b) 400 °C, (c) 425 °C and (d) 450 °C on glass substrates (bar size = 10μm except (d) indicating 1 μm).

There was little or no deposition observed on glass substrates below 350 °C. Deposition occurred at growth temperatures above 375 °C. The thickness of films varied with deposition temperatures. SEM images are shown in Figure 1. The FeS$_2$ films deposited from $[Fe(S_2CNMe^iPr)_3]$ on glass substrates tended to be denser at higher temperatures. The highest growth rate was *ca.* 2 μm/hr at 425 °C although between 375 and 425 °C we are probably in a region where diffusion limits the rate of film growth.

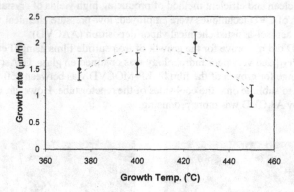

Fig. 2 Growth rates *vs* deposition temperatures for [Fe(S$_2$CNMe^{i}Pr)$_3$] by AACVD.

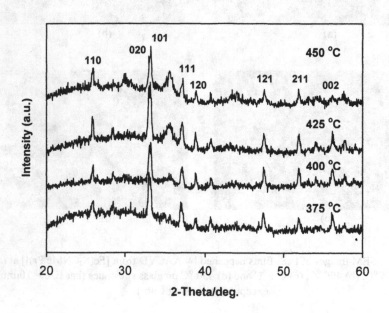

Fig. 3 X-ray diffraction patterns for FeS$_2$ grown on glass at various deposition temperatures from [Fe(S$_2$CNMe^{i}Pr)$_3$].

Table 1 X-ray diffraction data for FeS_2 grown from [Fe(S$_2$CNMe^{i}Pr)$_3$].

JCDPS 37-475 d-spacing (Å), (Intensity: %), (hkl)		375 °C	400 °C	425 °C	450 °C
3.439 (60)	(110)	3.445 (27)	3.439 (41)	3.444 (39)	3.440 (43)
2.712 (35)	(020)	2.707 (100)	2.709 (90)	2.706 (93)	2.706 (87)
2.693 (100)	(101)	2.699 (75)	2.695 (100)	2.696 (100)	2.695 (100)
2.413 (45)	(111)	2.425 (53)	2.418 (47)	2.417 (35)	2.410 (35)
2.315 (40)	(120)	2.323 (25)	2.314 (35)	2.319 (33)	2.314 (25)
2.221 (1)	(200)	2.214 (31)	2.211 (23)	2.213 (17)	2.212 (17)
1.912 (30)	(121)	1.918 (40)	1.915 (47)	1.913 (28)	1.913 (28)
1.757 (70)	(211)	1.758 (33)	1.759 (45)	1.759 (31)	1.757 (31)
1.693 (20)	(002)	n/a	1.692 (21)	1.687 (9)	n/a
1.675 (16)	(130)	n/a	1.673 (12)	n/a	n/a
1.595 (25)	(031)	1.596 (14)	1.567 (21)	1.593 (11)	1.596 (18)

Figure 2 shows the growth rates *vs* deposition temperatures in this work. The sulfur to iron ratio in the deposited films has a strong influence on the optoelectronic properties of pyrite. Therefore it is necessary to control the stoichiometry. EDAX studies indicated that a film deposited at 400 °C from [Fe(S$_2$CNMe^{i}Pr)$_3$] had only iron and sulfur in a close to one:two ratio. However, films grown at higher temperatures tended to be slightly iron rich.

The films grown at 400 – 500 °C from [Fe(S$_2$CN^{n}Bu$_2$)$_3$] were sulfur rich whilst FeS_2 films grown from the asymmetric iron precursor, [Fe(S$_2$CNMe^{i}Pr)$_3$] were slightly sulfur deficient. All the films grown from [Fe(S$_2$CNMe^{i}Pr)$_3$] gave reasonable peaks in the XRD patterns (Fig. 3) indicating monoclinic FeS_2 (JCPDS 37-475) whilst films from [Fe(S$_2$CN^{n}Bu$_2$)$_3$] gave amorphous XRD patterns. The diffraction patterns observed are shown in Fig. 2 and the data collected from the films are summarized in Table 1. It can be seen that the growth of thin films using [Fe(S$_2$CNMe^{i}Pr)$_3$] at various temperatures gave only the monoclinic FeS_2 phase. However, at higher deposition temperatures, the most intense peak was found at the (101) plane whilst at the lowest temperature (375 °C), the highest peak was at the (020) plane.

CONCLUSIONS

Good quality thin films of FeS_2 have been prepared by AACVD using [Fe(S$_2$CNRR')$_3$]$_n$ as single source precursors. By changing the alkyl groups in the precursor it is possible to deposit stoichiometrical crystalline iron sulfide (FeS_2) thin films. SEM images showed that the films grown at lower deposition temperatures form as hexagonal phases and at higher temperatures tended to form various morphology. Further work will be focus on dialkyl and mixed alkyl-monothiocarbamato iron(III) precursors for the deposition of FeS_2. They may also be useful as precursors for the growth of TOPO capped (tri-*n*-octylphosphineoxide) nanoparticles of FeS_2.

ACKNOWLEDGEMENTS

This work has been funded by EPSRC grants to P.O.B. We would like to thank Mr. K. Pell (QMW) for EDAX and SEM and Dr. M. Odlyha (Birkbeck) for TGA. J.H.P thanks the Rotary International Foundation for a studentship. P.O.B is the Sumitomo/STS Visiting Professor of Materials at IC and Professor of Inorganic Materials at The Manchester Materials Science Centre and The Chemistry Department, Manchester University.

REFERENCES

1. A. Ennaoui, S. Fiechter, C. Pettenkofer, N. Alonso-Vante, K. Buker, M. Bronold, C. Hopfner, and H. Tributsch, Sol. Energy Mater. Sol. Cells **29**, 289 (1993).
2. D.M. Schleigh and H.S.W. Chang, J. Crystal Growth **112**, 737 (1991).
3. B. Thomas, C. Hoepfner, K. Ellmer, S. Fiechter and H. Tributsch, J. Crystal Growth **146**, 630 (1995).
4. C. Hoepfner, K. Ellmer, A. Ennaoui, C. Pettenkofer, S. Fiechter and H. Tributsch, J. Crystal. Growth, **151**, 325 (1995).
5. B. Thomas, T. Cibik, C. Hoepfner, K. Diesner, G. Ehlers, S. Fiechter and K. Ellmer, J. Mat. Sci. **9**, 61 (1998).
6. B. Meester, L. Reijnen, A. Goossens and J. Schoonman, J. Phys. IV France **9**, Pr8-613 (1999).
7. G. Smestad, E. Ennaoui, S. Fiechter, H. Tributsch, W. K. Hofman and M. Birkholz, Sol. Energy Mater. **20**, 149, 1990.
8. I.J. Ferrer and C. Sanchez, J. Appl. Phys. **70**, 2641 (1991).
9. M. Birkholz, D. Lichtenberger, C. Hoepfner and S. Fiechter, Sol. Energy Mater. Sol. Cells **27**, 243 (1992).
10. S. Bausch, B. Sailer, H. Keppner, G. Willeke, E. Bucher and G. Frommeyer Appl. Phys. Lett. **57**, 25 (1990).
11. A. Ennaoui, G. Schlichtlorel, S. Fiechter and H. Tributch, Sol. Energy Mater. Sol. Cells **25**, 169 (1992).
12. G. Smestad, A. Da Silva, H. Tributsch, S. Fiechter, M. Kunst, N. Meziani and M. Birkholz, Sol. Energy Mater. **18**, 299 (1989).
13. B. Rezig, H. Dalman, M. Kanzai, Renewable Energy **2**, 125 (1992).
14. P. O'Brien, D.J. Otway and J.R. Walsh, Thin Solid Films **315**, 57 (1998).

VOLATILE LIQUID PRECURSORS FOR THE CHEMICAL VAPOR DEPOSITION (CVD) OF THIN FILMS CONTAINING ALKALI METALS

Randy N. R. Broomhall-Dillard, Roy G. Gordon and Valerie A. Wagner
Harvard University Chemical Laboratories, Cambridge, MA 02138

ABSTRACT

The first volatile, liquid compounds of alkali metals were synthesized and used for the CVD of materials containing alkali metals. Amides of the type $MNR^1(SiMe_2R^2)$ and $MN(SiMe_2R^2)_2$ [M = Li, Na, K; R^1 = t-butyl, t-amyl; R^2 = ethyl, n-propyl, i-propyl, n-butyl, i-butyl, n-hexyl, n-octyl] were made and characterized. The lithium amides were prepared via the deprotonation of the parent amine using butyl lithium. The sodium and potassium amides were formed by transamination of sodium amide and potassium bis(trimethylsilyl)amide with the parent amines. For example, lithium bis(ethyldimethylsilyl)amide was prepared from butyl lithium and bis(ethyldimethylsilyl)amine and was distilled as a clear, colorless liquid at 122 °C (0.2 Torr) having a viscosity of 37 cP at 40 °C. These alkali metal amides can be used as convenient liquid sources for CVD of mixed metal oxides containing alkali metals, such as the non-linear optical material lithium niobate, lithium-containing materials for battery electrodes, electrochromic tungsten bronzes, and the pyroelectric and ferroelectric material potassium tantalate.

INTRODUCTION

Chemical vapor deposition (CVD) is a versatile method for preparing solid materials in the form of films, powders, and fibers. Successful use of CVD requires reliable sources of reactant vapors, which are most easily generated from liquids. Solids are less convenient as sources of vapors for a variety of reasons. Solids often have low vapor pressures, and the kinetics of vaporization of solids is usually slow. The surface area from which solids evaporate changes as a function of time, causing a non-reproducible vapor flux. Decomposition products and impurities can segregate on the surface of a solid, further degrading their reproducibility as vapor sources. In fact, most practical applications of CVD use liquid sources, rather than solids.

Solid sources can also be dissolved in a liquid solvent, and the liquid solution can subsequently be flash-vaporized. This approach, however, introduces large amounts of solvent vapors into the CVD reactor. The solvent vapors can be hazardous because of flammability or toxicity and may introduce impurities, such as carbon, into the deposited material. Solvents also increase the effort needed to dispose of wastes from the process.

Alkali metals are essential components of many important materials, such as those listed in the abstract. Unfortunately, volatile liquid compounds have not been available for the alkali metals, so CVD of these materials has had to rely on sublimation of inconvenient solid sources, like lithium bis(trimethylsilyl)amide.[1]

In the present paper, the first room-temperature liquid, volatile compounds are reported for lithium, sodium, and potassium, and methods for their synthesis are given. These liquid compounds can be distilled or flash-vaporized to form vapors suitable for CVD of materials containing alkali metals, particularly lithium, sodium, and potassium. Because they are miscible with each other and with many organic solvents, they may also be useful for forming liquid mixtures or solutions for spray coating, spin coating or sol-gel deposition.

These amine and alkali amide compounds have a general formula as given below, where M is an alkali metal, E^1 and E^2 are silicon or carbon, R^1 and R^2 are alkyl groups, and n is 2 or 3.

Parent amine   Alkali amide

Table 1. Amine ligands

Amine	E	R^1	R^2	t
bis(n-octyldimethylsilyl)amine	Si	n-Oct	n-Oct	14
bis(n-hexyldimethylsilyl)amine	Si	n-Hex	n-Hex	10
bis(n-butyldimethylsilyl)amine	Si	n-Bu	n-Bu	6
bis(i-butyldimethylsilyl)amine	Si	i-Bu	i-Bu	4
bis(n-propyldimethylsilyl)amine	Si	n-Pr	n-Pr	4
tert-amyl(n-butyldimethylsilyl)amine	C	Et	n-Bu	4
tert-amyl(i-butyldimethylsilyl)amine	C	Et	i-Bu	3
tert-amyl(n-propyldimethylsilyl)amine	C	Et	n-Pr	3
tert-butyl(n-butyldimethylsilyl)amine	C	Me	n-Bu	3
tert-amyl(i-propyldimethylsilyl)amine	C	Et	i-Pr	2
bis(ethyldimethylsilyl)amine	Si	Et	Et	2
tert-amyl(ethyldimethylsilyl)amine	C	Et	Et	2
tert-butyl(n-propyldimethylsilyl)amine	C	Me	n-Pr	2
tert-amyl(trimethylsilyl)amine	C	Et	Me	1
tert-butyl(ethyldimethylsilyl)amine	C	Me	Et	1

Table I identifies a list of some suitable amine ligands. The number t in this table is the number of angular variables (torsion angles corresponding to rotation around C-C single bonds) in excess of those present in the reference compound bis(trimethylsilyl)amine, whose alkali salts are solid at room temperature. Methyl rotations about their three-fold axes were not counted, since these motions do not change the intermolecular interactions as much as the other torsions do. As t increases, the number configurations available to the ligand increases, and thus its ability to impede crystallization increases. Hence, larger t is, the greater is the ability of the ligand to keep the corresponding metal-ligand compounds in liquid form at room temperature.

The large alkyl groups (R^1 and R^2 larger than methyl) adopt multiple molecular conformations that frustrate the crystallization of the compounds and keep them in liquid form.

SYNTHESIS OF LIQUID ALKALI COMPOUNDS

<u>Synthesis of Bis(Trialkylsilyl)Amide Ligands.</u> The preferred bis(trialkylsilyl)amines may be prepared by known methods, such as the condensation of ammonia with a trialkylchlorosilane.[2]

As a specific example of this method, the synthesis of bis(n-propyldimethylsilyl)amine is given as follows: All experimental manipulations were carried out using standard Schlenk techniques under dry nitrogen either in a glove box or on a Schlenk line unless otherwise stated. Commercial (Gelest or United Chemical Technologies) n-propyldimethylchlorosilane (25.0 g, 0.183 mmol) was dissolved in 150 mL of dry ether. Ammonia gas was bubbled into the solution until it was no longer absorbed and continued for an additional hour in order to ensure the completion of the reaction. The solution was refluxed for one hour, and the solid byproduct NH$_4$Cl was removed by filtration. Distillation was used to remove the ether and excess ammonia, yielding a colorless liquid (17.2 g, 86 %) which was shown to be the desired product, bis(n-propyldimethylsilyl)amine, by NMR analysis, and was used without further purification.

Other bis(trialkylsilyl)amines were made in a similar manner, by substituting other trialkylchlorosilanes for n-propyldimethylchlorosilane.

<u>Synthesis of Alkyl(Trialkylsilyl)Amide Ligands.</u> Alkyl(trialkylsilyl)amines may be synthesized by condensation of primary amines with trialkylchlorosilanes. For example, n-butyldimethylchlorosilane reacts with tert-amylamine (Acros) to form tert-amyl(n-butyldimethylsilyl)amine.

<u>Synthesis of Alkali Amides.</u> Alkali metal compounds can be formed between these amide ligands and the alkali metals in various ways. For lithium, it is convenient to react a solution of butyl lithium with the amine. The sodium and potassium amides were most readily formed by the transamination of the parent amine with sodium amide or potassium bis(trimethylsilyl)amide, respectively.

Lithium bis(n-propyldimethylsilyl)amide was prepared by the slow addition via syringe of a hexane solution of butyl lithium (11.7 mL of 2.73 M solution, 31.9 mmol) to a stirred hexane solution (75 mL) of bis(n-propyldimethylsilyl)amine (6.93 g, 31.9 mmol) at room temperature. Stirring was continued for several hours, and the solution was then refluxed for one hour. The hexane was evaporated under vacuum, leaving 5.8 g of a pale yellow liquid. It was distilled at a temperature of 130 °C and a pressure of 0.15 torr to yield 5.25 g (74%) of clear liquid lithium bis(n-propyldimethylsilyl)amide. Its viscosity was measured to be 23.3 centipoise at 40 °C.

Similar methods were used to prepare other distillable liquid lithium compounds having the properties listed in Table 2. The molecular masses of these new compounds were determined by cryoscopy in p-xylene solution. Their "molecular complexities," defined as the ratio of the cryoscopic molecular mass to the theoretical monomeric value, fall between 2 and 3. Thus the solutions are most likely to contain dimers and trimers.

Table 2. Liquid lithium amides

Lithium amide	E	R^1	R^2	Viscosity (centipoise @ 40°C)	Molecular Complexity	Vapor Pressure (°C/Torr)
bis(ethyldimethylsilyl)amide	Si	Et	Et	37	2.45	122/0.2
tert-amyl(i-propyldimethylsilyl)amide	C	Et	i-Pr	409	2.41	137/0.2
bis(3,3-dimethylbutyldimethylsilyl)amide	Si	Z^i	Z^i	247	2.02	225/0.9
tert-amyl(i-butyldimethylsilyl)amide	C	Et	i-Bu	497	2.65	145/0.1
tert-amyl(n-propyldimethylsilyl)amide	C	Et	n-Pr	810	2.74	171/0.3
bis(n-propyldimethylsilyl)amide	Si	n-Pr	n-Pr	23.3	2.16	130/0.15
bis(i-butyldimethylsilyl)amide	Si	i-Bu	i-Bu	32.9	1.97	145/0.05
tert-amyl(triethylsilyl)amide	C	Et	Et^{ii}	162	2.20	157/0.095
bis(n-butyldimethylsilyl)amide	Si	n-Bu	n-Bu	22.4	2.28	145/0.085

$^i Z = (CH_2)_2C(CH_3)_3$; $^{ii} Me_2$ replaced by Et_2

Sodium bis(n-propyldimethylsilyl)amide was prepared as follows: Sodium amide (1.26 g, 0.0322 mol) was placed in dry benzene and bis(n-propyldimethylsilyl)amine (7.00 g, 0.0322 mol) was added. The mixture was stirred and refluxed for several hours. The benzene solution was filtered through celite, and then the benzene was evaporated under vacuum, leaving 6.31 g (82 %) of a yellow liquid product, sodium bis(n-propyldimethylsilyl)amide. Its viscosity was measured to be $7.1x10^4$ centipoise at 40 °C. It was distilled at a temperature of 213 °C and a pressure of 0.3 torr.

Similar methods were used to prepare other distillable liquid sodium compounds having the properties listed in Table 3.

Table 3. Liquid sodium amides

Sodium amide	E	R^1	R^2	Viscosity (centipoise @ 40°C)	Molecular Complexity	Vapor Pressure (°C/Torr)
bis(n-propyldimethylsilyl)amide	Si	n-Pr	n-Pr	$7.1x10^4$	2.26	213/0.3
bis(i-butyldimethylsilyl)amide	Si	i-Bu	i-Bu	$2.8x10^4$	2.08	189/0.08
bis(n-butyldimethylsilyl)amide	Si	n-Bu	n-Bu	$>10^7$	1.89	231/0.5
bis(n-hexyldimethylsilyl)amide	Si	n-Hex	n-Hex	$1.5x10^4$	2.01	265/0.3

Potassium bis(n-hexyldimethylsilyl)amide was prepared as follows: Potassium bis(trimethylsilyl)amide (5.07 g, 25.6 mmol) and bis(n-hexyldimethylsilyl)amine (7.66 g, 25.6 mmol) were added to a flask and 50 mL toluene was added. The clear yellow solution was stirred at room temperature for 18 hours and then refluxed for two hours. The toluene and hexamethyldisilazane byproduct were removed from the brown toluene solution under vacuum with heating to 150°C to yield a brown oil (5.40 g, 63%). Its viscosity was measured to be 271 centipoise at 40 °C. It may be flash vaporized from a heated nozzle for CVD applications.

Alternatively, the potassium amide may be dissolved in small amounts of organic solvents to form concentrated solutions that may be flash vaporized.

Similar methods were used to prepare other vaporizable liquid potassium compounds having the properties listed in Table 4.

Table 4. Potassium precursors

Potassium amide	E	R^1	R^2	Viscosity (centipoise @ 40°C)	Molecular Complexity	Melting Point (°C)
bis(i-butyldimethylsilyl)amide	Si	i-Bu	i-Bu	205	2.02	53
bis(n-butyldimethylsilyl)amide	Si	n-Bu	n-Bu	230	1.90	45
bis(n-hexyldimethylsilyl)amide	Si	n-Hex	n-Hex	271	1.80	<20
bis(n-octyldimethylsilyl)amide	Si	n-Oct	n-Oct	183	1.63	<20

These liquid alkali metal amides are generally completely miscible with organic solvents, including hydrocarbons, such as dodecane, tetradecane, xylene and mesitylene, and with ethers, esters, ketones and chlorinated hydrocarbons. These solutions generally have lower viscosities than the pure liquids, so that in some cases it may be preferable to nebulize and evaporate the solutions rather than the pure liquids. In these instances, however, very concentrated solutions, e.g. greater than one molar, may be obtained. The liquids or solutions can also be evaporated with thin-film evaporators or by direct injection of the liquids into a heated zone.

The liquids and solutions all appeared to be non-pyrophoric. The precursors generally react with moisture in the ambient air, and should be stored under an inert, dry atmosphere such as pure nitrogen gas.

CVD EXPERIMENT

Liquid lithium bis(n-ethyldimethylsilyl)amide was mixed with mesitylene to lower the viscosity below 5 centipoise so that the precursor could be nebulized into tiny droplets (about 20 microns in diameter) by a high-frequency (1.4 MHz) ultrasonic system.[3] The resulting fog was carried by nitrogen into the deposition zone where it mixed with O_2 at 200 °C. The precursor concentration in the deposition gas stream was 0.36 mol%, the oxygen concentration was 17 mol%, and the total flow rate was 0.60 L/min. A thin film was deposited on a silicon substrate placed on the bottom of the tube. The lithium-containing film was easily dissolved in water. The refractive index was determined to be 1.48-1.49, by using drops of Cargille certified index of refraction fluids, which more closely resembles lithium hydroxide (1.45-1.46) than lithium oxide (1.64).

Films were also produced containing both lithium and niobium by mixing liquid lithium bis(n-ethyldimethylsilyl)amide and liquid niobium (V) diethylamide with mesitylene in a 1:1:5 mole ratio. The mixed precursor films were deposited by the same method used for the single precursor lithium films at 250°C. The concentration of the lithium precursor in the deposition gas stream was 0.27 mol%, the niobium precursor concentration was 0.27 mol%, the oxygen concentration was 16 mol%, and the total flow rate was 0.60 L/min. The silicon substrates were coated with an iridescent film. The lithium to niobium ratio in the resulting film was determined by ablating the film with a 193 nm argon fluoride excimer laser and analyzing the ablated atoms by quadrupole mass spectrometry, Figure 1. The mole ratio of lithium to niobium was found to vary along the length of the silicon substrate. An approximate ratio of 0.3:1 Li:Nb was observed after several centimeters in the reaction zone, indicating that the niobium amide was more reactive than the lithium amide under these conditions. Interestingly, these silicon containing precursors do not deposit a silicon impurity in the film.

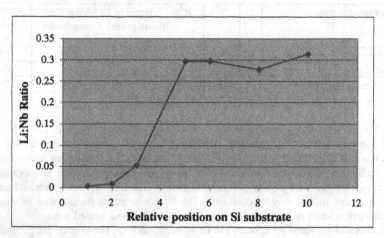

Figure 1. Mole Ratio of Li to Nb as Determined by Laser Ablation-Mass Spectrometry

CONCLUSIONS

The first volatile liquid compounds of lithium, sodium and potassium have been synthesized using easily prepared amines from commercially available reagents. Films containing alkali metals can be deposited from vapors of the precursor compounds and oxygen or other sources of oxygen. As an example, lithium niobate films were prepared by CVD using one of the new liquid lithium compounds as a precursor. Similarly, a liquid precursor for potassium may be combined with a tantalum precursor to provide a process for depositing potassium tantalate films having nonlinear optical properties.

Many other alkali-containing compounds might be deposited using these volatile liquid precursors. For example, one might deposit lithium phosphorus oxide nitride materials for use as solid electrolytes in batteries.[4] Similarly, sodium-potassium niobate (NKN) films may be formed with low loss tangent and a strong electric field dependence of rather low dielectric permittivity and used in tunable microwave devices.[5] $Na_{0.1}K_{0.9}Ta_{0.55}Nb_{0.45}O_3$ films with high pyroelectric sensitivity may be made and fabricated into night-vision devices.[6] By using a tungsten-containing precursor along with these precursors, tungsten bronzes having useful optical, electrical and electrochromic properties may be made and formed into electrochromic windows, mirrors and displays.[7] Similarly, by combining liquid lithium precursors with precursors for cobalt, nickel, vanadium, and/or other transition metals, electrochromic materials[8] or electrodes for lithium batteries[9] may be deposited. $LiAlO_2$ buffer layers could be deposited as substrates for growth of GaN.[10] Vapors of a sodium-containing precursor can be used to supply a sodium dopant for copper indium diselenide solar cells.[11]

The liquid compounds may also be used for spray coating, spin coating and sol-gel deposition of materials containing alkali metals. The high solubility and miscibility of these precursors is an advantage in forming the required solutions needed for such applications.

ACKNOWLEDGMENTS

This work was supported in part by the National Science Foundation grant CHE 95-10245.

REFERENCES

1. For reviews, see L. G. Hubert-Pfalzgraf, Electrochemical Soc. Proc. **97-25**, 824 (1998); W. A. Wojtczak, P. F. Fleig and M. J. Hampden-Smith, Adv. Organometallic Chem. **40**, 215 (1996); A. A. Drozdov and S. I. Troyanov, Main Group Met. Chem. **XIX**, 547 (1996); R. E. Sievers, S. B. Turnipseed, L. Huang and A. F. Lagolate, Coord. Chem. Rev. **128**, 285 (1993).
2. R. C. Osthoff and S. W. Kantor, Inorg. Syntheses **5**, pp. 55-64 (1957).
3. R. G. Gordon, F. Chen, N. J DiCeglie, Jr., A. Kenigsberg, X. Liu, D. J. Teff and J. Thornton, Mater. Res. Soc. Symp. Proc. **495**, pp. 63-68 (1998).
4. J. B. Bates, N. J. Dudney, G. R. Gruzalski, R. A. Zuhr, A. Choudhury, C. F. Luck and J. D. Robertson, J. Power Sources **43-44**, pp. 103-110 (1993).
5. C.-R. Cho, S. I. Khartsev, A. M. Grishin and T. Lindback, Mater. Res. Soc. Proc. **574**, pp. 249-254 (1999).
6. H. R. Beratan, K. R. Udayakumar, C. M. Hanson, J. F. Belcher and K. Soch, Mater. Res. Soc. Symp. Proc., in press (Paper BB8.4, Spring 1999 MRS Meeting).
7. M. Rubin, K. von Rottkay, S.-J. Wen, N. Ozer and J. Slack, Sol. Energy Mater. Sol.Cells **54**, pp. 49-57 (1998).
8. A. Talledo and C. G. Granqvist, J. Appl. Phys. **77**, 4655 (1995).
9. C. O. Kelly, H. D. Friend and R. Higgins, Proc. 13th Annual Battery Conf. On Applications and Advances (California State University – Long Beach, 1999), p. 335.
10. E. S. Hellman, Z. Lilienthal-Weber and D. N. E. Buchanan, MRS Internet J. Nitride Semiconductor Res. **2**, 30 (1997); W. Koh, S.-J. Ku and Y. Kim, Mater. Res. Soc. Symp. Proc. **495**, pp. 69-72 (1998).
11. T. Nakada, T. Kume and A. Kunioka, Sol. Energy Mater. Sol. Cells **50**, pp. 97-103 (1998).

MOCVD OF CuInE$_2$ (WHERE E = S or Se) AND RELATED MATERIALS FOR SOLAR CELL DEVICES

MICHAEL KEMMLER[a], MICHAEL LAZELL,[b] PAUL O'BRIEN,[b*] and DAVID J. OTWAY[a]

a. *Department of Chemistry, Imperial College of Science, Technology and Medicine, Exhibition Road, London, SW7 2AZ, UK.*
b. *Department of Chemistry, and The Manchester Materials Science Centre, Manchester University, Oxford Rd, Manchester, M13 9PL. UK.*
Email addresses: p.obrien@ic.ac.uk; d.j.otway@ic.ac.uk.

ABSTRACT

Thin film(s) of chalcopyrite CuInE$_2$ (where E = S or Se) have been grown by low-pressure metal-organic chemical vapour deposition (LP-MOCVD) using the precursors [In(E$_2$CNMe^{n}Hexyl)$_3$] and [Cu(E$_2$CNMe^{n}Hexyl)$_2$]. Similarly, thin films of ME (where M = Zn, Cd; E = S, Se) have been deposited from precursors of general formula [M(E$_2$CNMe^{n}Hex)$_2$]$_x$. Films were grown on glass between 400 - 500 °C, and characterized by X-ray diffraction, optical spectroscopy (UV/Vis), EDAX and scanning electron microscopy.

INTRODUCTION

Ternary compound semiconductors such as copper indium disulfide/diselenide (CuInS$_2$ or CuInSe$_2$) are promising materials for use in high efficiency, radiation hard, solar cells.[1] There have been only a few reports of the deposition of CuInE$_2$ by CVD methods. A halogen transport VPE method[2] has been used to grow single crystals. CuInSe$_2$ films contaminated with In$_2$Se$_3$ have been deposited by MOCVD using copper(II) hexafluoroacetylacetonate mixed with trimethylamine, triethyl indium and hydrogen selenide.[3,4] A plasma enhanced process using both hexafluoroacetylacetonate copper and indium complexes and a novel selenium source 4-methyl-1,2,3-selenadiazole has also been used.[5] Chichibu has reported[6] the growth of heteroepitaxial layers of CuInSe$_2$ using cyclopentadienylcoppertriethylphosphine, trimethyl indium and diethylselenide as the precursors; the first successful MOVPE results.

There are problems in developing a simple thermal system for the deposition of these and related materials that centre on the difficulty in finding thermally compatible precursors for the elements involved. The authors of the above papers have sought to overcome such problems in a variety of ways *e.g.* by the introduction of new S/Se sources or modifications to well known copper sources. There are problems and even serious potential hazards associated with systems which allow copper(II) complexes and metal alkyls to mix in the effluent from the reactor. We have been developing a range of dithio- and diseleno-carbamato complexes of various metals which have been successfully used to deposit a wide range of semiconductors.[7] One particularly successful modification to the sulfur/selenium containing ligands has been to develop compounds in which the parent amine is asymmetrically substituted and involves a bulky or extended alkyl substituent.[8,9] Compounds with these ligands are air stable and sufficiently volatile for the deposition of thin films of materials such as Cu$_2$E, In$_2$E$_3$, Ga$_2$E$_3$, ZnE and CdE (E = S or Se). Success with the binary parents of CuInE$_2$ has encouraged us to deposit the ternary phase. In this paper we report on a simple thermal MOCVD process for copper indium disulfide/diselenide; preliminary results have recently been reported.[10-12]

EXPERIMENTAL

The experimental method and apparatus used for the low pressure MOCVD of $CuInE_2$ was as described in earlier papers.[13] The growth of the thin films was carried out in a low-pressure ($\approx 10^{-2}$ Torr) MOCVD reactor tube which has been described elsewhere.[13] A graphite susceptor held the substrate (dimensions 10 mm x 15 mm) which was heated by a tungsten halogen lamp. A typical growth run (in temperature range 450 - 500 °C) involved the use of approximately 100 mg of stoichiometrically mixed sample, and lasted for 1- 2 hours.

Films were deposited on glass microscope slides. The precursors used were respectively the *bis*- and *tris*- complexes of methyl,*n*-hexyl-diseleno- or -dithio-carbamate with copper (II) and indium, prepared as described in earlier papers.[11,14,15] In initial experiments, these were simply mixed in a 1:1 ratio in the evaporator. In subsequent experiments the effect of varying the ratio of Cu:In has been investigated, these results will be reported in a full paper. The compounds were prepared by methods detailed in earlier papers. Growth runs were typically for times between 30 minutes and 2 h. In 2 h. thick *ca.* 2 micron films were deposited (T source $\{T_s\}$ = 180-250°C, T growth $\{T_p\}$ = 400-450°C). X-ray diffraction studies were performed using secondary graphite monochromated CuK_α radiation on a Philips PW1700 series automated diffractometer. The sample was mounted flat and scanned from 10 - 90 ° in steps of 0.04 ° with a count time of 2 s. Samples were carbon coated before analysis. All EDAX and electron microscopy was then carried out in a Jeol Superprobe 733 microscope.

RESULTS AND DISCUSSION

The success of these compounds for the deposition of sulfides or selenides depends on quite subtle differences in both their thermal stability and mode of decomposition from the parent compounds such as the diethyldiselenocarbamates (see scheme below).[10]

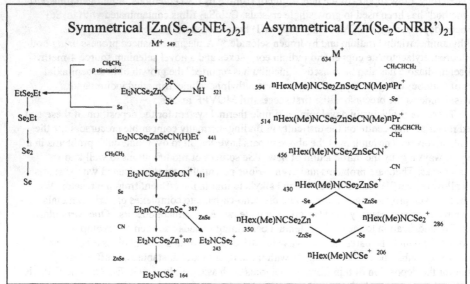

Figure 1. Decomposition Scheme for Symmetrical and Asymmetrical Precursors.

The combined GC/MS and HPLC results show that the symmetrical precursors such as [Zn(Se₂CNEt₂)₂] compounds decompose readily to give EtSe₂Et and eventually Se as a major product whereas the asymmetrical precursors such as [Zn(Se₂CNMe"Hex)₂] primarily give ZnSe and a ring closed organic fragment of formula SeCNMe"Hex⁺ that is volatile and removed *in vacuo.*

The structure of one of the precursors used for deposition, [Zn(Se₂CNMe"Hex)₂], can be seen in figure 2 below. The zinc metal center is tetrahedral in geometry. The analogous cadmium compound is actually a dimer with pseudo five coordinate geometry; see figure 3.[15]

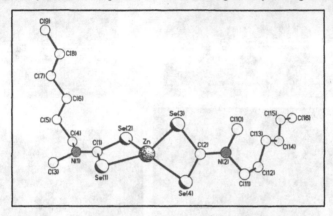

Figure 2. Molecular Structure of [Zn(Se₂CNMe"Hex)₂].

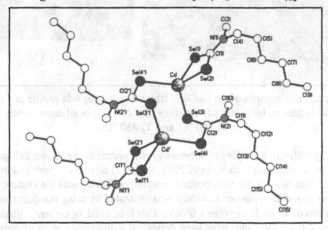

Figure 3. Molecular Structure of [Cd(Se₂CNMe"Hex)₂]₂.

Films deposited using the Cu/In precursors in a 1:1 ratio in the evaporator were specular, black and reasonably adherrent. Growth rates were estimated as approximately of the order of 1 micron/h. Typical scanning electron micrographs of some of the materials deposited are shown in Figure 4. In general the films are composed of a mixture of acicular and hexagonal platelets typical of Cu(In/Ga)E$_2$. The platelets in particular appear orientated orthogonal or at an acute angle to the substrate surface.

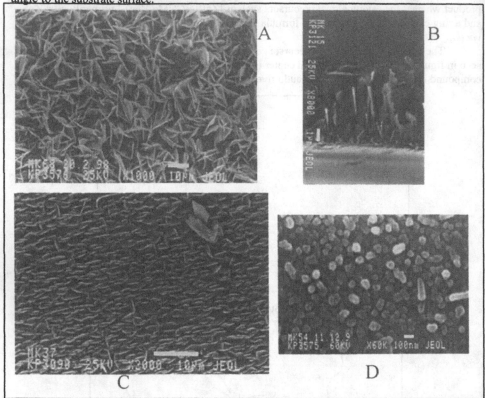

Figure 4. Electron micrographs of **(a)** CuGaS$_2$ film on glass; **(b)** side profile of CuInSe$_2$ film on glass; **(c)** CuInSe$_2$ film on InP(100); **(d)** CuInSe$_2$ film on ITO; in all cases grown over 1 h. with T$_s$ 250 °C, and T$_p$ 450 °C.

The X-ray diffraction profiles of some as grown materials are shown in Figure 5.The indices are assigned according to those in JCPDS, the (112) diffraction peak is dominant but the other main peaks indicate that the as-deposited material is CuInE$_2$ with the chalcopyrite structure. The optical band gap of the as grown CuInSe$_2$ was estimated by using the direct band gap method (from plots of α^2 vs. Energy) as 1.08 eV. This is in good agreement with the accepted value of between 1.0-1.1 eV. Films have been deposited with various ratios of copper to indium precursor and show little deviation from the 1:1 stoichiometry films. Increasing the length of the growth run leads to densification of the films.

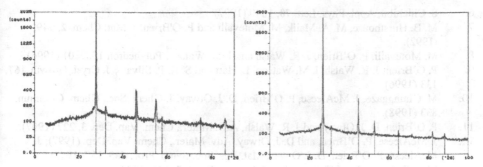

Figure 5. XRD Patterns of **(a)** CuInS$_2$ film on glass; **(b)** CuInSe$_2$ film on glass. Growth details as for Figure 4.

CONCLUSIONS

In this paper we have demonstrated that by making a judicious choice of precursors the low-temperature growth of the potentially important ternary material CuInSe$_2$ can be achieved by MOCVD. Similarly, a number of other important optoelectronic materials can also be grown. The route represents an advance over those reported in the literature which have involved plasma assisted growth and/or inappropriate combinations of precursors. We are at present developing this approach with the aim of depositing epitaxial layers and the deposition of related quaternary materials.

ACKNOWLEDEMENTS

We would like to thank the sponsors of this work, the EPSRC, the Royal Society and the Leverhulme Foundation. POB is the Royal Society Amersham International Research Fellow and the visiting Sumitomo/STS Professor of Materials Chemistry at IC and is Professor of Inorganic Materials at the University of Manchester. We would also like to thank Mr. Michael Kemmler for assistance with growth work and Mr. Richard Sweeney (IC) for XRD results and Mr. Keith Pell for SEM (QMW).

REFERENCES

1. See for example : V. Nadenau, D. Braunger, D. Hariskos, M. Kaiser, Ch. Koble, M. Ruckh, U. Ruhle, R. Schaffer, D. Schmid, T. Walter, S. Zwergart and H. W. Schock, Prog. Photo. Res. Appl. **3**, 363 (1995).
2. O. Igarashi, J. Cryst. Growth **130**, 343 (1993).
3. V. Sagnes, A. Salesse, M. C. Artaud, S. Duchemin, J. Bougnot and G. Bougnot, J. Cryst. Growth **124**, 620 (1992).
4. F. Ouchin, P. Gallon, M. C. Artaud, J. Bougnot and S. Duchemin, Crystal Res. Technol. **31**, S513 (1996).
5. P. A. Jones, A. D. Jackson, P. D. Lickiss, R. D. Pilkington and R. D. Tomlinson, Thin Solid Films **238**, 4 (1994).

6.   S. Chichibu, Appl. Phys. Lett. **70**, 1840 (1997).
7.   M. B. Hursthouse, M. A. Malik, M. Motevalli and P. O'Brien, J. Mat. Chem. **2**, 949 (1992).
8.   M. Motevalli, P. O'Brien, J. R. Walsh and I. M. Watson, Polyhedron **15**, 2801 (1996).
9.   P. O'Brien, J. R. Walsh, I. M. Watson, L. Hart and S. R. P. Silva , J. Cryst. Growth **167**, 133 (1996).
10.   M. Chunggaze, J. McAleese, P. O'Brien, D. J. Otway, J. Chem. Soc., Chem. Commun. 833 (1998).
11.   P. O'Brien, D. J. Otway and J. R. Walsh, Adv. Mater., Chem. Vap. Dep. **3**, 227 (1997).
12.   J. McAleese, P. O'Brien, and D. J. Otway, Adv. Mater., Chem. Vap. Dep. (1997); J. McAleese, P. O'Brien, D. J. Otway, Mat. Res. Soc. Symp. Proc. **485**, 157 (1998).
13.   M. A. Malik and P. O'Brien, Adv. Mater. Opt. Elect. **3**, 171 (1994).
14.   M. R. Lazell, P. O'Brien, D. J. Otway and J. R. Walsh, Adv. Mater. Opt. Elect accepted for publication (1999).
15.   M. R. Lazell, P. O'Brien, D. J. Otway, J.-H. Park, Chem. Mater. accepted for publication (1999).
16.   P. O'Brien, M. Chunggaze, D. J. Otway unpublished results (1999).

Solution Deposition of
Electronic Ceramics

WET-CHEMICAL SYNTHESIS OF THIN-FILM SOLAR CELLS

R.P Raffaelle[*], W. Junek[*], J. Gorse[**], T. Thompson[**], J.D Harris[***], J. Cowen[***],
D. Hehemann[***], G. Rybicki[****], and A.F. Hepp[****]
[*]Rochester Institute of Technology, Rochester, NY 14623
[**]Baldwin-Wallace College, Berea, OH 44017
[***]Kent State University, Kent, OH 44242
[***]NASA Glenn Research Center, Cleveland, OH 44135

ABSTRACT

We have been working on the development of wet-chemical processing methods that can be used to create thin film photovoltaic solar cells. Electrochemically deposition methods have been used to produce copper indium diselenide (CIS) thin films on molybdenum coated polymer substrates. CIS has an extremely high optical absorption coefficient, excellent radiation resistance, and good electrical conductivity and thus has proved to be an ideal absorber material for thin film solar cells. A series of compositionally different p-type CIS films were produced by using different electrochemical deposition potentials. Cadmium sulfide (CdS) window layers were deposited directly on these CIS films using a chemical bath process. CdS is a naturally n-type wide-bandgap semiconductor which has good transparency and is well lattice-matched to CIS. Zinc oxide thin films were grown by electrochemical deposition directly on the CdS films. ZnO is a transparent and conductive thin film that serves as the top contact of the cells. The structural and elemental properties of the individual ZnO, CdS and CIS films were characterized by x-ray diffraction and energy dispersive spectroscopy. The electrical behavior of the CdS on CIS junctions was determined using current versus voltage measurements. We will discuss the performance of these devices based on the physical properties of the component films and the processing methods employed in their fabrication.

INTRODUCTION

The development of high power-to-weight ratio photovoltaic solar cells for use in space applications is essential to several proposed NASA programs. These cells should be flexible for easy deployment and show good radiation resistance. Thin-film solar cells based on copper indium diselenide grown on polymer substrates are an excellent candidate for such applications [1]. It is also desirable that the methods used to produce these cells are cost-effective, easily scalable, and involve low temperatures, as to be compatible with polymer substrates. Wet-chemical methods of producing thin-film photovoltaic materials have the potential to meet these goals.

$CuInSe_2$ (CIS) is a leading candidate for use as a thin-film solar cell absorber layer. CIS has an extremely high optical absorption coefficient, good electrical conductivity, and good radiation resistance [2]. CIS has an optical bandgap of around 1.1 eV, which although not ideal, can be improved by the substitutional doping of Ga for In and/or S for Se [3]. Thin-film solar cells utilizing $CuIn_{1-x}Ga_xSe_2$ (CIGS) absorber layers have reached efficiencies of nearly 18% [4].

The most commonly used junction material for CIS-based solar cells is cadmium sulfide (CdS). The hexagonal for of CdS is well lattice matched to CIS. CdS is naturally n-type and has a good optical transparency and an optical bandgap of approximately 2.4 eV [5]. The CdS films are most commonly deposited using a chemical bath deposition technique [6].

155

Zinc oxide (ZnO) is a transparent conducting oxide that is used as the top contact for thin-film solar cells (see Figure 1). ZnO has an optical bandgap of around 3.3 eV and conductivity as high as 200 $\Omega^{-1}cm^{-1}$[7]. Electrochemical deposition and chemical bath deposition techniques have been used to deposit ZnO thin films [8,9].

EXPERIMENT

$CuInSe_2$ was electrochemically deposited on Mo foil, Mo coated poly(benzobisiazole) or PBO, and indium tin oxide (ITO) coated glass from a solution containing 1 mM $CuSO_4$, 10 mM $In_2(SO_4)_3$, 5 mM SeO_2, 25 mM citric acid, and 10% ethanol by volume [10]. The films were deposited at room temperature using a deposition potentials from −1.1 to −1.3 V versus a saturated calomel electrode (SCE) in 0.05 V increments. The substrates were mounted in a rotating disk electrode and rotated at 500 rpm. The active electrode area had a radius of 1.27 cm. In order to deposit a 1 μm thick film, a deposition time of 600 s was used. The as-deposited films were then annealed in a flowing Argon atmosphere at 600 °C for 1 hour.

CdS buffer layers were grown on the CIS-based absorber layers and ITO coated glass slides using chemical bath deposition. The substrates were immersed for 600 s in a 60 °C solution of 1mM Cd-acetate, 1 M NaOH, and 60 mM thiourea. The substrates were rotated at 500 rpm during the CdS deposition.

The top contacts to the CdS on CIS cells were made by electrochemical deposition of ZnO. The ZnO was deposited on the CdS on CIS junctions and on ITO coated glass at −1.0 V vs. SCE for 300 s using a solution of 100 mM $Zn(NO_3)_2$ in deionized water held at 60 °C. The substrates were rotated at 500 rpm during the ZnO deposition.

The surface morphology of the films were examined in a Hitachi S-4700 FE-SEM. Energy dispersive spectroscopy was also performed in the SEM using an EDAX DX prime system utilizing ZAF standardless correction. Film thickness was determined using a Dektak II profilometer. X-ray diffraction on the films was performed using a Phillips PW 3710 diffractometer. Transmission spectrophotometry was performed on the films deposited on ITO coated glass using a Perkin Elmer Lamba 19 spectrophotometer. The resulting absorption coefficients versus photon energy plots were used to determine the optical bandgaps of the films.

An Alessi four-point probe was used to monitor the conductivity of the different films deposited in this study. The four-point probe was also used to type the different semiconductors deposited via the Seebeck effect. Current versus voltage measurements on the ZnO/CdS/CIS junctions were performed using a computer controlled Keithley 236 source/measure unit. Contacts were made to the device using evaporated Al contacts and an Alessi Rel-2100 wafer probing station.

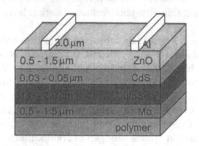

Figure 1. Thin-film photovoltaic solar cell.

RESULTS

The CIS films deposited at –1.1 V vs. SCE were imaged in a scanning electron microscope before and after annealing. The as-deposited CIS film is smooth and continuous (see Figure 2(a). A coalescing of the individual grains is seen upon annealing in flowing argon (see Figures 2(a) and 2(b)). Examining this behavior under high magnification shows that as the grain coalesce the volume of voids and the size of the grain boundaries also increase (see Figure 2(c) and 2(d)).

Figure 3 shows a comparison of the morphology of annealed CIS films as a function their deposition voltages. An increase in the homogeneity of the grain is seen as the deposition voltage is made more negative. Energy dispersive spectroscopy on these samples shows that this trend also corresponds to a trend towards stoichiometry. The [Cu] to [In] ratio varies from 1.15 to 0.95 as the deposition potential is reduced from –1.1 V to –1.3 V vs. SCE.

The surface morphology of chemically bath deposited CdS on CIS deposited at –1.2 V vs. SCE is shown in Figure 4(a). Although the annealed CIS had voids and wide grain boundaries, the CdS over-layer is smooth and continuous. The chemical bath process does a good job of filling the voids and buffering the polycrystalline absorber film. Figure 4(b) shows the electrodeposited ZnO top contact. The film shows continuous but dendritic growth.

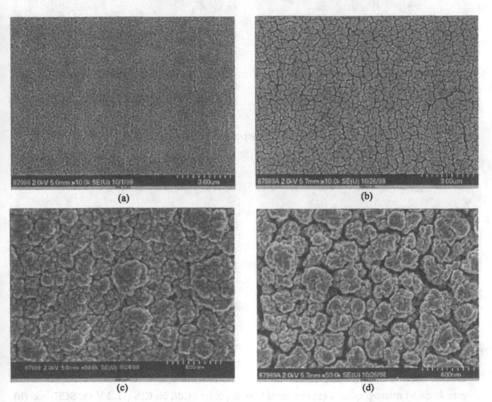

(a)

(b)

(c)

(d)

Figure 2. SEM micrographs of CIS deposited at –1.1 V vs. SCE (a) before and (b) after annealing. High magnification SEM images of the same CIS (c) before and (d) after annealing.

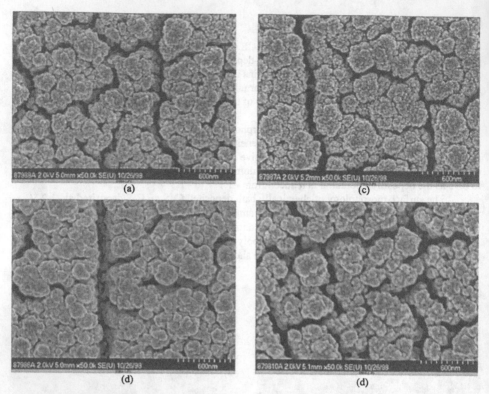

Figure 3. SEM micrographs of annealed CIS deposited on Mo at (a) –1.15 V; (b) –1.20 V; (c) –1.25 V; and (d) –1.30 V vs. SCE.

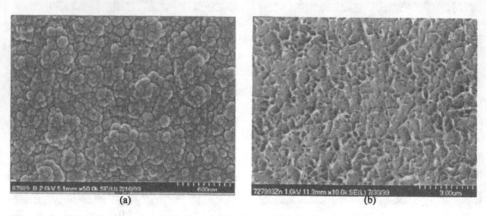

Figure 4. SEM micrographs of (a) chemical bath deposited CdS on CIS (-1.2 V vs. SCE) and (b) electrodeposited ZnO on CdS.

The crystal structure of CIS deposited on the Mo coated PBO was examined by XRD both before and after annealing (see Figure 5). The as-deposited film had poor crystallinity as seen in the poor intesity and broadness of the Bragg peaks. However, upon annealing at 600 °C in flowing argon for 1 hour a dramatic increase in the crystallininty is evident. The peak indices shown correspond to the chalcopyrite crystal structure. The PBO substrate became discolored and brittle due to the annealing process. The CdS was found to be hexagonal and the ZnO was shown to have the zincite structure.

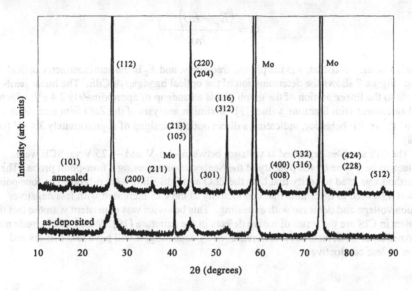

Figure 5. XRD pattern of as-deposited and annealed CIS on Mo on PBO.

The thickness as function of time for the ZnO films was determined using a profilometer (see Figure 6). The thickness of the electrodeposited ZnO was shown to increase linearly with deposition time. This behavior is consistent with Faraday's law of electrochemistry.

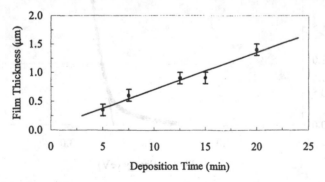

Figure 6. Thickness of ZnO electrodeposited at −1.0 V vs. SCE on ITO coated glass as a function of time. The error bars represent thickness variation across the film.

Transmission spectroscopy on the ZnO and CdS window layers was performed using the films deposited on ITO coated glass. Both the ZnO and CdS films exibited good transparancy throughout the entire visible spectrum and in the near IR. The transmission data was converted to absorption coefficients using the film thickness determined from profilometry. The absorption coefficients (α) were used to find the optical bandgaps using the following direct bandgap relation

$$\alpha = \frac{\left(E_g - h\upsilon\right)^{1/2}}{h\upsilon} \tag{1}$$

where h is Planck's constant, ν is the photon frequency, and E_g is the semiconductor optical bandgap. Figure 7 shows the determination of the optical bandgap of CdS. The linear least-squares fit to the linear portion of the graph yields a bandgap of approximately 2.4 eV, which is in good agreement with literature values [5]. A similar analysis of the ZnO film also yielded a linear $(\alpha h\nu)^2$ vs. $h\nu$ behavior, indicating a direct optical bandgap of approximately 3.4 eV (see figure 8).

The CIS samples deposited at voltages between –1.1 V and –1.25 V vs. SCE were determined to be p-type by the polarity of the Seebeck voltage using a four-point probe. The same method was used to verify that both the ZnO and CdS films were n-type. The four-point probe was also used to determine that conductivity of the CIS increased with less negative deposition voltage and decrease with annealing. This behavior was consistent with the fact that the carriers in CIS are the result of native defects in the structure [10]. When CIS is made more stoichiometric, or its defects have been removed by annealing, it will have less carriers and therefore be less conductive.

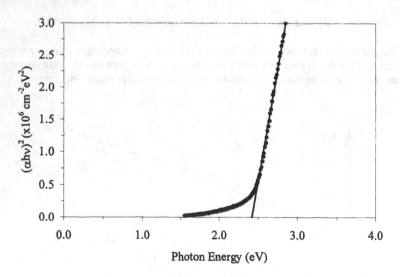

Figure 7. Optical bandgap determination of CdS.

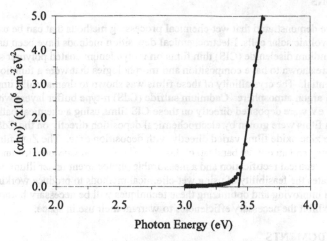

Figure 8. Optical bandgap determination of ZnO.

The current versus forward bias voltage behavior of a device that consisted of electrodeposited ZnO on chemical bath deposited CdS on electrodeposited CuInSe$_2$ is shown in Figure 9. The device exhibited good rectification with a measurable short-circuit photocurrent. The device had an open-circuit voltage of 230 mV and short-circuit current of approximately 10 mA/cm^2 under ambient room lighting. The small open circuit voltage and reverse bias leakage current indicate point defects near the junction and/or grain boundary shunting. The shallow slope of the forward-bias current versus voltage behavior also indicates a large series resistance.

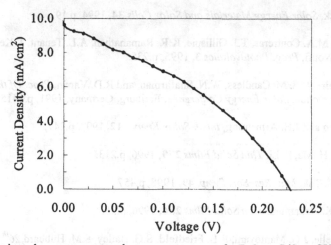

Figure 9. Illuminated current versus voltage measurement of a ZnO on CdS on CIS thin-film solar cell.

CONCLUSIONS

We have demonstrated that wet-chemical processing methods that can be used to create thin film photovoltaic solar cells. Electrochemical deposition methods have been used to produce p-type copper indium diselenide (CIS) thin films on molybdenum coated polymer substrates. These films were shown to have composition and morphologies that were a function of their deposition potential. The crystallinity of these films was shown to dramatically improve with annealing in an argon atmosphere. Cadmium sulfide (CdS) n-type buffer layers with and optical bandgap of 2.4 eV were deposited directly on these CIS films using a chemical bath process. Zinc oxide thin films were grown by electrochemical deposition directly on the CdS films. The thickness of the zinc oxide films varied directly with deposition time. The ZnO films showed a good transparency and an optical bandgap of 3.4 eV. The ZnO on CdS on CIS junctions exhibited good electrical rectification and a measurable photocurrent under illumination. This work demonstrates the feasibility of using wet-chemical methods to produce working solar cells. Future work on improving and optimizing these techniques will be necessary before the cells of this type will exhibit the necessary efficiencies to warrant their use in space.

ACKNOWLEDGMENTS

The authors would also like to acknowledge the support of the NASA grants NCC3-710, NCC3-563, and the NASA Glenn Research Center Directors Discretionary Fund.

REFERENCES

1. D.J. Flood, *Int'l Photovolt. Solar Energy Conf.*, **5**, Japan, 1990, p.551.

2. K. Zweibel, H.S. Ullal, and B. von Roedem, *25th IEEE Photovoltaic Specialists Conf.*, Washington, 1996, p.159.

3. H.W. Schock, *Solar Energy Materials and Solar Cells* **34**, 1994, p.19.

4. J.R. Tuttle, M.A. Contreras, T.J. Gillispie, K.R. Ramanathan, A.L. Tenant, J. Keane, A.M. Gabor, and R, Noufi, *Prog. Photovoltaics* **3**, 1995, p.235.

5. R.W. Birkmire, B.E. McCandless, W.N. Shafarman, and R.D. Varrin, *Proc. Of the 9th European Photovoltaic Solar Energy Conference*, Freiburg, Germany, 1991, p.1415.

6. B.R. Lanning and J.H. Armstrong, *Int. J. Solar Energy* **12**, 1992, p.247.

7. J. Ma, F. Ji, H. Ma, S. Li, *Thin Solid Films* **279**, 1996, p.213.

8. J. Lee and Y. Tak, *Mat. Res. Soc. Symp.* **49**, 1998, p.457.

9. K. Ito and K. Nakamura, *Thin Solid Films* **286**, 1996, p.35.

10. R.P. Raffaelle, J.G. Mantovani, R.B. Friedfeld, S.G. Bailey, S.M. Hubbard, *26th IEEE Photovoltaic Specialists Conf.*, Anaheim, 1997, p.559.

SYNTHESIS OF HIGH-K TITANIUM OXIDE THIN FILMS FORMED BY METALORGANIC DECOMPOSITION

Hisashi Fukuda, Yoshihiro Ishikawa, Seiogo Namioka and Shigeru Nomura
Department of Electrical and Electronic Engineering, Faculty of Engineering, Muroran Institute
of Technology, 27-1 Mizumoto-cho, Muroran-shi, Hokkaido 050-8585, Japan
E-mail address: fukuda@mmm.muroran-it.ac.jp

ABSTRACT

Titanium oxide (TiO_2) thin films were formed on a Si substrate by metalorganic decomposition (MOD) at temperatures ranging from 600 to $1100\,°C$. As-deposited films were in the amorphous state and were completely transformed after annealing at $600\,°C$ to a crystalline structure with anatase as its main component. During crystallization, a reaction between TiO_2 and Si occurred at the interface, which resulted in the formation of a thin interfacial SiO_2 layer. Capacitance-voltage measurement showed good dielectric properties with a maximum dielectric constant of 76 for films annealed at $700\,°C$. For the crystallized TiO_2 films, the interface trap density was 1×10^{11} $cm^{-2}eV^{-1}$, and the leakage current was 1×10^{-8} A/cm^2 at 0.2 MV/cm. The modified structure of $TiO_2/SiO_2/Si$ is expected to be suitable for the dielectric layer in an integrated circuit in place of conventional SiO_2 films.

INTRODUCTION

Thin silicon dioxide (SiO_2) films have been commonly used as a gate insulator in ultralarge-scale integrated (ULSI) circuits. With SiO_2 film as the gate oxide in metal-oxide-semiconductor field-effect transistors (MOSFETs), the transconductance and charge storage capacity increase upon decreasing the thickness of the SiO_2 film, resulting in high device performance. For a given bias voltage, there are two approaches to increasing the charge capacity. One is to reduce the thickness of gate oxide film. Another approach is to increase the dielectric constant of the insulator. ULSIs are made as thin as possible by reducing the thickness of SiO_2 layers. In fact, the gate oxide film thickness must be decreased to below 2 nm for 0.1-μm-rule ULSIs owing to the downscaling of devices. However, in ultrathin SiO_2 films below 3 nm, leakage current arising from direct tunneling and defects increases, resulting in poor yields.

Recently, there have been a number of studies on using insulator with high dielectric constant as an alternative dielectric to SiO_2 film in ULSIs. Among these insulators, titanium oxide (TiO_2) is very promising as a possible replacement for SiO_2 [1-10]. The TiO_2 films have been proposed for variety applications in optics, electronics and chemistry because of physical and chemical stability, high refractive index, excellent transmittance in the visible and near-infra-red region. In addition, the dielectric constant of some dielectric films can also be modified by adding a small amount of TiO_2 into the films [11,12]. In ULSIs, TiO_2 films have the following properties: a dielectric constant which is up to 10 times higher than that of SiO_2, low leakage current for a given gate voltage and high breakdown strength as compared with SiO_2. Moreover, it is expected to have smaller fixed, trapped, and mobile charge densities, both at the interface and in the bulk. TiO_2 films can be formed by a number of methods, such as reactive sputtering [1,2], plasma-enhanced chemical vapor deposition (PECVD) [3,4], low-pressure chemical vapor deposition (LPCVD) [5], and metalorganic chemical vapor deposition (MOCVD) [6,7,10]. Recently, metalorganic decomposition (MOD) processing has been extensively used in thin-film technology because of easy composition control, good homogeneity, and uniform deposition over a large substrate surface area. Some crystallized and multicomponent thin films (PZT, $SrBi_2Ta_2O_9$, $Bi_4Ti_3O_{12}$, and $Ta_2O_5-Al_2O_3$ etc.) fabricated by MOD have been reported [13-17].

In this paper, we report on the fabrication of thin TiO_2 films on Si by MOD, and investigated the TiO_2 film structure, crystallization behavior and dielectric properties as a function of annealing temperature and ambient.

Mat. Res. Soc. Symp. Proc. Vol. 606 © 2000 Materials Research Society

EXPERIMENTAL

The starting materials were p-type (1-2 Ω cm) silicon (100)-oriented wafers, which were cut into 1.0 cm $\times$ 1.0 cm squares. These wafers were first rinsed in deionized water and methyl alcohol and then cleaned by a standard RCA method. TiO_2 thin films were fabricated by MOD technique using titanium tetrakis isopropoxide (TTIP: $Ti(i\text{-}OC_3H_7)_4$) as a precursor. Isopropyl acetate ($CH_3CO\text{-}O\text{-}iPr$) was used as a solvent. The precursor films was directly deposited onto the Si wafers by the spin-coating technique. The thickness of the film was controlled by the spin-coating conditions. After deposition, the films were dried in air on a hot plate at 110°C for 10 min, then preannealed in nitrogen at 400°C for 30 min in a furnace tube. Crystallization of the films was carried out in nitrogen and/or oxygen atmosphere at temperatures ranging from 600 to 1000°C for 30 min. The crystallinity of the films was examined using a Raman spectrometer excited with the 514.5 nm line of an Ar^+ laser (100 mW) as the excitation source, and a X-ray diffractometer (XRD) using Cu Kα radiation at 40 kV. The TiO_2 and TiO_2/Si interface structure was analyzed using a variable-angle spectroscopic ellipsometer at wavelengths ranging from 350 to 1250 nm at 10 nm intervals. The directly measured values are ellipsometric angles Ψ and Δ as a function of wavelength. Unknown model parameters, such as layer thickness and volume fraction of each constituent in the composite layer, were designated as fitting parameters. Electrical measurements were conducted on films in a metal-insulator-semiconductor (MIS) configuration. Several aluminum dot electrodes of 0.5 mm diameter were evaporated over the area of the TiO_2 films. Capacitance-voltage (C-V) and current-voltage (I-V) measurements were carried out using an electrical analyzer.

RESULTS AND DISCUSSION

Figure 1 shows typical Ψ and Δ curves obtained for TiO_2 films after annealing at 700°C. The broken lines represent measured values and the solid lines are curves simulated using the mathematical model called the Cauchy dispersion model [19]. The ellipsometric parameters, Ψ and Δ, are defined by the relation

$$R_p/R_s = \tan[\Psi(\lambda)]\exp[i\,\Delta(\lambda)], \qquad (1)$$

where R_p and R_s are the parallel and perpendicular reflection coefficients of incident light on the sample as a function of wavelength λ. Ellipsometry is a technique based on measurements of the state of polarization of the incident and reflected light, leading to the determination of the ratio R_p/R_s. In Fig. 1, Ψ and Δ are modelled by calculating reflectance spectra based on trial values of thickness d_i using the complex index of reflection N ($=n-ik$) where n is the real part and k is the imaginary part of the complex refractive index for a particular layer. The quantity used to describe the agreement between measured and calculated values is given by the mean square error (MSE) defined as

$$MSE = (1/N)\sum_{i=1}[\,\Psi_i^m-\Psi_i^c)^2+(\Delta_i^m-\Delta_i^c)^2\,], \qquad (2)$$

where N is the number of data points, Ψ_i^m and Δ_i^m are the measured values at λ_i, and Ψ_i^c and Δ_i^c are the calculated values. The MSE was 2.538. The calculated TiO_2 and SiO_2 film thicknesses in this experiment were 66.0 nm and 11.0 nm, respectively. As shown in Fig. 2, the thickness of the interfacial SiO_2 layer increased with increasing temperature. In contrast, the thickness of the TiO_2 layer was fairly constant. This behavior is due to the fact that during annealing, oxidant (O_2 and/or H_2O) diffuse through the TiO_2 layer into the Si substrate, causing a reaction that forms the $TiO_2/SiO_2/Si$ structure. A schematic illustration of the interfactial reaction is shown in Fig. 3. As-deposited sample may form a TiO_{2-x}/Si structure. During the crystal growth of TiO_2, the reaction starts at the TiO_2/Si interface. After the annealing, the structure changes to $TiO_2/TiO_{2-x}/SiO_2/Si$, and finally to $TiO_2/SiO_2/Si$ multilayered structure.

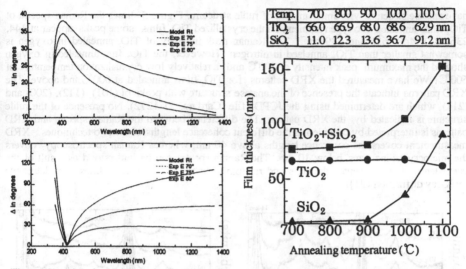

Temp.	700	800	900	1000	1100°C
TiO₂	66.0	66.9	68.0	68.6	62.9 nm
SiO₂	11.0	12.3	13.6	36.7	91.2 nm

Fig.1 Ψ and Δ dependences on wavelength for the TiO₂/SiO₂/Si structure.

Fig.2 TiO₂ and SiO₂ thicknesses as a function of annealing temperature.

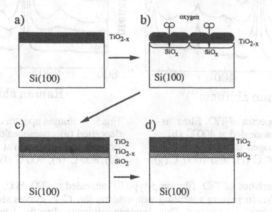

Fig.3 Schematic illustration of the interfacial reaction. (a) as-deposited, (b) after preannealed at 400°C, (c) after crystallization in oxygen ambient, (d) final structure.

Figure 4 shows the Raman spectra of TiO₂ films as-deposited (after being dried), preannealed at 400°C, and crystallized at temperatures from 600 to 1000°C in nitrogen ambient. The peaks at 300 cm⁻¹ and 520 cm⁻¹ are assigned to the Si substrate. No significant peaks due to the crystalline TiO₂ phase can be seen in the spectra of as-deposited films. In the crystallized TiO₂ films, some peaks appear at 144, 236, 400, 448, 612 and 639 cm⁻¹. The peaks at 144, 400 and 639 cm⁻¹ are assignable to, respectively, Raman tensors E_g, B_{1g} and E_g of the anatase structure. In contrast, the bands at 236, 448 and 612 cm⁻¹ are identified as $B_{1g}+B_{1g}$, E_g and A_{1g} of the rutile structure; respectively [4,10]. The peaks of the Raman spectra are reasonably good agreement with the spectra of anatase and rutile TiO₂ powder samples (not shown). The peak intensity of the 448 cm⁻¹ of rutile structure seems to be enhanced by annealing at 800 to 900°C; however, it decreases above 1000°C. The results indicate that the TiO₂ films prepared from MOD have anatase structure with small amount of rutile structure. Figure 5 shows the Raman spectra of TiO$_2$ films annealed in oxygen ambient. In the crystallized TiO$_2$ films, some peaks appear at 144, 236, 400, 448, 612 and 639 cm^{-1}. The Raman peak intensity of TiO$_2$ annealed in oxygen is somewhat smaller than TiO$_2$ annelaed in nitrogen. However, the TiO$_2$ film annealed in oxygen showed the maximum peak intensity at 700℃ and a relatively low crystallization temperature of 600℃. We have measured the XRD patterns for TiO$_2$ films annealed at 600℃ and above. The XRD patterns indicate the presence of the anatase structure with peaks of (101), (112), (200) and (211), which are determined using the JCPDS file (Card no. 21-1272). No presence of the rutile structure is indicated by the XRD pattern. The difference between the Raman spectra and XRD patterns is interpreted by considering the different coherence lengths of the two techniques. XRD measurement covers the coherence lengths above 40 nm, whereas Raman spectroscopy covers the microcrystalline sizes below 10 nm. The Raman spectra can be understood as result of the coexistence of well-oriented domains and randomly distributed microcrystallites, not detectable by X-ray diffraction [21].

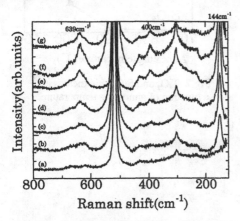
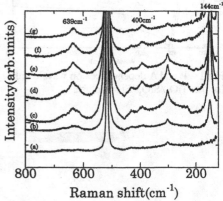

Fig.4　Raman spectra of TiO$_2$ films as-deposited (a), preannealed at 400℃ (b), crystallized in nitrogen at 600℃ (c), 700℃ (d), 800℃ (e), 900℃ (f), and 1000 ℃(g).

Fig.5　Raman spectra of TiO$_2$ films as-deposited (a), preannealed at 400℃ (b), crystallized in oxygen at 600℃ (c), 700 ℃ (d), 800℃ (e), 900℃ (f), and 1000 ℃ (g).

The C-V characteristics of TiO$_2$ film on p-type Si annealed at 700, 800, 900 and 1000℃ are shown in Fig. 6. With increasing annealing temperature, the C-V curves show a decrease in the maximum (accumulation) capacitance. This decrease originates directly from the increase in the thickness of the interfacial SiO$_2$ layer. It is considered that C_{max} measured in the accumulation region is the total series capacitance of TiO$_2$ and underlying SiO$_2$ films. Therefore, the total capacitance C_{tot} is given by

$$1/C_{tot} = 1/C_{TiO2} + 1/C_{SiO2}, \qquad (3)$$

where C_{TiO2} and C_{SiO2} are the capacitances of TiO$_2$ and SiO$_2$, respectively. The capacitances are calculated from

$$C_i = \varepsilon_0 \varepsilon_i A/d_i, \qquad (4)$$

where ε_0 and ε_i and are the permittivity of free space and the dielectric constant of the i-th layer with a thickness of d_i, respectively, and A is the area of the electrode spot (=1.96×10^{-3}cm^2). In the calculation of the dielectric constant of the TiO$_2$ layer, spectroscopic ellipsometry data are used in which the thickness of the interfacial SiO_2 layer is 5.0, 11.0, 12.3, 13.6, 36.7 and 91.2 nm for annealing temperatures of 600, 700, 800, 900, 1000 and 1100℃ in oxygen ambient, respectively, whereas the thickness of the TiO_2 layer is fairly constant in the range of 63-68 nm. For the sample annealed at 700℃, C_{TiO2} and C_{SiO2} are calculated as 1.99 nF and 615 pF respectively; then, the dielectric constant of TiO_2, ε_{TiO2}, is calculated as 76. The dielectric constant of TiO_2 films was calculated as a function of annealing temperature. The dielectric constant of TiO_2 annealed at 600℃ remains at 24, which is nearly the same as those of as-deposited and preannealed samples. On increasing the temperature to 700℃, the dielectric constant increases sharply and a maximum (ε_{TiO2} =76) appears in the samples, which is exactly the crystallization temperature detected from Raman spectra and XRD measurements. However, the dielectric constant decreases rapidly when annealing temperature increases above 800℃. This is due to some stoichiometric change with increasing temperature. In fact, the maximum dielectric constant is obtained for the samples annealed at 700℃ in which the Raman peak intensity of the rutile structure is maximum at 700℃ and disappears above 800℃ owing to the growth of the SiO_2 layer. The higher dielectric constant in the annealed film is assumed to result from some film densification and crystallization of the minor amorphous phase in the as-deposited films that occurred during the heat treatment.

The C-V curves move parallel to the negative bias direction with increasing annealing temperature, owing to SiO_2 growth at the TiO_2/Si interface, as shown in Fig.2. The flat-band voltage, V_{FB}, lies on the slope of the C-V curve between the accumulation and the inversion region. V_{FB} is defined as

$$V_{FB} = \phi_{ms} - (Q_f + Q_m + Q_{ot})/C_{ox}, \qquad (5)$$

where the difference in work function between the metal and the semiconductor, Q_f, is the oxide fixed charge, Q_m is the mobile ionic charge, Q_{ot} is the interface trap charge, and C_{ox} is the capacitance of the oxide. The value of V_{FB} in the sample annealed at 700℃ was + 0.5 V, suggesting the presence of positive charge in the oxide film. The positive charge could result from oxygen vacancies due to dangling and/or broken bonds present at the TiO_2/SiO_2 interface. The samples annealed above 700℃ showed a further shift of the C-V curve in the negative bias direction. Afer annealing at 1000℃, V_{FB} shifted to –0.8 V. The V_{FB} shift toward the negative direction reflects an increase in the density of defect states in the insulator. This results from the formation of distinct grain boundaries, thus creating more dangling bonds. The formation of a new interface between the TiO_2 layer and the SiO_2 layer could provide more interfacial oxide traps. The interface trap density, D_{it}, was typically about 1×10^{11} $eV^{-1}cm^{-2}$, which was calculated from the C-V characteristics at a frequency of 1 MHz.

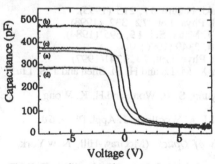

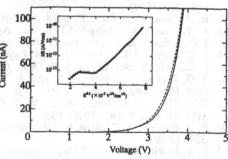

Fig.6 Capacitance-voltage characteristics of TiO_2-based MIS structures annealed at 600℃ (a), 700℃ (b), 800℃ (c), and 900℃ (d).

Fig.7 Current-voltage characteristics of TiO_2 film annealed at 700℃ in oxygen ambient. The inset shows the Poole-Frenkel plots derived from I-V curve.

Figure 7 shows the leakage current (I) versus applied voltage (V) for the annealed TiO_2 films. The I-V curve indicates that the leakage current density is 4×10^{-9} A/cm^2 below 0.1 MV/cm, and increases to 1×10^{-8} A/cm^2 at 0.2 MV/cm. The plot of I-V curve indicates that the current increases exponenitally with the sqare root of the field, implying the Poole-Frenkel type and the dominant conduction is bulk-limited through the TiO_2 rather than electrode-limited. Another conduction due to thermionic emission has been also reported previously [6].

CONCLUSION

In summary, TiO_2 thin films that have a crystalline structure in which anatase is the main component, were successfully prepared on Si substrates by metalorganic decomposition. A maximum dielectric constant of 76 was obtained for TiO_2 films annealed at 700°C in oxygen ambient, and the film had a low leakage current and a low interface trap density. The results indicate that TiO_2 thin films may be used as a suitable dielectric layer in fully integrated circuits.

ACKNOWLEDGMENTS

We would like to thank Michio Suzuki of J. A. Woollam Japan Co., Ltd. for VASE measurement and fruitful discussion. This work was supported by the Foundation for the Promotion of Material Science and Technology of Japan (MST Foundation).

REFERENCES

1. P. Lobl, M. Huppertz and D. Mergel, Thin Solid Films **251**, 72 (1994).
2. J. -Y. Gan, Y. C. Chang and T. B. Wu, Appl. Phys. Lett. **72**, 332 (1998).
3. W. Heitmann, Appl. Opt. **10**, 2414 (1971).
4. H. -K. Ha, M. Yoshimoto, H. Koinuma, B. -K. Moon and H. Ishiwara, Appl. Phys. Lett. **68**, 2965 (1996).
5. N. Rausch and E. P. Burte, J. Electrochem. Soc. **140**, 145 (1993).
6. H. -S. Kim, D. C. Gilmer, S. A. Campbell and D. L. Polla, Appl. Phys. Lett. **69**, 3996 (1996).
7. S. A. Campbell, D. C. Gilmer, X. -C. Wang, M. -T. Hsieh, H. -S. Kim, W. L. Gladfelter and J. Yan, IEEE Trans. Electron Devices **44**, 104 (1997).
8. H. Shin, M. R. De Guire and A. H. Heuer, J. Appl. Phys. **83**, 3311 (1998).
9. H. -S. Kim, S. A. Campbell, D. C. Gilmer, V. Kaushik, J. Conner, L. Prabhu and A. Anderson, J. Appl. Phys. **85**, 3278 (1999).
10. A. Turkovic, M. Ivanda, A. Drasner, V. Vranesa and M. Persin, Thin Solid Films **198**, 199 (1991).
11. R. F. Cava, W. F. Peck, Jr. and J. J. Krajewski, Nature **37**, 215 (1995).
12. J. -Y. Gan, Y. C. Chang and T. B. Wu, Appl. Phys. Lett. **72**, 332 (1998).
13. J. Fukushima, K. Kodaira and T. Matsushita, J. Mater. Sci. **19**, 595 (1984).
14. P. C. Joshi and S. B. Desu: J. Appl. Phys. **80**, 2349 (1996).
15. P. C. Joshi, S. Stowell and S. B. Desu, Appl. Phys. Lett. **71**, 1341(1997).
16. Z. G. Zhang, Y. N. Wang, J. S. Zhu, F. Yan, X. M. Lu and H. M. Shen and J. S. Liu, Appl. Phys. Lett. **73**, 3674 (1998).
17. G. D. Hu, I. H. Wilson, J. B. Xu, W. Y. Cheung, S. P. Wong and H. K. Wong, Appl. Phys. Lett. **74**, 1221(1999).
18. P. G. Snyder, M. C. Rost, H. Bu-Abbud and J. A. Woollam, J. Appl. Phys. **60** 3293 (1986).
19. F. A. Jenkins and H. E. White, *Fundamentals of Optics* (McGraw-Hill, New York, 1957) pp. 468.
20. R. M. A. Azzam and N. M. Bashara, *Ellipsometry and Polarized Light* (Elsevier, New York, 1977) Chap.4.
21. J. Kreisel, S. Pignard, H. Vincent and J. P. Senateur, Appl. Phys. Lett. **73**, 1194 (1998).

Sol-Gel Synthesis of BaTiO₃ Based Films for Photonic Applications

V.Fuflyigin, H.Jiang, F.Wang, P.Yip, P.Vakhutinsky and J.Zhao

NZ Applied Technologies, Woburn, MA 1801, USA

ABSTRACT

High quality barium titanate films were grown on (001)LaAlO$_3$ and r-Al$_2$O$_3$ in a wide thickness range of 0.5-10 μm by sol-gel technique. Significant improvement of the films' crystallinity and optical quality was observed if acetate precursors are used vs. alkoxide precursors, and in the presence of lead oxide. The material is transparent at 350-2000 nm, indicating the possibility of its application in light controlling devices at wavelengths used in optical communication: 1300 and 1500 nm. Maximum field induced relative phase shift of 0.22 radian was measured in the film with composition of Ba$_{0.9}$Pb$_{0.1}$TiO$_3$ under a field strength of 3·10^6 V/cm.

INTRODUCTION

Barium titanate is a promising material in many aspects, i.e. for electro-optic and microwave devices and DRAMs. This is due to the almost unique combination of its ferroelectric, photorefractive and electro-optic properties [1,2]. High-quality films of BaTiO$_3$ with a wide range of thickness are required for electro-optical applications in wave-guide and transmission modes because of the possibility to attain a large phase shift at low optical loss in BaTiO₃. The growth of thin epitaxial films of BaTiO₃ has been demonstrated by several techniques: chemical vapor deposition (CVD), rf-sputtering, sol-gel and dipping-pyrolisis [3-5]. However, all mentioned techniques failed to produce sufficiently thick films of high optical quality. These films are required for certain photonic applications in which electro-optic films are used in transmission or reflection modes [6].

It should be mentioned that there are a number of difficulties in fabrication of BaTiO$_3$ films related to the BaTiO$_3$ phase itself. The main difficulty is the slow kinetics of BaTiO$_3$ phase formation. This results in a low growth rate and requires a growth at elevated temperatures to produce high quality films of barium titanate. High processing temperature is an apparent obstacle in integrating of BaTiO$_3$ films if other device components can not sustain high temperatures. The goal of this work was to discuss some approaches to the sol-gel synthesis of the barium titanate films for photonic and microelectronic applications. In general, these approaches can be also useful for the synthesis of other high temperature oxides especially in the form of thin and thick films.

EXPERIMENTAL

Metal alkoxides are traditionally used as precursors in sol-gel process. These substances are very sensitive to ambient moisture and carbon dioxide. Also, these solutions are usually stable for rather short time because of the fast gelation process. This results in poor reproducibility of the films properties. Mixed barium-titanium alkoxides recently became popular sol-gel precursors because of their improved stability against ambient moisture and carbon dioxide, and also because they yielded better quality films at lower temperature [7].

169

We have used two types of precursors for growing barium titanate films: alkoxide and acetate based precursors. Barium-titanium alkoxide precursor was prepared by dissolving metal barium in absolute isopropanol. Titanium isopropoxide (Strem Chemicals) was then added followed by addition of acetic acid ($CH_3COOH/Ti=0.025/1$) to promote formation of the mixed alkoxide. Mixture was refluxed for 24 hours. Addition of water ($H_2O/Ti=0.1-0.25/1$) was followed by 2 hours reflux. The prepared solution was diluted with absolute isopropanol to adjust the concentration and viscosity. Solution was stable for 7-10 days.

The mixed acetate-alkoxide precursor solution was prepared by dissolving barium acetate (99.999%, Aldrich Chemical Co.) in deionized water. This solution was then mixed with a solution of titanium (diisopoxide)bis(acetylacetonate) (Strem Chemicals) in methanol. Acetic acid was added to adjust pH which was kept in the range of 5-6. Precursor solution was then diluted with methanol and was used to grow barium titanate films. The solution thus prepared was stable for 4-6 months. Strontium acetate and lead acetate adduct with acetic acid were used as strontium and lead precursors when these element were added to the precursor solutions.

Deposition was performed using standard dip-coating technique. Details of the deposition procedure are described elsewhere [9]. Deposition temperature was in the range of 450-750°C.

XRD spectra were recorded on "Rigaku" diffractometer using $Cu_{K\alpha}$ radiation. Phi-scanning and rocking curve measurements were taken using "Phillips" four-circle diffractometer. Films thickness was measured using profilometer "Dektak".

RESULTS AND DISCUSSION

Precursor chemistry and films' composition.

We have deposited films of barium titanate on (001) lanthanum aluminate substrates using both alkoxide and acetate based precursor solutions. Crystallization temperature of barium titanate was strongly dependent on the precursor used. If an alkoxide precursor is used, the

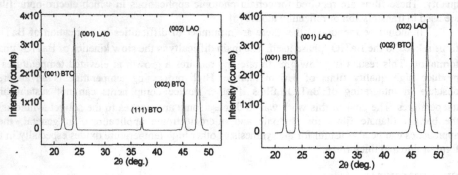

Fig.1 XRD patterns of BTO films deposited on (001) LAO using alkoxide (a) and acetate (b) precursors.

presence of barium titanate phase was detected only at 650°C. If an acetate based precursor is employed, the $BaTiO_3$ phase occurs already at 500°C. XRD spectrum of $BaTiO_3$ film deposited on $LaAlO_3$ at 650°C from alkoxide precursor reveals presence of (001) and (111) reflections as shown in (Fig.1a). This speaks out for its polycrystalline nature.

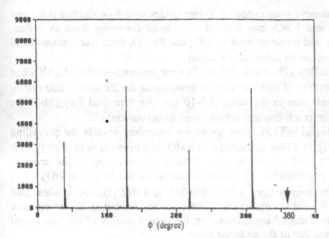

Fig.2. φ–scan of (111) reflection of BTO film grown on (001) LAO.

At the time only peaks corresponding to (00*l*) reflections are present in θ–2θ scan of the barium titanate film deposited from acetate based solution (Fig1b). This difference can be understood by considering processes taking place during decomposition of the oxide precursors. Barium acetate melts at 490°C, and simultaneously it starts decomposing. It is well known that solid state reactions proceed faster in the presence of liquid phase. Therefore, one can speculate that presence of the molten barium acetate promotes

nucleation and crystallization of $BaTiO_3$ even at a relatively low temperature. Liquid phase does not form during the decomposition of alkoxide precursor [9]. This results in a higher crystallization temperature for $BaTiO_3$ formed from alkoxide solution. Further characterization

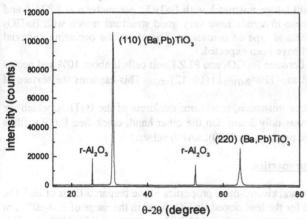

Fig.3 XRD pattern of the $Ba_{0.9}Pb_{0.1}TiO_3$ film films deposited on r-sapphire.

of the $BaTiO_3$ film deposited on (001)$LaAlO_3$ at 500°C revealed a perfect in-plane alignment of $BaTiO_3$ crystallites as shown in Fig.2, which presents φ-scan of (111) reflection of $BaTiO_3$. Additional proof of the high quality of $BaTiO_3$ layer comes from the rocking curve measurements performed for (002) reflection of $BaTiO_3$ film. Full Width at Half Maximum (FWHM) was 0.15°.

Other approach to promoting crystallization of $BaTiO_3$ consists of using additives having a low melting point or forming

eutectics with film's components (fluxes). We have chosen lead (II) oxide for this purpose. Lead oxide has relatively low melting point (886°C) and might form eutectics with other oxide components of the film promoting diffusion and, thereby, enhancing the phase formation process. As a result of more intense phase formation and sintering processes, $BaTiO_3$ films with better crystallinity (and transparency) were expected. Other oxides with low melting point and high volatility, such as Bi_2O_3 and MoO_3 can also be considered for using them as a flux. However we have preference to lead oxide because $BaTiO_3$ and $PbTiO_3$ form a continuous solid solution, hence phase segregation can be potentially avoided.

Films with a composition $Ba_{1-x}Pb_xTiO_3$ (x=0-0.25) were prepared on (001) $LaAlO_3$ and r-sapphire substrates. A 2 at.% excess of lead was used to compensate for the loss of lead during thermal treatment. Film thickness was in the range 0.5-10 μm. As deposited $Ba_{1-x}Pb_xTiO_3$ (x=0-0.2) were highly transparent, crack free and adhered well to the substrate.

The X-ray pattern of $Ba_{1-x}Pb_xTiO_3$ films grown on r-sapphire reveals the prevailing (110) orientation of the films (Fig.3). Films deposited on $LaAlO_3$ had orientation of (00l) type. It is to be noted that (101) and (110) peaks are not resolved. That allows us to suggest the structure of $Ba_{1-x}Pb_xTiO_3$ films to be pseudocubic. The lattice constant of the $Ba_{1-x}Pb_xTiO_3$ films deposited on r-sapphire changes linearly from 3.999(±0.003)Å to 4.092 (±0.004)Å when lead content changes from 0 to 0.2. We did not observe a reduction of the crystallization temperature of lead doped films compared to undoped ones. However, the optical quality of the lead doped films was significantly higher then that of the undoped ones.

Role of the substrate material.

In order to take advantage of this anisotropy to fit specific applications, control of the crystallographic orientation during the materials growth is critical. This is particularly important for the applications of these materials in light-controlling devices since devices with different operational modes would require material with different crystallographic orientation relative to the incident light beam.

$LaAlO_3$ has a relatively small lattice mismatch with $BaTiO_3$: parameter **a** is 3.874 Å and 3.994 Å respectively. Moreover, these materials have very good structural match with $BaTiO_3$ since they belong to the same structural type of perovskite. Therefore, the occurring epitaxial growth of $BaTiO_3$ on $LaAlO_3$ could have been expected.

Minimum lattice mismatch between $BaTiO_3$ and PLZT unit cells is about 10% and occurs for the relative orientation defined as $<110>_{BaTiO3}||(10\ \overline{1}2)_{Al2O3}$. This explains the texture of (110) type developed in $BaTiO_3$

As a result of the large lattice mismatch, maximum thickness of the $BaTiO_3$, which we were able to attain on r-sapphire, was only 3 μm. On the other hand, crack free $BaTiO_3$ films with the thickness up to 10 μm were prepared on (100)$LaAlO_3$ substrates.

Optical and electro-optic properties.

We have characterized optic and electro-optic properties of the prepared films of $BaTiO_3$. Optical transmission was measured for the lead doped $BaTiO_3$ films in the range of 320-2000 nm using an optical spectrometer. The material was found to be transparent at 350-2000 nm. This indicates the possibility of their application in light controlling devices at wavelengths used in optical communication systems, that is 1300 and 1500 nm. To accurately measure the optical loss in barium titanate films on r-sapphire, prism-coupling experiments were performed. The optical loss coefficient of 1.5-3 μm thick barium titanate films measured at 633 nm was in the range of 2-2.5 dB/cm. This is significantly lower than the value of optical loss in undoped barium titanate films of comparable thickness grown by MOCVD or by sol-gel techniques, which is usually in the range of 20-30 dB/cm [10].

A transmission differential ellipsometer was employed to measure the electro-optic properties of the prepared films of lead doped barium titanate. Details of this method are described elsewhere [11]. Coplanar interdigitated aluminum electrodes were fabricated on dip-coated films by a standard lift-off process. Electro-optic measurements were conducted on films with a thickness of 0.9-1.1 μm. The material exhibited a strong electro-optic effect with electro-optic hysteresis loops characteristic for ferroelectric materials (Fig.4). A maximum relative phase shift of 0.22 radian (normalized to thickness of 1 μm) was achieved in transmission-mode under a moderate electric field for $Ba_{0.9}Pb_{0.1}TiO_3$ film.

The value of the second-order electro-optic coefficient (R) was calculated as follows:

$$R = \lambda \Delta\Gamma / (\pi n^3 E^2 L) \qquad (1)$$

where $\Delta\Gamma$ is the relative phase shift in radians, λ is a wavelength in nm, n is the refractive index of the film, E is the density of the electric field in V/m and L is the film thickness in nm. The value of R for $Ba_{0.9}Pb_{0.1}TiO_3$ film was calculated to be $0.65 \cdot 10^{-16}$ $(m/V)^2$. This number is higher than that measured for $(Pb,La)(Ti,Zr)O_3$ (PLZT) film with composition of 8/65/35, which is one of the most commonly used electro-optic materials, grown on r-sapphire. 1-2 μm thick PLZT films exhibited a phase shift of 0.15 radian (recalculated for 1 μm film). This translates into value of $R = 0.45 \cdot 10^{-16}$ $(m/V)^2$ [11].

It should be mentioned that the value of the phase shift measured for $BaTiO_3$ films grown on r-sapphire by MOCVD was 0.23 rad for 1 μm film. This is very close to phase

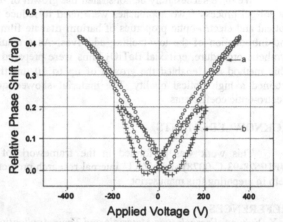

Fig.4. Electro-optic hysteresis loops for the films with compositions of $Ba_{0.9}Pb_{0.1}TiO_3$ (a) and $Ba_{0.8}Pb_{0.2}TiO_3$ (b).

shift observed in the lead doped barium titanate. However, the optical quality of $Ba_{1-x}Pb_xTiO_3$ films is significantly better than that of undoped films grown by MOCVD or sol-gel.

A higher lead content leads to the decrease the electro-optic effects (Fig.4). This can be explained by the fact that the magnitude of the electro-optic effect is believed to be directly connected with the value of the dielectric constant of electro-optic material [12]. The Curie temperature of $Ba_{1-x}Pb_xTiO_3$ solid solution changes from 120°C for x=0, to 490°C for x=1. Hence, the dielectric constant of $Ba_{1-x}Pb_xTiO_3$ at room temperature decreases with the increase of the lead content. This leads to the decrease of the electro-optic effect. At the same time the shape of the electro-optic hysteresis loop became more square at a higher lead content, indicating well-developed memory effect. Apparently, domain switching becomes more difficult as the Curie temperature moves further from the room temperature.

$BaTiO_3$ films deposited on (001)$LaAlO_3$ demonstrated substantially lower electro-optic effect ($R=0.15 \cdot 10^{-16}$ $(m/V)^2$) compared to $BaTiO_3$ and $Ba_{1-x}Pb_xTiO_3$ films deposited on r-sapphire. Orientation of the ferroelectric films appears to be an important factor affecting electro-optical properties of the $BaTiO_3$ films. $BaTiO_3$ films with orientation of (110) type appear to have a larger EO effect compared to those having (001) orientation. This observation for barium titanate films can not be interpreted, however, solely by considering crystallographic factor. The [111] direction lies in the substrate plane if film has (110) orientation and is, therefore, perpendicular to the incident light beam. In this arrangement, a maximum phase retardation between ordinary and extraordinary components of the light beam can be reached. The ferroelectric $BaTiO_3$ phase, which exists at 0-120°C has a tetragonal symmetry (*4mm*) and the spontaneous polarization parallel to [001] direction. The rhombohedral (*3m*) phase exists only at the temperatures lower than -90°C [1]. In our opinion, different stress conditions and their implications for domain switching in $BaTiO_3$ films deposited on different substrates should be taken into account in this case. This problem certainly deserves a separate detailed study.

CONCLUSIONS

We have successfully demonstrated the growth of thin and thick films of barium titanate by sol-gel process. Two approaches were used to reduce processing temperature and improve optical and electro-optic properties of barium titanate films: use of the acetate based precursor and introduction of the low-melting additives. As a result of the optimization of the sol-gel synthetic procedure, epitaxial $BaTiO_3$ films were prepared at the temperature as low as 500°C. This method also enabled the growth of up to 10 μm thick films of barium titanate, which retained a high optical quality. The material showed quadratic electro-optic behavior with electro-optic coefficients.

ACKNOWLEDGMENTS

This work was performed in the framework of USAF Rome Laboratory Contract (#F19628-93-C-0078) and NZAT' internal research project. Authors also thank E.Salley for his help in preparing this manuscript.

REFERENCES
1. X. Yuhuan, *Ferroelectric Materials and Their Applications* (Elsevier Science Publishers B.V., Amsterdam, 1991), p.101-159.
2. B.Hoerman, G.M.Ford, L.D.Kaufmann, B.Wessels Appl.Phys.Lett., **73**, (1998)
3. J. Zhao, V.Fuflyigin, F.Wang, P.Norris, L. Bouthilette, C. Woods, J.Mater.Chem **7**, 933 (1997)
4. K.Takemura, T.Sauma, Y.Miyasaka, Appl.Phys.Lett., **64**, 2967 (1994)
5. T.Hayashi, T.Tanaka, Jap.J.Appl.Phys, **34**, 5100 (1995).
6. F.Wang, K.K.Li, V.Fuflyigin, Appl.Optics, **37**, 7490 (1998)
7. T.Hayashi, N.Ohji, K.Hiohara, T.Fugunaga, H.Maiwa, Jap.J.Appl.Phys, **32**, 4092 (1993).
8. V.Fuflyigin, K.K.Li, F.Wang, H.Jiang, S.Liu, J.Zhao, P.Norris, P.Yip, *in High-Temperature Superconductors and Novel Inorganic Materials*, edited by G.Van Tendeloo, (Kluwer Academic.Publ. 1999), p.279-284.
9.V.Fuflyigin unpublished data
10.F.Wang unpublished data
11. F.Wang, E.Furman, G.H. Haertling, J.Appl.Phys. **78**, 9 (1995)
12. F. Wang, Phys.Rev. B, **59**, pp.9733-9736 (1999).

FERROELECTRIC COMPOSITE OF $Ba_{1-x}Sr_xTiO_3$ WITH Al_2O_3 AND MgO SYNTHESIZED BY SOL-GEL METHOD

Pramod K. Sharma*, K. A. Jose, V. V. Varadan and V . K. Varadan
Center for the Engineering of Electronics and Acoustic Materials,
Department of Engineering Science and Mechanics, The Pennsylvania State University,
University Park, PA 16802, USA

ABSTRACT

Barium strontium titanate with low dielectric constant, low loss tangent and high tunability, has unique application as phase shifter in radar and antenna. This work presents an alternative fabrication of the composite of barium strontium titanate with metal oxides such as Al_2O_3 and MgO. $Ba_{1-x}Sr_xTiO_3$ (BST), xergels of various compositions (x = 0.2, 0.4, 0.5 and 0.6) were prepared by sol-gel method. Xerogels were calcined at 800 °C. The structure of the final powders was investigated by x-ray diffraction (XRD). The resultant BST was then mixed with metal oxides in the desired weight percentage and ball milled in an organic solvent (ethyl alcohol) with a binder (acrylic polymer). The final samples were sintered at 1250 °C in air. The dielectric properties of the composites were determined by using impedance analyzer at the frequency of 1MHz. This work has reported the effect of Al_2O_3 and MgO, on the electronic behavior of the final BST-metal oxide composites.

INTRODUCTION

Recent advances in ceramic composites and thin film deposition technologies have resulted in new ferroelectrics with voltage-tunable dielectric constants at room temperature and improved dielectric loss. On going research in ferroelectric materials renewed interest in developing phase shifters at microwave frequencies [1]. The most commonly used phase shifters are ferrite and diode phase shifters [2]. The phased shifter antennas transmitted and received signals without mechanical rotating the antenna. These antennas are currently constructed using ferrite phase shifting elements. The ferrite phase shifting elements are relatively large and heavy. Barium strontium titanate (BST) exhibits a large tunability (variation of dielectric constant with applied DC bias fields) and low loss tangent. It also has an insensitivity of di- electronic properties to change in environmental conditions. So it is well suitable for dielectric phase shifters [3-5].

Sol-gel technique has obtained much interest because of its many advantages such as easier composition control, better homogeneity, lower processing temperature, lower cost [6-9]. This work presents a brief study on the synthesis of $Ba_{1-x}Sr_xTiO_3$ (x=0.2, 0.4,

0.5, 0.6) by sol-gel method and its composite fabrication with metal oxides e.g. Al_2O_3 and MgO. This study also illustrates a novel method to alter electronic properties of the $Ba_{1-x}Sr_xTiO_3$ so that they can be adjusted for a particular application.

EXPERIMENTAL

1. Synthesis of $Ba_{1-x}Sr_xTiO_3$

BST was synthesized by an alternate sol-gel method. Titanium tetra iso-propoxide $(Ti(O-C_3H_7)_4)$ and catalyst, were mixed in the appropriate molar ratio with 2-methoxyethanol solvent and refluxed for 2 hrs at 80 °C. Separate solutions of Ba and Sr were prepared by dissolving 2,4-pentadionate salts of Ba and Sr, in methoxyethanol. Mild heating was required for a complete dissolution of salts. The metal salt solution was then transferred to the prepared titania sol slowly. The solution was refluxed for another 6 hrs. The sol was then hydrolyzed by a particular concentration of water. It is important to note

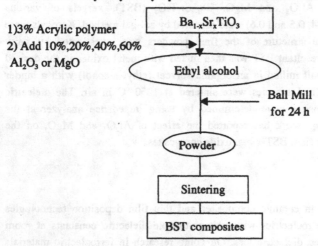

Fig. 1 Schematic diagram of fabrication of BST composite with metal oxides.

that direct addition of water leads to precipitation in the sol. Therefore a mixture of water/solvent has to be prepared, and then added to the sol drop by drop. The resultant sol was refluxed for 2 hrs to complete hydrolysis. This sol was kept in an oven at 90 °C to obtain the xerogel. Xerogels with various compositions (x = 0.2, 0.4, 0.5 and 0.6) were then calcined at 800 °C. Structure of the $Ba_{1-x}Sr_xTiO_3$ samples after calcination were investigated by x-ray diffraction (XRD) using a Scintag diffractometer (DMC 105) with copper Kα radiation for 2θ from 25° to 70°.

2. Fabrication of BST composite with metal oxides:

The resultant BST powder in appropriate weight percent (40% wt. to 90% wt.), was mixed with metal oxide e.g. Al_2O_3 or MgO. Ethyl alcohol and 3 wt. % of acrylic polymer, were used as solvent and binder respectively. This slurry was ball milled for 24 hours. The final mixture was then air-dried and dry-pressed at approximately 8 tons. These composites were then fired under air atmosphere initially at 300 °C for 2 hrs and finally 1250 °C for 5 hours. Heating and cooling rate of the furnace was 1 °C per minute. Fig. 1 shows the schematic diagram of the fabrication process.

3. Electrical measurement of BST composite:

Proper electroding of the composite ceramics must be done. Ag paste was painted on both sides of the pellets. These pellets were dried at 70 °C for 24 hours to form ohmic contact electrodes. Permittivity values at 1 MHz were measured at room temperature by an impedance analyzer (HP 4192A).

RESULTS AND DISCUSSION

Fig. 2 shows the XRD pattern of BST (x=0.2, 0.4, 0.6 mole) and confirms the formation of crystalline material in these powders is $Ba_{1-x}Sr_xTiO_3$. It is seen in the figure that pure $Ba_{1-x}Sr_xTiO_3$ was observed only x=0.4. At x=0.2, the amount of strontium was

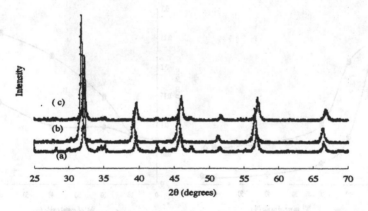

Fig. 2 XRD patterns of BST with x (a) 0.2, (b) 0.4 and (c) 0.6 mole

small and excess barium forms the barium carbonate. The Sr rich compound along with barium carbonate, was appeared at x=0.6 due to incomplete dissolution of Sr with Ba at 800 °C.

The use of Al_2O_3 and MgO in the composite fabrication of $Ba_{1-x}Sr_xTiO_3$ was made in order to change the electronic properties of $Ba_{1-x}Sr_xTiO_3$. The variation in the dielectric constant of $Ba_{1-x}Sr_xTiO_3$ (at x=0.4) with respect of addition of Al_2O_3 and MgO are shown in Fig. 3. It is apparent from the Fig.3 that the dielectric constant (the dielectric constant which is a complex quantity i.e. $\varepsilon = \varepsilon' - \varepsilon''$ where ε' is real permittivity, ε'' is imaginary permittivity) decreases with increasing wt. % of metal oxide. The dielectric constant of BST composite with Al_2O_3 is observed to be lower than BST composite with MgO at a particular wt. % of any metal oxide. The decrease in dielectric constant with metal oxides can be attributed due to the increase in loss tangent. It is well known that the dielectric constant is related to the energy stored in the material where as the dielectric loss is related to the power dissipation in the same material.

The low dielectric constant of BST is desirable for its application in phase shifting device since the impedance matching for these becomes easier. It is important to note that the low dielectric constant does not affect the phase shifting ability provided the dimension of the material is sufficient. Fig. 4 indicates that the tangent loss (tan δ = $\varepsilon''/\varepsilon'$) increases when Al_2O_3 increases. The trend is similar for the $Ba_{1-x}Sr_xTiO_3$ (at x=0.4) composite with MgO. Therefore, it is assumed that addition of metal oxides plays an important role in affecting the grain boundary of $Ba_{1-x}Sr_xTiO_3$ which leads to an increase in dielectric loss [10,11].

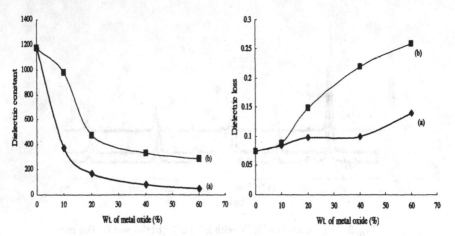

Fig. 3 Variation in dielectric constant
of BST (x=0.4) with (a) Al_2O_3 (b) MgO.

Fig. 4 Dielectric loss of BST with
(a) Al_2O_3 and (b) MgO.

The electronic properties of the material influenced by the tunability i.e. how much the dielectric constant changes with applied voltage. The amount of the phase shifting ability is directly related to the tunability. Higher tunabilities of this material is desirable. Hence tunability is an important parameter while considering the application of BST for antenna and radar. Tunability can derived by the following equation for a ferroelectric material [12].

$$\text{Tunablity}(\%) = \frac{\varepsilon_{V_0} - \varepsilon_V}{\varepsilon_{V_0}} \times 100$$

where
ε_{V_0} = Dielectric constant at zero voltage,
ε_V = Dielectric constant at applied voltage

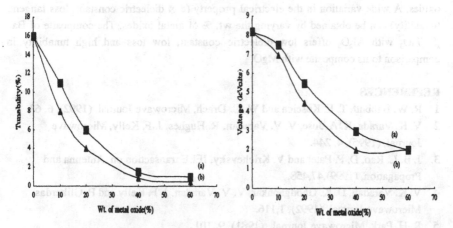

Fig. 5 Tunability of BST composite with   Fig. 6 Voltage of BST composite (a) Al_2O_3
(a) Al_2O_3 and (b)MgO.   and (b)MgO.

Fig. 5 shows the tunablity of $Ba_{1-x}Sr_xTiO_3$ (at x=0.4) composite with Al_2O_3 or MgO at 1 MHz. Measurements indicated that the maximum tunability could be obtained at 10 wt. % of metal oxides. It decreases with increasing concentration of metal oxides. As expected tunability shows similar trend with metal oxides as dielectric constant (refer to Fig.2). The dielectric loss is inversely related to the tunability and it is apparent in the Figs. 3 and 5 respectively. At lower wt. % of the metal oxide, the dielectric loss is low but the tunability is high. At higher wt. % of the metal oxide, the tunability is lower due to high dielectric loss. This behavior is similar for both the $Ba_{1-x}Sr_xTiO_3$ composites derived from Al_2O_3 and MgO.

Bias voltage decreases with increase in weight percent of metal oxides as shown in Fig. 6. The maximum voltage used here was 8 kV however it is still lower than it reported elsewhere [13]. At lower level of metal oxide (0-20 wt.%), higher values of electronic properties were obtained in comparison to higher level of metal oxide (30-60 wt.%). Hence electronic properties of the BST composite can be adjusted with the help of formulation of metal oxide in BST.

CONCLUSION

This study offers a new method of fabrication of the $Ba_{1-x}Sr_xTiO_3$ composite with metal oxide. Value of x=0.4 was chosen for the purity of initial material. The tunability at this concentration of x in pure $Ba_{1-x}Sr_xTiO_3$ was higher than its composite with any metal oxides. A wide variation in the electrical property (e.g. dielectric constant, loss tangent, tunability) can be obtained by varying the wt. % of metal oxides. The composite of $Ba_{1-x}Sr_xTiO_3$ with Al_2O_3 offers low dielectric constant, low loss and high tunability in comparison to its composite with MgO.

REFERENCES

1. R. W. Babbutt, T. E. Koscica and W. C. Drach, Microwave Journal, (1992), 6, 63.
2. V. K. Varada, K. A. Jose, V. V. Varadan, R. Hughes, J. F. Kelly, Microwave Journal, (1995), 4, 244.
3. J. B. L. Rao, D. P. Patel and V. Krichevsky, IEEE transaction on Antenna and Propagation, (1999),47,458.
4. V. K. Varadan, D. K. Ghodgaonkar, V. V. Varadan, J. F. Kelly and P. Glikerdas, Microwave Journal, (1992), 1,116.
5. R. H. Park, Microwave Journal, (1981), 9, 101.
6. C. J. Brinker and G. W. Scherer, Sol-gel science: The physics and chemistry of sol gel processing ; Academic Press, Boston (1990).
7. R. C. Mehrotra. Structure and Bonding, (1992), 77, 153.
8. S. Sakka and K. Kamiya, J. Non-Cryst. Solids, (1982), 48, 31.
9. Pramod K. Sharma, K. A. Jose, V. V. Varadan and V. K. Varadan, Mater. Res. Soc. Symp. (1999) 606.
10. S. B. Herner, F. A. Selmi, V. V. Varadan and V. K. Varadan, Mater. Lett. (1993), 15, 317.
11. N. H. Chan, R. K. Sharma and D. M. Smyth, J. Am. Cer. Soc. (1982), 65, 165.
12. L. Sengupta, E. Ngo, S. Stowell, M. O'Day, R. Lancio, U.S. Patent, (1996) 5486491.
13. L. Sengupta, E. Ngo, S. Stowell, M. O'Day, R. Lancio, U.S. Patent, (1996) 5312790.

EPITAXIAL GROWTH OF $Sr_{0.3}Ba_{0.7}Nb_2O_6$ THIN FILMS PREPARED BY SOL-GEL PROCESS

Keishi Nishio, Jirawat Thongrueng, Yuichi Watanabe and Toshio Tsuchiya
Dep. of Materials Sci. and Tech. Science University of Tokyo, Noda Chiba, 278-8510, JAPAN

ABSTRUCT

We succeeded in the preparation of strontium-barium niobate ($Sr_{0.3}Ba_{0.7}Nb_2O_6$: SBN30) that have a tetragonal tungsten bronze type structure thin films on $SrTiO_3$ (100), STO, or La doped $SrTiO_3$ (100), LSTO, single crystal substrates by a spin coating process. LSTO substrate can be used for electrode. A homogeneous coating solution was prepared with Sr and Ba acetates and $Nb(OEt)_5$ as raw materials, and acetic acid and diethylene glycol monomethyl ether as solvents. The coating thin films were sintered at temperature from 700 to 1000°C for 10 min in air. It was confirmed that the thin films on STO substrate sintered above 700°C were in the epitaxial growth because the 16 diffraction spots were observed on the pole figure using (121) reflection. The <130> and <310> direction of the thin film on STO were oriented with the c-axis in parallel to the substrate surface. However, the diffraction spots of thin film on LSTO substrate sintered at 700°C were corresponds to the expected pattern for (110).

INTRODUCTION

Ferroelectric materials with chemical formula such as $Sr_xBa_{1-x}Nb_2O_6(0.25 \leq x \leq 0.75)$ (SBN) have a tetragonal type tungstenbronze structure and have been studied for various devices [1-6]. In particular, $Sr_{0.3}Ba_{0.7}Nb_2O_6$ (SBN30) has an attractive effect on the ferroelectric and optoelectric properties (e. g., piezoelectric effect, linear optoelectric effect and high refraction index). The SBN30 ceramic is expected to be applied for a piezo type infrared sensor [4,5], an optical modulation device [7,8], an elastic surface wave filter [8,9] and a hologram device [10]. The SBN30 ceramic with a tetragonal tungsten bronze type structure belongs to the point group of 4mm symmetry at room temperature[6]. Especially, the c-axis (00n) oriented SBN thin films are suitable as the optical modulation device and the elastic surface wave filter device.

The sol-gel process has received much interest due to a potential of precise control of chemical stoichiometry, homogeneity, low temperature processing and decreasing costs [11-15]. We have already reported that $Sr_{0.3}Ba_{0.7}Nb_2O_6$ thin films were prepared on $SrTiO_3(100)$ single crystal substrate and MgO(100) single crystal substrate [16]. In this study, an attempt is made to prepare highly c-axis oriented and epitaxial SBN30 thin films on $SrTiO_3(100)$ and 0.75w% La doped $SrTiO_3$ (100) single crystal substrates by sol-gel process, and the dielectric properties of the obtained films were measured.

EXPERIMENT

A coating solution and thin films of $Sr_{0.3}Ba_{0.7}Nb_2O_6$ were prepared as shown in Figure 1. Strontium acetate $[Sr(OCOCH_3)_2 \cdot 0.5H_2O]$, bariun acetate $[Ba(OCOCH_3)_2]$ and pentaethoxyniobium $[Nb(OC_2H_5)_5]$ were used as raw materials. Acetic acid and diethylene glycol monoethylether (DGME) were used as solvents. An alkoxide $(M(OR)_n)$ is stabilized by DGME, because DGME $(C_2H_5OCH_2CH_2OCH_2CH_2OH)$ has high ability of coordination to metal alkoxides [17]. Firstly, $Sr(OCOCH_3)_2 \cdot 0.5H_2O$ and $Ba(OCOCH_3)_2$ were dissolved in the acetic acid at room temperature and stirred at 110°C for 4 h. Then, it was cooled down to room temperature. $Nb(OC_2H_5)_5$ mixed with DGME were mixed to the solution in a dry box filled with N_2 gas. The thin films were formed on $SrTiO_3$ (100), STO, and La doped $SrTiO_3$ (100) (La : 3.73w% or 0.35w%), LSTO, single crystal substrates. The coating films were deposited by spin-coating method at 1000 rpm for 5 s and 4500 rpm for 40 s. After dried the coating film in air at 200℃ for 10 min, the thin films were sintered in air at temperatures from 700 to 1000℃ for 10min.

Using an XRD diffractometer (Rigaku model CN4148) with a thin film attachment, crystal phases of the thin films were identified. X-ray pole figures of the thin films were investigated using X-ray diffraction MXP system of MAC Science. The surface of the thin film was investigated with atomic force microscopy (AFM) images (SII SPA300). The *P-E* hysteresis loops of the thin film were investigated using a Sawyer-Tower bridge with a triangular field of 50 Hz.

RESULTS

Figure 2 shows XRD patterns of the SBN30 thin film on STO and LSTO substrate prepared at the temperature from 700 to 1000℃ for 10 min. The only diffraction pattern of $SrTiO_3$ (100) substrate was observed.

Figure 3 shows (a) and (b) the pole figures of SBN30 thin films on STO (100) sintered at 700 and 1000℃, (c) and (d) LSTO (100) substrates sintered at 700 and 900℃ using (121) reflection. It was confirmed that the obtained thin films were in the epitaxial growth because the diffraction spots were observed. Figure 3 - (e) and (f) show a pole figures which corresponds to the expected pattern for (310) and (440) oriented SBN30 thin film. Figure 3 - (a) and (b) are in a good agreement with summation of (130) pole figure (Figure 3 - (d)) and rotated (130) by 90° of pole figure. In other words, <130> and <310> directions of the thin film prepared on $SrTiO_3$ (100) were oriented with the c-axis in parallel to the substrate surface. Figure 3 - (c) is in a good agreement with (440) pole figure. Figure 3 - (d) is in a good agreement with summation of (130) pole figure and (440) pole figure. The thin film sintered at 900℃ oriented with two directions.

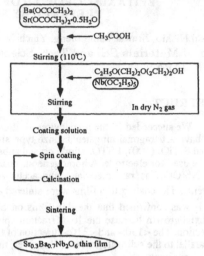

Figure 1　Preparation procedure of SBN30 thin film.

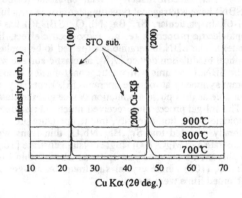

Figure 2　XRD patterns for SBN30 thin film on 0.75w%La doped $SrTiO_3$(100) by heat treatment for 10min.

Figure 4 shows AFM images of the thin film prepared on STO (100) and LSTO (100) single crystal substrates, which were heat treated at the temperature of 700, 900 and 1000℃, respectively. The grains of the thin film prepared on STO substrate were pillar-shaped and arranged in right angle to each other. However, the grains of thin film prepared on LSTO by heat treatment at 700℃ were cubic, and those of the thin film prepared by heat treatment at 900℃ were mixed pillar-shape and cubic. From the results of AFM images and that of pole figures, it is thought that the pillar-shape grain is (130) oriented and cubic grain is (110) oriented.

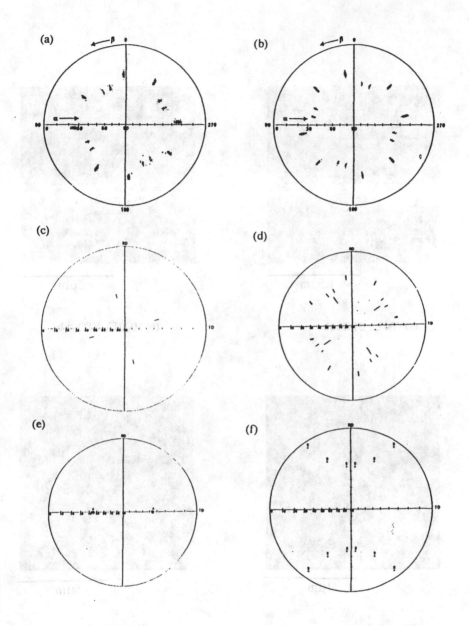

Figure 3 Pole figures for SBN30 thin films and expected pattern using (121) reflection.
(a) prepared on SrTiO₃ single crystal substrate by heat treatment at 700℃, (b) prepared on SrTiO₃
single crystal substrate by heat treatment at 1000℃, (c) prepared on La doped SrTiO₃ single
crystal substrate by heat treatment at 700℃, (d) prepared on SrTiO₃ single crystal substrate by
heat treatment at 1000℃, (e) expected pattern (440) and (f) expected pattern (310).

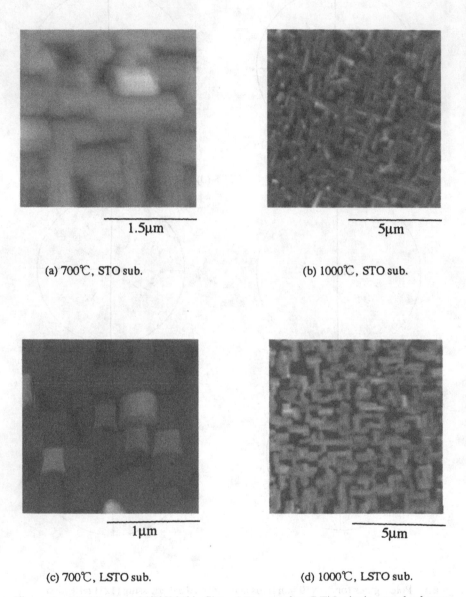

(a) 700℃, STO sub.

(b) 1000℃, STO sub.

(c) 700℃, LSTO sub.

(d) 1000℃, LSTO sub.

Figure 4   AFM images of SBN30 thin films. (a) prepared on $SrTiO_3$ single crystal substrate by heat treatment at 700℃, (b) prepared on $SrTiO_3$ single crystal substrate by heat treatment at 1000℃, (c) prepared on La doped $SrTiO_3$ single crystal substrate by heat treatment at 700℃, (d) prepared on $SrTiO_3$ single crystal substrate by heat treatment at 1000℃.

The thin film prepared on SrTiO₃(100) substrate was found to exhibit enhanced orientation for the c-axis in parallel to the substrate surface. Therefore, the grains of the thin film were pillar-shaped and aligned right angle each other at random. The grains of the thin films were influenced from lattice of the substrate. The mismatching between SrTiO₃(100) and the SBN30(100) are Ma = Mb = 5.75% and Mc = 1.39%. The thin film prepared on SrTiO₃(100) substrate orientated in the c-axis direction, because the thin film affected by the SrTiO₃ lattice. The lattice constant of a-axis is equal to that of b-axis. As the result, the grains of the thin film were pillar-shaped and rivaled right angle at random. To the epitaxyal growth of SBN30 crystal in the direction of <130> or <310> on SrTiO₃ substrate, SBN30 (100) phase must be rotate 18.4° to SrTiO₃ substrate. However, the thin film prepared on La doped SrTiO₃ by heat treatment at 700 ℃ showed epitaxial grouth in the direction of <110>. It is thought that this phenomenon is caused by La doped into SrTiO₃. However its has not been clarified, yet.

To measure the ferroelectric properties, the thin films (film thickness 0.09μm) were prepared on LSTO substrates. Figure 5 shows P-E hysteresis loops (measured at room temperature, 50 Hz) of the SBN30 thin films prepared on LSTO substrate by heat treatment at 700℃ and 900℃ (a) and that of the thin film prepared on Pt poly crystal substrate by heat treatment at 1000 ℃ (b). The remanent polarization (P_r) of the thin film on Pt substrate increased with increasing heat treatment temperature, P_r (heat treated at 1000℃) = 0.4 μC/cm². The coercive

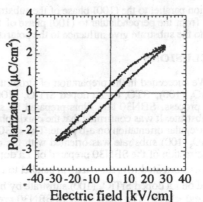

(a) prepared on Pt polycrystal substrate by heat treatment at 1000℃

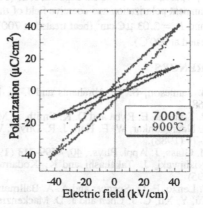

(b) prepared on La doped SrTiO₃ single crystal substrate.

Figure 5 P-E hysteresis loops of SBN30 thin films measurement at room temperature, 50Hz.

field (E_c) decreased with increasing heat treatment temperature, E_c (heat treated at 1000℃) = 3.80 kV/cm. It was observed that the grain size and density of the thin film increased with increasing heat treatment temperature.The remanent polarization and coercive field of the thin films on LSTO substrate increased with increasing heat treatment temperature,e. g., E_c = 6.55 kV/cm and P_r = 2.03 μC/cm² (heat treated at 700℃), E_c = 4.80 kV/cm and P_r = 4.42 μC/cm² (heat treated at 900℃). It was considered that the ferroelectric properties were improved as the orientation of the crystal phases. It is well known that there is the polarization direction of SBN30 in the a-b phase. The thin film prepared by heat treatment at 700℃ was (110) orientation parallel to the (100) phase of the substrate. a-axis and b-axis of the SBN30 slanted 45° from the perpendicular to (100) phase of the substrate. The thin film prepared by heat treatment at 900℃ was (110) and (310) orientation parallel to the (100) phase of the substrate. a-axis and b-axis of the SBN30 slanted 45° and 18° from the perpendicular to (100) phase of the substrate. It is thought that the angle of the phases to the substrate give influence to the polarization.

CONCLUSIONS

We succeeded in the preparation of strontium-barium niobate, $Sr_{0.3}Ba_{0.7}Nb_2O_6$, thin films on $SrTiO_3$ (100), STO, or La doped $SrTiO_3$ (100), LSTO, single crystal substrates by a spin coating process. SBN30 thin films prepared by sol-gel process received the strongly influence from substrate. It was confirmed that the c-axis phase of SBN30 was a one axis orientation films due to irregular orientation on a-b plane. The <130> and <310> direction of the SBN30 prepared on $SrTiO_3$ (100) substrate was oriented with the c-axis in parallel to the substrate surface. The <110> direction of the SBN30 prepared on La doped $SrTiO_3$ (100) substrate by heat treatment at 700℃ was oriented with the c-axis in parallel to the substrate surface. The orientation of thin film prepared on La doped $SrTiO_3$ (100) substrate by heat treatment at 900℃ shows two direction for <130> and <110>. The grain shape of SBN30 prepared on $SrTiO_3$ (100) by heat treatment above 700℃ and that of La doped $SrTiO_3$ (100) substrate by heat treatment at 1000℃ substrate were pillar. These pillar-shaped grains were aligned right angle each other. However the grain shape of SBN30 prepared on 0.75w% La doped $SrTiO_3$ (100) by heat treatment at 700℃ was cubic.

The remanent polarization and coercive field of the thin films on LSTO substrate were $E_c = 6.55$ kV/cm and $P_r = 2.03$ $\mu C/cm^2$ (heat treated at 700℃), $E_c = 4.80$ kV/cm and $P_r = 4.42$ $\mu C/cm^2$ (heat treated at 900℃).

REFERENCES

1. P. B. Jamieson, S. C. Abrahams and J. L. Bernstein, J. Chem.Phys.,48 [11], 5048-57 (1968).
2. M. P. Trubelja, E. Ryba and D. K. Smith, J. Mater. Sci., 31, 1435-43 (1996).
3. R. R. Neurgaonkar, W. F. Hall, J. R. Oliver, W. W. Ho and W. K. Cory, Ferroelectrics, 87, 167-79 (1988).
4. A. M. Glass, J. Appl. Phys., 40, 4699-713 (1969); 41,2268(E) (1970).
5. S. Nishiwaki, J. Takahashi and K. Kodaira, J. Ceram. Soc. Jpn., 103 [12], 1246-50 (1995).
6. P. V. Lenzo, E. G. Spencer and A. A. Ballman, Appl. Phys. Lett., 11 [1], 23-24 (1967).
7. R. Xu, Y. Xu, C. J. Chen and J. D. Mackenzie, J. Mater. Res., 5 [5], 916-18 (1990).
8. W. Sakamoto, T. Yogo, K. Kikuta, K. Ogiso, A. Kawase and S. Hirano, J. Am. Ceram. Soc., 79 [9], 2283-88 (1996).
9. R. R. Neurgaonka, M. H. Kalisher, T. C. Cim, E. J. Staples and K. L. Keester, Mater. Res. Bull., 15, 1235-40 (1980).
10. J. B. Thaxter, Appl. Phys. Lett., 15 [7], 210-12 (1969)
11. S. Sakka, Science of Sol-Gel method, in Japanese, Agne Shoufu-sha (1988).
12. H. Yanagida and M. Nagai, Science of Ceramics, in Japanese, Gihoudou Shyuppan (1993).
13. S. Hirano, T. Yogo, K. Kikuta,H. Urahata, Y. Isobe, T. Morishita, K. Ogiso and Y. Ito, Mat. Res. Soc. Symp. Proc., Vol. 271, 331-38 (1992)
14. YU-FU KUO and TSEUNG-YUEN TSENG, J. Mat. Sci., 31, 6361-68 (1996)
15. YI HU, J. Mat. Sci., 31, 4255-59 (1996)
16. KEISHI NISHIO, NOBUHIRO SEKI, JIRAWAT THONGRUENG, YUICHI WATANABE AND TOSHIO TSUCHIYA, J. Sol-Gel Sci. and Tech., 16, 37-45 (1999)
17. T. Nakano, M. Kumagai and T. Funahashi, Proceedings of Fall Meeting The Ceramic Society of Japan, 242-243 (1990)

DEPOSITION OF CRACK-FREE BaTiO₃ AND Pb(Zr,Ti)O₃ FILMS OVER 1 μm THICK VIA SINGLE-STEP DIP-COATING

H. KOZUKA, M. KAJIMURA, K. KATAYAMA, Y. ISOTA, T. HIRANO
MSE Department, Kansai University, Suita 564-8680, Japan, kozuka@ipcku.kansai-u.ac.jp

ABSTRACT

BaTiO₃ and PZT films were prepared by single-step dip-coating from alkoxide-acetate solutions containing polyvinylpyrrolidone (PVP). Crack-free BaTiO₃ and PZT films over 1 μm in thickness were obtained via single-step deposition. Stepwise heating of the gel films was found to improve densification of BaTiO₃ films, reducing the thickness and increasing the optical transmittance, which was not, however, the case with PZT films, where the stepwise heating rather induced crack formation, leading to degraded transmittance. Residual stress was evaluated on spin-coating BaTiO₃ films by measuring the substrate curvature, where a significant reduction in tensile stress was found to be caused by PVP.

INTRODUCTION

Alkoxide-based sol-gel coating is already a technique widespread in laboratories for preparing ceramic coatings of a variety of functions. People in industries, however, are not encouraged enough to utilize this technique in mass production especially of polycrystalline ceramic coatings. One of the factors discouraging people is the low value of the critical thickness, the maximum thickness achievable without crack formation via non-repetitive, single-step deposition, which is often less than 0.1 μm for non-silicate polycrystalline films.

Film thickness over submicrometer is often favored in ceramic coatings. For example, as addressed by Tu and Milne, ferroelectric coatings thicker than 1 μm are useful as pyroelectric infrared detectors and bulk acoustic wave bandpass filters as well as piezoelectric actuators [1]. Although cycles of gel film deposition and firing can be employed in laboratories, such repetitive coating is impracticable in industrial production.

Chelating agents [2] and diols [1,3] have been proposed to be effective in increasing the critical thickness of sol-gel coatings. Crack-free, single-layer PZT of 1 μm thickness has already been achieved by Tu et al., who employed diols as the solvent [1,3]. Recently the present authors have reported that incorporation of polyvinylpyrrolidone (PVP) in precursor solutions can increase the critical thickness of TiO₂ [4] and BaTiO₃ [5,6] films. PVP was thought to block the condensation sites of metalloxane polymers in films and promote stress relaxation. In the present paper, formation of BaTiO₃ and PZT films from PVP-containing solutions is described, on the basis of film thickness, optical transmittance and X-ray diffraction measurements and SEM observation. Effect of PVP on the residual stress in films was also studied.

EXPERIMENT

PVP powders of average molecular weight of 630000 were employed. Starting solutions of molar compositions, Ba(CH₃COO)₂ : Ti(OC₃H₇^i)₄ : PVP : CH₃COOH : H₂O : i-C₃H₇OH = 1 : 1 : 0 - 1 : 20 : 20, were prepared, following the procedure illustrated in Fig. 1 (a). The moles of PVP represent those of the monomer (polymerization repeating unit) of PVP. The solutions were kept standing at 25°C in a sealed glass container for various periods of time and served as coating solutions for BaTiO₃ films.

A starting solution of molar composition, Pb(CH₃COO)₂·3H₂O : Zr(OC₃H₇^i)₄ : Ti(OC₃H₇^i)₄ : PVP : CH₃COOH : H₂O : CH₃OCH₂CH₂OH : n-C₃H₇OH = 1.18 : 0.53 : 0.47 : 1 : 10 : 5 : 10 : 1, was prepared, following the procedure shown in Fig. 1 (b). The solution was kept standing at 25°C in a sealed glass container for 64 h and served as coating solutions for PZT films.

Dip-coating was conducted on silica glass substrates (ca. 20 x 40 x 1.2 mm³). Substrate withdrawal speeds of 3 and 1 - 3 cm min⁻¹ were employed for preparing BaTiO₃ and PZT films,

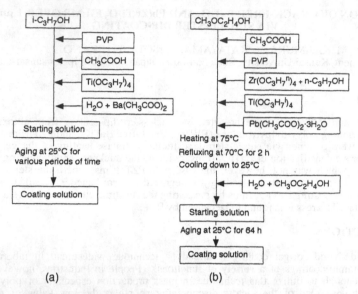

Fig. 1. Flow chart showing the procedure for preparing (a) $BaTiO_3$ and (b) PZT films.

respectively. Gel films were converted to ceramic thin films either by transferring into an electric furnace of 700°C and heating there for 10 min or by stepwisely heating at 300°, 500° and 700°C where a 10 min heat-treatment was conducted at each temperature. Dip-coating and heat-treatment were always performed just one time, not repeated.

Viscosity of the sols was measured using a Brookfield DI+ viscometer. Thickness of the films was measured by a Kosaka Laboratory SE-3400 contact probe surface profilometer. A part of the gel films was scraped off with a surgical knife before heat-treatment, and the level difference between the coated and the scraped parts was measured after heat-treatment. Crack formation was examined by a JEOL JSM-T330 scanning electron microscope and by the contact probe surface profilometer. Optical transmission spectra were measured on films by a Shimadzu UV-2400PC spectrophotometer using a silica glass substrate as the reference. X-ray diffraction patterns were measured by a Rigaku RTP300 X-ray diffractometer with Cu Kα radiation operating at 50 kV and 150 mA.

The residual stress in films was determined by substrate curvature measurement, where gel films were deposited on one side of a silica glass substrate 0.3 mm in thickness (20 x 50 mm²) by spin-coating at a rotation speed of 3440 rpm, followed by heat-treatment. Before and after the deposition the curvature of the back side of the substrate was measured three times along the central line of the substrate using the contact probe surface profilometer. Film stress, σ_s, was determined from the curvature data by using modified Stoney's formula [7];

$$\sigma_f = \frac{1}{3} \frac{E_s}{1-\nu_s} \frac{t_s^2}{t_f} \frac{\delta}{r^2} \qquad (1)$$

where t_s, E_s and ν_s are the thickness, Young's modulus and Poisson's ratio of the substrate, respectively, t_f is the film thickness, and δ the deflection of the substrate at a distance r from the center of the substrate. For this measurement, the curvature of the substrate was measured three times, and the film thickness was measured at three different positions.

RESULTS AND DISCUSSION

BaTiO₃ Films

The thickness of the BaTiO₃ films as well as the sol viscosity are plotted against the sol reaction time in Fig. 2. The films were prepared from the solutions of $PVP/Ti(OC_3H_7^i)_4 = 0, 0.5$ and 1.0, and were directly heated at 700°C. The film thickness increased with sol reaction time due to the increased sol viscosity, and the films finally cracked on heating as denoted by crosses in the figure. As seen in the figure, the maximum film thickness achieved without crack formation was significantly increased by introducing PVP in the sol. As suggested by Saegusa and Chujo in their work on organic-inorganic composite materials, the C=O groups of PVP can make strong hydrogen bonds with the OH groups of inorganic polymers [8]. Such hydrogen bonds can block the polymerizable sites of the inorganic polymers in gel films, which can retard the condensation reaction and promote the plastic deformation or structural relaxation, leading to suppression of the stress generation in films.

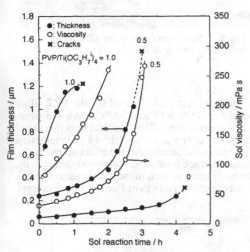

Fig. 2. Dependence of the film thickness and sol viscosity on sol reaction time for BaTiO₃ films and sols. The films were heated at 700°C. Crosses denote the films with cracks and the other marks represent the crack-free films.

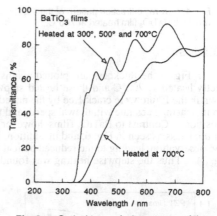

Fig. 3. Optical transmission spectra of the BaTiO₃ film directly heated at 700°C and that stepwisely heated at 300°, 500° and 700°C. The films were prepared from the sol of $PVP/Ti(OC_3H_7^i)_4 = 1.0$ with a viscosity of 150 mPa s.

Heat-treatment conditions affected the final thickness and transparency of the BaTiO₃ films. BaTiO₃ films were prepared under different heating conditions from a sol of $PVP/Ti(OC_3H_7^i)_4 = 1$, having a viscosity of 150 mPa s, by single-step dip-coating. When the gel film was directly heated at 700°C, a crack-free, 1.2 μm thick BaTiO₃ film was obtained. When the gel film was heated stepwisely at 300°, 500° and 700°C, on the other hand, the film thickness decreased and a crack-free, 0.7 μm thick film resulted. Fig. 3 shows the optical transmission spectra of these two films. The 1.2 μm thick film was very slightly milky in appearance and transmitted the visible light as seen in the figure, while the stepwisely heated 0.7 μm thick film exhibited higher transmittance. Plastic deformation or structural relaxation is thought to be promoted in gel films by the stepwise heating.

Figs. 4 and 5 show X-ray diffraction pattern and SEM micrograph of the crack-free 1.2 μm thick BaTiO₃ film. The film was perovskite BaTiO₃ without detectable voids in microstructure. The crystallite size determined with Scherrer's equation was 18 nm.

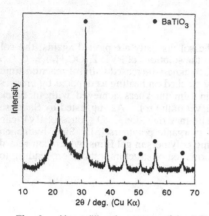

Fig. 4. X-ray diffraction pattern of the 1.2 μm thick BaTiO₃ film heated at 700°C.

Fig. 5. SEM micrograph of the 1.2 μm thick BaTiO₃ film heated at 700°C.

PZT Films

In Fig. 6 the thickness are plotted against substrate withdrawal speed for the PZT films directly heated at 700°C and those heated stepwisely at 300°, 500° and 700°C. All the films shown in the figure were crack-free by the naked eye observation. The film thickness increased with increasing substrate withdrawal speed, and thickness over 1 μm was achieved by single-step deposition. Contrast to BaTiO₃ films, however, the stepwise heating did not affect so much the film thickness as seen in Fig. 6, and in addition, induced crack formation, which was detected by SEM observation, leading to the reduced optical transmittance as seen in the transmission spectra (Fig. 7). Thus the stepwise heating was found to be ineffective in densification in the present

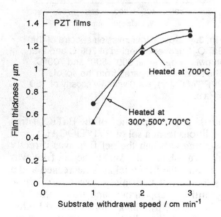

Fig. 6. Film thickness plotted against the substrate withdrawal speed for the PZT film directly heated at 700°C and that step-wisely heated at 300°, 500° and 700°C.

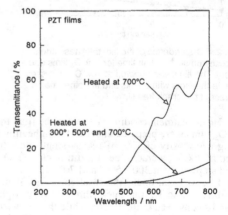

Fig. 7. Optical transmission spectra of the PZT films directly heated at 700°C and those stepwisely heated at 300°, 500° and 700°C. The substrate with-drawal speed was 2 cm min⁻¹.

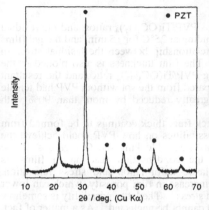

Fig. 8. X-ray diffraction pattern of the PZT film prepared with a substrate withdrawal speed of 2 cm min⁻¹ and directly heated at 700°C.

Fig. 9. SEM micrograph of the PZT film prepared with a substrate withdrawal speed of 2 cm min⁻¹ and directly heated at 700°C.

PZT films. The stepwise heating brings about some cause that hinders the promotion of plastic deformation or structural relaxation in PZT films, which is now under consideration.

The PZT films heated at 700°C were milky in appearance but transmitted light as seen in the spectra (Fig. 6) The film was perovskite PZT without pyrochlore as seen in the X-ray diffraction pattern (Fig. 8), and was crack-free, but rather porous as revealed in the SEM micrograph (Fig. 9). The crystallite size determined with Scherrer's equation was 13 nm.

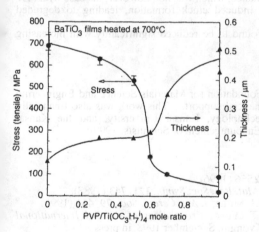

Fig. 10. Dependence of the residual stress and thickness on the PVP/Ti(OC₃H₇ⁱ)₄ ratio for spin-coating BaTiO₃ films. The error bars represent the maximum and minimum stress values calculated from thickness and curvature data obtained through three time measurements.

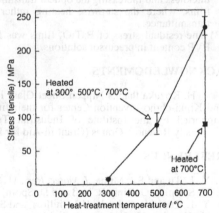

Fig. 11. Stress as a function of heat-treatment temperature for spin-coating BaTiO₃ film directly heated at 700°C and that stepwisely heated at 300°, 500° and 700°C. The sol aged for 15 min was served as the coating solution. For the stepwisely heated sample, the stress measure-ment was made after each heat-treatment.

Residual Stress in Films

BaTiO$_3$ films were prepared from sols of various PVP/Ti(OC$_3$H$_7^i$)$_4$ ratios, and the residual stress in films was measured. The sols were kept standing at 25°C for 15 min, and the gel films were directly heated at 700°C. Fig. 10 shows the relationship between the residual stress in BaTiO$_3$ films and the PVP/Ti(OC$_3$H$_7^i$)$_4$ mole ratio. The film thickness is also plotted in the figure. The film thickness increased with increasing PVP/Ti(OC$_3$H$_7^i$)$_4$ ratio, and the resultant films were all crack-free. It is seen that the film prepared from the sol without PVP had tensile stress as high as 700 MPa, while the stress was greatly reduced by more than 90% with increasing PVP/Ti(OC$_3$H$_7^i$)$_4$ ratio.

It would be the reduced stress that allowed crack-free, thick coatings to be formed from PVP-containing solutions. There would be two possibilities on how PVP could achieve the reduction in film stress. First, PVP can act as plasticizer, blocking the OH groups of the metalloxane polymers, suppressing and retarding the condensation reaction in films as mentioned above, which can lead to promoted plastic deformation and stress relaxation. Second, thermal decomposition of PVP can result in increase in film porosity, which can lower the elastic modulus of the film, leading to reduction in stress. The latter possibility is somehow regrettable when dense film formation is aimed at, but cannot be neglected. As a matter of fact, it was found that the stepwise heating of the gel film, which promoted film densification as described in the previous section, resulted in higher film stress than the direct heating as shown in Fig. 11.

CONCLUSIONS

BaTiO$_3$ and PZT films were prepared by single-step dip-coating from alkoxide-acetate solutions containing polyvinylpyrrolidone (PVP).
(1) Crack-free BaTiO$_3$ and PZT films over 1 μm in thickness could be obtained, respectively, via single-step deposition.
(2) Stepwise heating of the gel films improved the densification of BaTiO$_3$ films, reducing the thickness and increasing the optical transmittance, which was not, however, the case with PZT films, where the stepwise heating rather induced crack formation, leading to degraded transmittance.
(3) The residual stress of BaTiO$_3$ films was found to be reduced significantly with increasing PVP content in precursor solutions.

ACKNOWLEDGMENTS

H. Kozuka thanks Nippon Sheet Glass Foundation for Materials Science and Engineering and Kinki-chiho Invention Center for their financial support. This work was also financially supported by the Institute of Industrial Technology, Kansai University, and the Kansai University Research Grants (Grant-in-Aid for Encouragement of Scientists, 1999).

REFERENCES

1. Y.-L. Tu and S.J. Milne, *J. Mater. Res.*, **11**, 2556 (1996).
2. H. Schmidt, G. Rinn, R. Naβ and D. Sporn, *Mat. Res. Soc. Symp.*, **121**, 743 (1988).
3 Y.-L. Tu, M.L. Calzada, N.J. Phillips, and S.J. Milne, *J. Am. Ceram. Soc.*, **79**, 441 (1996).
4. H. Kozuka, K.-S. Jang, C.-S. Mun, and T. Yoko, *Proceedings of the 3rd International Meeting of Pacific Rim Ceramic Societies*, Kyongju, Spetember 1998, in press.
5. H. Kozuka and M. Kajimura, *Chem. Lett.*, 1029 (1999).
5. H. Kozuka and M. Kajimura, *J. Am. Ceram. Soc.*, in press.
7. R. Glang, R.A. Holmwood and R.K. Rosenfeld, *Rev. Sci. Instrum.*, **36**, 7 (1965).
8 T. Saegusa and Y. Chujo, *Makromol. Chem., Macromol. Symp.*, **64**, 1 (1992).

NEAR-STOICHIOMETRIC BARIUM TITANATE SYNTHESIS
BY LOW TEMPERATURE HYDROTHERMAL REACTION

Kyoungja Woo*, Guang J. Choi, Young S. Cho
Clean Technology Research Center, Korea Institute of Science and Technology,
P.O. Box 131, Cheongryang, Seoul 130-650, Korea, kjwoo@kist.re.kr

ABSTRACT

Barium-deficiency of barium titanate particles prepared by low temperature hydrothermal reaction has been notorious. It has been believed that barium-deficiency is caused by the high solubility of barium source compared with titanium. Here is reported the synthesis of near-stoichiometric barium titanate powders with ultrafine particle size and high crystallinity by low temperature hydrothermal reaction from barium acetate and titanium tetra(methoxyethoxide). Barium titanate particles were synthesized in the spherical, metastable cubic crystalline grains with size distribution between 60 ~ 90 nm in diameter. Ultrafine particle size was resulted from the control of the hydration rate and the decrease of Ti-O-Ti cross-linking extent of titanium precursor. Increasing barium to titanium molar ratio in reactant could not overcome the notorious barium-deficiency but, improved stoichiometry and produced finer and less agglomerated particles. Interestingly, adding a slight pressure to autogeneous one to make total 4 ~ 10 atm has yielded near-stoichiometric, highly crystalline, and less agglomerated barium titanate particles. It seems like that the total pressure around 4 ~ 10 atm provides strong force enough to push barium ions into the interstitial points of perovskite structure and stabilize it. These particles, which were in metastable cubic form as synthesized, initiated phasetransition to tetragonal form by calcination at 400 °C.

INTRODUCTION

Stoichiometric(Ba/Ti=1) $BaTiO_3$ powder with ultrafine and monodispersed morphology has been of great concern due to its high electronic applications. Commercial $BaTiO_3$ powder synthesized by solid state reaction or by chemical precipitation processes limits the current trend of device miniaturization in the electronics industry because of limited size scale, microstructural variations, non-stoichiometric compositions, and poor electrical reproducibility. $BaTiO_3$ synthesis by low temperature hydrothermal reaction has been developed as a plausible candidate for chemical industry [1-3]. This method uses low temperature, short period of time, autogeneous pressure, and reactive Ti-precursor. Low reaction temperature less than 200 °C is one of the prominent merits to utilize the conventional reactor technology. The major difficulties in this process are to slow down hydrolysis rate and to avoid very fast and spontaneous self-condensation between Ti-OH groups. Failure in controlling these phenomena leads to large and polydispersed particles in size and very rapidly forming Ti-O-Ti cross-linking may result in the phase segregation of TiO_2 and $BaTiO_3$. Another notorious difficulty of low temperature hydrothermal reaction is to control Ba/Ti stoichiometry, that is, to overcome Ba-deficiency in barium titanate. Barium-deficiency has been believed to be caused by the high solubility of barium source compared with titanium.

In this study is reported near-stoichiometric, highly crystalline, and ultrafine $BaTiO_3$ preparation by low temperature hydrothermal reaction using $Ba(OAc)_2$ and $Ti(OCH_2CH_2OCH_3)_4$.

EXPERIMENT

Hydrolysis of Ti-Precursor

Each Ti-precursor was added to large excess of water with vigorous stirring. This solution was stirred for 30 more minutes before the separation of the solid residue by centrifugation. To investigate the hydrolysis behavior of $Ti(OCH_2CH_2OCH_3)_4$ precursor in NaOH solution, the solid residue was separated from the mixed solution just before hydrothermal treatment in the synthetic process of $BaTiO_3$. The solid residues were dried at 60 °C in a vacuum oven overnight and sampled as disk specimen mixed with KBr to obtain FT-IR spectra (Nicolet Magna 750).

Synthesis and Characterization of BaTiO₃

$Ti(OCH_2CH_2OCH_3)_4$ in 4eq. 2-propanol was added to 1M $Ba(OAc)_2$ aqueous solution with vigorous stirring. The combined solution was stirred for 30 more minutes for homogeneous mixing. Meanwhile, NaOH solution (pH>13.5 in the final mixture [4]) was prepared in a PTFE lined stainless steel autoclave with stirring and heating equipment. The mixed solution was poured into the autoclave and underwent hydrothermal reaction under the condition of 80 °C, 1 hr period, 0.20 M [5]. The reaction variables were the pressure (autogeneous ~ 10 atm) and the Ba/Ti mole ratio (1.0 ~ 2.5) of the reactants. After the hydrothermal reaction was done, the reacted solution was cooled down in a cold water bath and washed with ammonia water three times using centrifugation. Then the solid product was dried at 80 °C in a vacuum oven for 12 hours. Some particles were calcined for 1 hr at various temperatures (400 ~ 1200 °C) under moisture free air using a box furnace.

Surface morphology of $BaTiO_3$ particles was measured by FE-SEM (Hitachi S-4200). Each specimen was sensitized with Au-sputtering to avoid charging during SEM analysis. The crystalline structure of $BaTiO_3$ particles was determined by XRD (Shimadzu XRD-6000 or Rigaku Miniflex) analysis using Cu K_α radiation. Raman spectra were obtained using Perkin-Elmer System 2000. Elemental analysis of $BaTiO_3$ powders was performed by ICP spectroscopy (Thermo Jariellash, Polyscom 61E).

RESULTS AND DISCUSSION

Hydrolysis Behavior and Particle Morphology

FT-IR spectra are shown in Figure 1 to compare the hydrolysis behaviors of Ti-precursors. Peaks at $1,020 \sim 1,125$ cm^{-1} correspond to Ti-O-C stretching vibrations. Intensity decrease and/or disappearance of these peaks is related to the partial and/or complete hydrolysis of titanium precursors. Because of methoxy substituent, methoxyethoxide can have electronic, steric, and weakly chelating effects around titanium ion so that it makes stronger bond with titanium ion than simple alkoxide. This has appeared as the partial hydrolysis of $Ti(OCH_2CH_2OCH_3)_4$ in (b) as a comparison with the case of $Ti(OCH(CH_3)_2)_4$ in (a), which underwent complete hydrolysis under the same condition. Methoxyethoxide groups were still partially retained even in as-precipitated powder (c), which was collected from KOH solution.

These results suggest that the partial and slow hydrolysis of $Ti(OCH_2CH_2OCH_3)_4$ precursor reduce Ti-O-Ti cross-linking extent greatly during $BaTiO_3$ synthesis. Conversely will increase the possibility of getting a homogeneous solution at atomic level by mixing $Ti(OCH_2CH_2OCH_3)_4$ and $Ba(OAc)_2$ solutions.

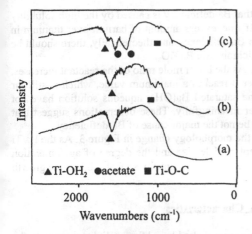

Figure 1. IR spectra of (a) completely hydrolyzed $Ti(OCH(CH_3)_2)_4$ in water, (b) and (c) partially hydrolyzed $Ti(OCH_2CH_2OCH_3)_4$ in water and in KOH solution, respectively.

▲Ti-OH₂ ●acetate ■Ti-O-C

2000   1000   0

Wavenumbers (cm⁻¹)

(a)

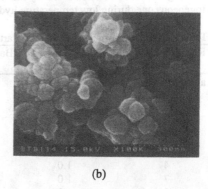
(b)

Figure 2. SEM pictures of $BaTiO_3$ particles prepared from different Ti-precursors (a) $Ti(OCH_2CH_2OCH_3)_4$, (b) $Ti(OCH(CH_3)_2)_4$.

For the reasons mentioned above, we reasoned that low temperature hydrothermal reaction using $Ti(OCH_2CH_2OCH_3)_4$ as a titanium precursor lead to homogeneous and pure $BaTiO_3$ crystalline powder with rather uniform and fine particle size. The case was found to be true as is shown in Figure 2. $BaTiO_3$ particles prepared from $Ti(OCH_2CH_2OCH_3)_4$ were finer, more uniform, and less agglomerated than those from $Ti(OCH(CH_3)_2)_4$.

Ba/Ti Ratio Effect of Reactants on Stoichiometry and Particle Characteristics

Generally, Ba-deficiency is notorious in $BaTiO_3$ particles synthesized by low temperature hydrothermal reaction. $BaTiO_3$ particles prepared at our reaction condition showed Ba-deficiency (typically, Ba/Ti = 0.87), too. Also existed Ba-deficiency (typically Ba/Ti = 0.83) in $BaTiO_3$ particles prepared by low temperature hydrothermal reaction using titanium acylate-based precursors [6], which were known to slow down the hydrolysis rate further than our Ti-precursor. Using microwave [7] as a heating source was reported to show improved stoichiometry but, still far from the ideal value 1. It has been believed that Ba-deficiency is caused by the high solubility of barium source compared with titanium [4]. If true, excess amount of barium over titanium in the reactant solution should be able to improve Ba deficiency and theoretically, there should be the optimal Ba/Ti ratio for the production of stoichiometric BaTiO$_3$.

Table I shows the elemental analysis results. As the Ba/Ti mole ratio of the reactant increases, that of the product increases but, decreases after it reaches a maximum value, which is still far less than 1. Washing the solid product with the saturated Ba(OH)$_2$ aqueous solution have not improved the stoichiometry of the final powder meaningfully. These observations suggest that the solubility difference of barium and titanium be not the major cause of Ba-deficiency.

Nonetheless, a consistency is illustrated for the morphology change in Figure 3. As the Ba/Ti mole ratio of the reactant increases, the resultant particle size and the degree of agglomeration decrease. It seems like that excess Ba(OAc)$_2$ acts as a kind of impurity so that inhibits the growth and the agglomeration of BaTiO$_3$ particles.

Pressure Effect on Stoichiometry and Particle Characteristics

The good control of stoichiometry has been accomplished by adding a slight pressure to the autogeneous one during low temperature hydrothermal treatment using 1:1 mole ratio of Ba and

Table I. Elemental Analysis of BaTiO$_3$ Particles Synthesized at Different Reaction Conditions

Reaction Pressure, atm	Ba/Ti mole ratio in reactant	Ba, weight %	Ti, weight %	Ba/Ti mole ratio in product
autogeneous (<2)	1.0	53.4	21.5	0.866
	1.3	54.5	20.9	0.910
	1.5	54.2	20.7	0.913
	2.0	56.9	20.9	0.950
	2.2	57.0	21.1	0.942
	2.5	56.6	21.5	0.916
4	1.0	56.4	20.1	0.979
7	1.0	56.2	20.3	0.966
10	1.0	56.5	20.1	0.980
Commercial (oxalate process)		55.9	20.5	0.951
Theoretically calculated for BaTiO$_3$		58.9	20.5	1.00

(a)   (b)   (c)

Figure 3. SEM pictures of BaTiO$_3$ particles prepared from different Ba/Ti mole ratios (a) 1.0 (b) 2.0 (c) 2.2.

Ti reactants. Table I shows that the typical Ba-deficiency was dramatically improved just by adding some external pressure to make total 4 ~ 10 atm for the system. It seems like that the total pressure around 4 ~ 10 atm provides strong force enough to push Ba ions into the interstitial points of perovskite structure and stabilize it. There was no distinguishable advantage of a higher pressure over 4 atm in terms of stoichiometry. The stoichiometry of our as-synthesized particles was found to be definitely better than that of the commercial ones.

XRD patterns of $BaTiO_3$ particles prepared at various pressures are compared in Figure 4. The crystallinity of $BaTiO_3$ particles synthesized under 4 ~ 10 atm was greater than that prepared under autogeneous pressure. No noticeable pressure effect on crystallinity was found in the range of 4 ~ 10 atm. In the pressurized reactions, the trend of increase in crystallinity coincides with that of improvement in stoichiometry.

A representative SEM micrograph of $BaTiO_3$ particles synthesized under 7 atm is shown in Figure 5. The resultant particles were spherical shaped, 60 ~ 90 nm in diameter. Interestingly, the degree of agglomeration in synthesized particles was greatly reduced as the pressure was raised to 4 ~ 10 atm, under which condition produced near-stoichiometric $BaTiO_3$. It has been reported that nanoparticles tend to make an agglomerate because of their high surface energy [8]. It seems like that agglomeration is somewhat related with Ba-deficiency in addition to the high surface energy. TiO_2-rich(Ba-deficient) phase of $BaTiO_3$ is known to show exaggerated grain growth during sintering [9]. It could be equally possible for TiO_2-rich phase to evolve exaggerated agglomeration during hydrothermal treatment for the powder synthesis.

Figure 6 shows Raman spectra of the particles calcined for 1 hr at various temperatures. A sharp absorption at 304 cm^{-1} is the characteristic peak of tetragonal $BaTiO_3$. As-synthesized particles appeared as a metastable cubic form on Raman spectrum as well as on XRD. Surprisingly, tetragonality appeared clearly at 400 °C. The intensity of this characteristic peak increased gradually according to temperature and showed a substantial increase when particles were calcined at 1200 °C.

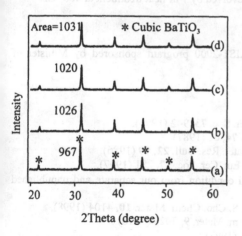

Figure 4. XRD patterns of $BaTiO_3$ prepared under various pressures (a) autogeneous (b) 4 atm (c) 7 atm (d) 10 atm.

Figure 5. SEM picture of $BaTiO_3$ prepared under 7 atm.

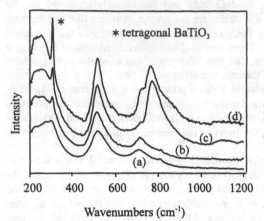

Figure 6. Raman spectra of BaTiO$_3$ particles calcined for 1 hr at various temperatures (a) as synthesized (b) 400 °C (c) 800 °C (d) 1200 °C.

CONCLUSIONS

Near-stoichiometric, highly crystalline, and ultrafine BaTiO$_3$ powders were synthesized from 1:1 ratio of 0.20 M Ba(OAc)$_2$ and Ti(OCH$_2$CH$_2$OCH$_3$)$_4$ by low temperature hydrothermal reaction under 4 ~ 10 atm, 80 °C, 1 hr period. The electronic, steric, and weakly chelating effect of methoxyethoxide ligand attributed to the evolution of ultrafine particles. The notorious Ba-deficiency has been overcome by adding a slight external pressure to the autogeneous one rather than by using a large excess amount of Ba over Ti. The slight external force seems likely to push Ba ions into the interstitial points of perovskite structure and stabilize it. Near-stoichiometric BaTiO$_3$ showed higher crystallinity than non-stoichiometric one. As-synthesized particles were in pure cubic crystalline phase. Tetragonality has evolved by 1 hr heat treatment at 400 °C.

ACKNOWLEDGMENT

This research has been conducted under KIST-2000 program sponsored by Minister of Science and Technology in Korea.

REFERENCES

1. F. Chaput, J. P. Boilot, A. Beauger, J. Am. Cer. Soc. **73**, 942 (1990).
2. A. K. Maurie, R. C. Buchanan, Ferroelectrics **74**, 61 (1987).
3. R. Vivekanadan, S. Philip, T. R. N. Kutty, Mater. Res. Bull. **22**, 99 (1986).
4. M. C. Blanco-Lopez, B. Rand, F. L. Riley, J. Eur. Cer. Soc. **17**, 281 (1997).
5. 80°C, 1hr, 0.20M was chosen as an optimal condition from our separate and unpublished result.
6. G. J. Choi, S. K. Lee, K. Woo, K. K. Koo, Y. S. Cho, Chem. Mater. **10**, 4104 (1998).
7. Y. Ma, E. Vileno, S. L. Suib, P. K. Dutta, Chem. Mater. **9**, 3023 (1998).
8. X. Li, W. -H. Shih, J. Am. Cer. Soc. **80**, 2844 (1997).
9. R. K. Sharma, N. -H. Chan, D. M. Smyth, J. Am. Cer. Soc. **64**, 448 (1981).

NEW APPROACHES TO CHEMICAL BATH DEPOSITION OF CHALCOGENIDES

PAUL O'BRIEN,[a*] MARKUS R. HEINRICH,[b] DAVID J. OTWAY,[b] ODILE ROBBE,[b] ALEXANDER BAYER,[b] and DAVID S. BOYLE.[c]

1. Department of Chemistry, and The Manchester Materials Science Centre, Manchester University, Oxford Rd, Manchester, M13 9P, UK.
2. Department of Chemistry, Imperial College of Science, Technology and Medicine, Exhibition Road, London, SW7 2AZ, UK
3. Department of Physics, Science Laboratories, University of Durham, South Road, Durham, DH1 3LE.
    Email addresses: p.obrien@ic.ac.uk; d.j.otway@ic.ac.uk; David.Smyth-Boyle@durham.ac.uk

ABSTRACT

We have been studying new approaches to conventional Chemical Bath Deposition (CBD) of chalcogenide containing materials, using continuous circulation and replenishment of CBD solution over a heated substrate. Crystalline thin films produced by this method offer potential for use in solar cell devices or other optoelectronic applications. Films of CdS, ZnS and the ternary material $Cd_xZn_{1-x}S$ have been deposited on TO-glass substrates. In this paper we demonstrate our approach for the deposition of CdS films. These have been characterized by XPS, SEM, XRD and UV/vis spectroscopy and shown to be good quality. The films have been used to fabricate Au/CdTe/CdS/TO-glass solar cells of efficiency 10.1% under AM1.5 illumination.

INTRODUCTION

Photovoltaic (PV) energy generation is currently dominated by crystalline Si cell technology. The market volume for PV modules in 1999 is expected to lie in the region of 200 MWp which accounts for the full world capacity for crystalline silicon production. The general market trend in the last 10 years for photovoltaics has consisted in a 25% rise per year. An increase in production capacity, in the field of crystalline silicon material, is not apparent at the present time. The lack of production facilities may be the main restraining factor in limiting the actual growth rate of this technology.[1] Thin film polycrystalline solar cells offer the potential for low cost solar energy conversion. Regarding CdTe-CdS thin film solar cells, four industrial enterprises are presently preparing to go into production, First Solar Inc. (capacity of 20 MWp p.a.) and BP-AMOCO (10 MWp p.a.) in the USA, ANTEC Solar in Germany (10 MWp pa.) and Matsushita in Japan (5 MWp p.a.).[2]

The large-scale exploitation of these devices is partly dependent on a reduction of the potential environmental impact of the technology. There is considerable potential for the development of cheaper and safer processes for their manufacture. Cadmium-containing compounds and wastes are highly regulated in the EU and elsewhere and the associated environmental legislation is subject to continuous review; trends for regulatory limits are increasing downwards.[3] The fabrication of CdS window layers by Chemical Bath Deposition (CBD) generates considerable Cd-containing waste. The chemical processes underlying CBD have been reviewed by ourselves and others.[4-6] Typical CBD processes for sulfides employ an aqueous alkaline solution (*ca.* 60-90 °C) containing the chalcogenide source, the metal ion and added base. A chelating agent is used to limit the hydrolysis of the metal ion and impart some stability to the bath, which would otherwise undergo rapid hydrolysis and precipitation. The technique under these conditions relies on the slow release of S^{2-} ions, *via* thermal decomposition of an organic precursor, into an alkaline solution in which the free metal ion is buffered at a low concentration.

A major limitation of current CBD methods is the inefficiency of batch processing techniques, in terms of the utilisation of starting materials and their conversion to thin films (*ca.* 2% of the cadmium source in CdS CBD is usefully converted. Also the process often employs high concentrations of ammonia, which is volatile and detrimental to the environment. An efficient system for recovering and recycling cadmium and other chemicals is required in order to minimise waste generation.

Our initial successful and reproducible attempts to lessen the environmental impact of CBD appear to be successful. Chemical modelling and speciation studies have enabled us to develop a novel high-efficiency CdS CBD process that utilizes low cadmium concentrations and provides a viable alternative to the use of volatile solutions of ammonia. Films have been deposited by conventional CBD from heated solutions and also in a novel reactor with a heated substrate.

RESULTS AND DISCUSSION

One major problem in the CBD of films of CdS and related materials is that the efficiency of the process is low due to homogeneous precipitation and deposition on the reactor walls. We have now sought to combine the use of dilute solutions with a novel approach of the heating of the substrate. Conventional CBD experiments involve heating of the whole deposition bath; we have modified the deposition system in such a way that the dilute solution is continuously circulated over a TO-glass substrate, in contact to a resistively heated immersion apparatus. In some ways this approach is analogous to the difference between a hot walled and a cold walled CVD reactor. A schematic representation of our simple design is shown in Figure 1.

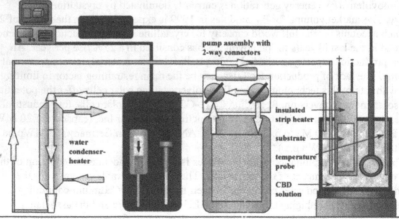

Figure 1. Schematic representation of high-efficiency CdS CBD system. Potential for continuous operation is achieved *via* alternate use of filters 1 and 2 (polypropylene 200 nm mesh) to facilitate recovery of particulate cadmium compounds. Organic by-products of the CdS CBD process are also removed during water treatment

Deposition of good quality CdS films on heated glass substrates was carried out using a low cadmium-ethylenediamine CBD solution ($[Cd^{2+}]$ = 0.001 mol dm^{-3}, [en] = 0.012 mol dm^{-3}, pH = 12.5). Films were also deposited from a conventional ammonia-containing bath and characterised in order to compare and contrast with those of the new process. Throughout the new deposition process the CBD solution remained optically clear, pale yellow in colour with no formation of CdS on reactor walls or homogeneous precipitate over 10 h. Films were removed and replaced periodically (*ca.* 45 min) with fresh TO-glass substrates. The films were characterized by spectroscopic methods (UV-vis and XPS), microscopy (TEM) and XRD. The as-deposited films were optically transparent, specular and yellow-orange in colour. The colour changed to pale yellow after annealing in air at 673 K. The bandgaps of films were determined before and after annealing, the latter produced a shift in the bandgap from 2.38 eV to 2.31 eV (Figure 2). The values were similar to those obtained for conventional CBD films. Glancing angle XRD measurements were recorded (Figure 3) and the diffractograms compared with standards in the JCPDS data files. Similar to observations made for conventional CBD films, the as-deposited films were composed of a mixture of cubic and hexagonal phases of CdS. Films were annealed in air for 30 min at 673 K, the process improved crystallinity and effected conversion to the hexagonal modification. Films deposited from baths containing ammonia appear to have a preferred [002] orientation, ethylenediamine baths deposit films that are in general more crystalline with alignment in the [101] direction. An approximate elemental ratio of Cd:S 1:1 was determined from EDAX measurements.

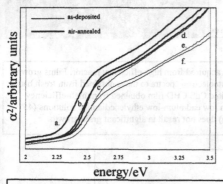

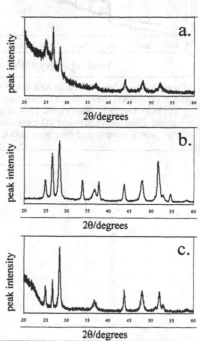

Figure 2 (above). UV-vis spectra of CdS CBD films on TO-glass, before and after annealing in air. Films deposited from 3 different chemical baths. Spectra *a* and *d* from Bath A (high cadmium-high ammonia bath; $[Cd^{2+}]$ = 0.01 mol dm^{-3}, [ammonia] = 5.02 mol dm^{-3}, pH = 13); *b* and *e* from Bath B (low cadmium-low ethylenediamine bath; ($[Cd^{2+}]$ = 0.001 mol dm^{-3}, [en] = 0.012 mol dm^{-3}, pH = 12.5); *c* and *f* from Bath C (low cadmium-low ammonia bath ($[Cd^{2+}]$ = 0.001 mol dm^{-3}, [ammonia] = 2.8 mol dm^{-3}, pH = 12.5.

Figure 3 (right). X-ray diffraction spectra of air-annealed CdS thin films on TO-glass deposited from different chemical bath solutions. Films grown from *a.* high cadmium-high ammonia baths ($[Cd^{2+}]$ = 0.01 mol dm^{-3}, [ammonia] = 5.02 mol dm^{-3}, pH = 13); *b.* low cadmium-low ethylenediamine baths ($[Cd^{2+}]$ = 0.001 mol dm^{-3}, [en] = 0.012 mol dm^{-3}, pH = 12.5); *c.* film deposited on heated TO-glass substrate from recirculating CBD solution as used for *b.*

No significant differences in XPS data were recorded for films from heated substrates in comparison with those obtained by conventional CBD. Photoelectron binding energies of 405.9 eV (Cd $3d_{5/2}$), 411.5 eV (Cd $3d_{3/2}$) and 162.5 eV (S 2p) were recorded for the major XPS peaks of our CBD CdS films. An important conclusion was the impurities in the as-deposited films did not increase for films grown from replenished baths (Figure 4). Minor concentrations of impurities assigned to carbon (C 1s 284.8 eV), nitrogen (N 1s 396.0eV) and oxygen (O 1s 532.1 eV) were present in all CBD films. These contaminants have been identified in CdS CBD films by other workers.[7] Annealing of films in air reduced the nitrogenous contaminant to a concentration below the detection limit of the instrument. The effect of air annealing was further investigated by TEM. The as-deposited films on heated substrates appeared to be composed of dense, asymmetric grains of diameter in the range 80-120 nm, no compelling evidence for grain growth with annealing was obtained (Figure 5). We have reported similar observations for conventional CBD of CdS from EN-buffered solutions.[8]

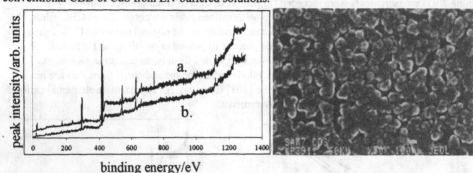

Figure 4 (above left). Wide scan XPS of CdS CBD films acquired from high efficiency reactor. Films grown from b. replenished baths after 6 h possess very similar photoelectron spectra to those obtained from fresh baths (45 min). **Figure 5 (above right).** TEM of an as-deposited CdS CBD film obtained from high-efficiency reactor. Films grown on heated TO-glass substrates from low cadmium-low ethylenediamine solutions (45 mins). Annealing of films in air (673 K for 30 min) does not result in significant grain growth.

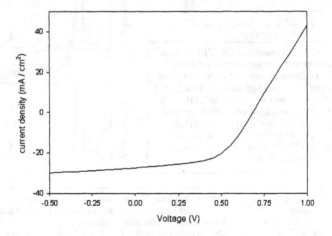

Figure 6. I-V characteristics of CBD-CdS/CSS-CdTe solar cell.

Solar cells were fabricated using CBD-CdS films from Bath B. CdTe layers (~5.5 μm) were deposited by close space sublimation.[9] After the $CdCl_2$ treatment, the cells were thoroughly rinsed to remove any remaining $CdCl_2$ and then etched in 0.03% Br - methanol for 10 - 15 s before gold contacts were deposited. Current-voltage (I-V) curves were recorded under illumination. Figure 6 shows the I-V curves recorded under AM1.5 illumination at room temperature. Values for the short circuit current (I_{sc}), open circuit voltage (V_{oc}), fill factor (FF), series resistance (R_s), shunt resistance (R_{sh}) and efficiency (η) were determined and are shown in Table 1.

J_{sc}(mA/cm^2)	V_{oc}(V)	FF%	η%	R_s (Ω.cm^2)	R_{sh}(Ω.cm^2)
27.3	0.68	54.4	10.1	6	196

Table 1. The PV characteristics of CBD-CdS/CCS CdTe solar cells.

The values were indicative of satisfactory solar cells, with short circuit current of 27.3 mA/cm^2, open circuit voltage of 684 mV and efficiency of 10.1%.

CONCLUSIONS

We have developed a novel high-efficiency CdS CBD process that utilizes low cadmium concentrations and provides a viable alternative to the use of volatile solutions of ammonia. Films have been deposited by conventional CBD from heated solutions, in a novel reactor with a heated substrate and a new geometry for deposition. The films obtained from the reactor have been characterised as good quality CdS by spectroscopic methods (UV-vis and XPS), microscopy (SEM and TEM) and powder XRD. No significant differences in the chemical or physical nature of films have been observed for films deposited from fresh and replenished baths. This suggests that continuous recycling may be possible. The approach may offer a viable route towards the high efficiency production of CdS layers for solar cell applications. Films have been used to fabricate CBD-CdS/CSS-CdTe solar cells with efficiencies of 10.1%.

ACKNOWLEDGEMENTS

We would like to thank the sponsors of this work, the EPSRC, the Royal Society and the Leverhulme Foundation. POB is the visiting Sumitomo/STS Professor of Materials Chemistry at IC and is Professor of Inorganic Materials at the University of Manchester. We would also like to thank Mr. Alexander Bayer for assistance with growth work and Mr. Richard Sweeney (IC) for XRD results, Dr Karl Senkiw (Department of Chemical Engineering, Imperial College) for the XPS analysis and Mr. Keith Pell for SEM (QMW).

REFERENCES

1.   K. Zweibel, Solar Energy Materials and Solar Cells **59**, 1 (1999).
2.   D. Bonnet, P. Meyers, Journal of Materials Research **13**, 2740 (1998).
3.   M. S. Patterson, A. K. Turner, M. Sadeghi, R. J. Marshall, Proc. 12th European Photovoltaic Solar Energy Conf. 950 (1994).
4.   P. O'Brien and J. McAleese, J. Mater. Chem **8**, 2309 (1998).

5.   A. C. Jones and P. O'Brien, CVD of Compound Semiconductors; Precursor Synthesis, Development and Applications, VCH, Weinheim, (1997).
6.   D. Lincot, Actualite Chimique, p.23 (1999).
7.   M. Stoev and K. Katerski, J. Mater. Chem. **6**, 377 (1996).
8.   P. O'Brien and T. Saeed, J. Crystal Growth **158**, 497 (1996).
9.   S. N. A. Alamri, PhD Thesis, University of Durham 46 (1999).

DIRECT FABRICATION OF LiCoO$_2$ FILM ELECTRODES USING SOFT SOLUTION-PROCESSING IN LiOH SOLUTION AT 20 - 200°C

S.W. SONG,* K.S. HAN,* M. YOSHIMURA,* Y. SATO,** A. TATSUHIRO **
*Center for Materials Design, Materials and Structures Laboratory, Tokyo Institute of Technology, 4259 Nagatsuta, Midori, Yokohama 226-8503, Japan, yoshimu1@rlem.titech.ac.jp
** Department of Applied Chemistry, Faculty of Engineering, Kanagawa University

ABSTRACT

Application of Soft Solution-Processing, which is defined by environmentally friendly processing using (aqueous) solution, to the field of rechargeable lithium microbattery has been demonstrated by fabricating LiCoO$_2$ film on the cobalt metal substrate in LiOH solution under hydrothermal condition. The film formation mecahnism could be interpreted in terms of chemical dissolution of cobalt metal plate in LiOH solution at fixed temperature of 20 - 200 °C, resulting in the formation of H$_{1-n}$CoO$_2$ where n value increases with fabrication temperature and precipitation as LiCoO$_2$ by heterogeneous nucleation on the substrate followed by crystal growth. The crystal structure and film properties have been characterized by X-ray diffraction, X-ray photoelectron, micro-Raman spectroscopic and scanning electron microscopic analyses. The films exhibited a good crystallinity despite the low reaction temperature without any post-synthesis annealing. The films prepared under different conditions showed different phase selection such as spinel (*Fd3m*) or hexagonal (*R3-m*), surface morphology and film thickness. An electrochemical activity of the LiCoO$_2$ films was evidenced by cyclic voltammogram revealing a good reversibility on Li-intercalation and -deintercalation.

INTRODUCTION

We have demonstrated a utility and advantage of Soft Solution-Processing by showing successful fabrication of multicomponent oxide films directly from solutions [1-3]. It has been achieved using interfacial reactions between reactive substrates of metal, alloy or oxide and species in an aqueous solution activated thermally (hydrothermal), and/or electrochemically. Other activation methods such as photo-, sono-, etc. will also be possible to accelerate the interfacial reactions. In addition, Soft Solution-Processing gives products with much higher homogeneity than solid state method and faster growth rates than that in gas or vacuum processing because concentration of components in solutions is generally higher than that in gas phase or vacuum. Liquids may be beneficial for charging, transportation, mixing, and/or separation of products, furthermore, give the possibility for acceleration of diffusion, adsorption, reaction rate, and crystallization, especially under hydrothermal condition. In reality, however, almost earlier works for fabricating such Li-M binary oxides films were done by highly sophisticated multistep processes including preparation of a bulk cathode material or its precursor and subsequent deposition from the gas phase using evaporation, sputtering or chemical vapor deposition [4-7]. It is clear that all those techniques are neither economic nor environmentally friendly. In this regard, we have intentionally attempted to apply Soft Solution-Processing to the preparation of cathode films of Li-M binary oxide for rechargeable lithium microbatteries. Recently, we have successfully fabricated Li-M binary oxide films, LiMO$_2$

(M = Ni, Co), using Soft Solution-Processing, hydrothermal/electrochemical treatment of metal plate with a concentrated LiOH solution at 20 - 200 °C [8-12]. In the present paper, we describe film fabrication of $LiCoO_2$ using hydrothermal method and investigate film properties such as crystal structure and morphology depending on reaction conditions.

EXPERIMENT

The $LiCoO_2$ films have been fabricated in a single synthetic step by hydrothermal treatment of a cobalt metal plate with 200 ml of 0.1 - 6M LiOH solution at 20 - 200 °C using autoclave as shown in Figure 1. The cobalt metal electrode potential with respect to Ag/AgCl reference electrode was simultaneously monitored during the hydrothermal reaction. The prepared samples were washed with deionized water several times to remove any residual LiOH solution, then air dried at room temperature. Note that no post-synthesis annealing was applied to these samples. The crystal structure of samples was characterized by an X-ray diffractometer (Model MXP^{3VA}, MAC Science Co.) with Ni filtered Cu Kα radiation between 10 ~°120 2θ. Raman scattering spectra were recorded using microprobe optics (Atago Bussan) at room temperature. All Raman spectra were excited with the 514.5 nm line of an Ar laser with 5 mW/cm^2 power. The scattered light was analyzed with a triple spectrometer (Model T64000, Jobin Yvon / Atago Bussan) and detected using liquid nitrogen cooled CCD (charge-coupled device) cameras. The film surface and thickness were examined using a scanning electron microscope (SEM, S-4500, Hitachi). The cyclic voltammogram of the films were recorded to check their electrochemical activity using EG&G potentiostat/galvanostat. It was carried out in 1.0M $LiClO_4$ propylene carbonate under and Ar atmosphere using Pt counter electrode and Ag/AgCl reference one, and potential was referred to a Li/Li$^+$ electrode.

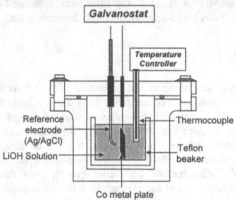

Figure 1. Diagram of autoclave.

RESULTS AND DISCUSSION

XRD patterns of the samples in Figure 2 exhibit the direct film formation of lithium cobalt oxide on the cobalt metal substrate. Note that $LiCoO_2$ powder is known to be crystallized in two phases depending on reaction temperature by the solid state method; hexagonal phase

(HT, *R-3m*) at high temperature (≥ 800 °C) or cubic spinel one (LT, *Fd3m*) at low temperature (400 °C). In this context, the peak at 19 ° corresponds to the (003) reflection for hexagonal layer structure or (111) one for spinel cubic structure of $LiCoO_2$ [13,14]. However, impurity phases such as $Co(OH)_2$ and $CoOOH$ were detected in lower LiOH concentration than 4M. The lithium cobalt oxide film would be exactly identified by the following micro-Raman spectroscopic and electrochemical analyses.

Film fabrication has been visualized more clearly in the SEM images as seen in Figure 3. It is recognized that well crystallized grains of lithium cobalt oxide form on the cobalt substrate in spite of low fabrication temperature without any post-synthesis annealing. Increasing of the fabrication temperature induced that lithium cobalt oxide crystallites became continuously bigger and film surface got rough and finally well developed layered grains were observed at higher temperature than 100 °C. The film thickness was 13 - 18 μm.

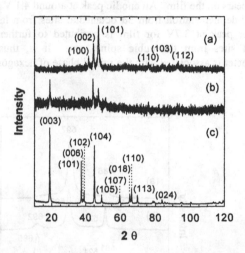

Figure 2. X-ray diffraction patterns for (a) cobalt metal plate, (b) lithium cobalt oxide film fabricated at 200 °C and (c) powder crystallized in a layered hexagonal structure.

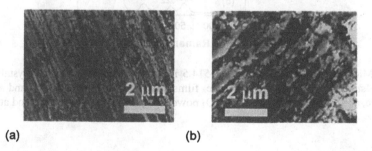

Figure 3. SEM images for (a) cobalt metal plate and (b) lithium cobalt oxide film fabricated at 150 °C.

According to the comparative micro-Raman spectroscopic analyses (Figure 4), the films obtained at higher temperature than 100 °C in 4M LiOH solution exhibit mainly two Raman active modes of A_{1g} and E_g from layered hexagonal structure with the trace of spinel one's, while the films fabricated at lower temperature than 100 °C and in lower LiOH concentration than 4M represent four modes A_{1g}, E_g and $2F_{2g}$ arising from cubic spinel one [15,16] and additional peaks from $Co(OH)_2$ and CoOOH. Therefore it becomes clear that films are crystallized in the mixture of hexagonal and cubic spinel phases of $LiCoO_2$ in 4M LiOH solution above 100 °C, whereas spinel one mainly forms at lower temperature than 100 °C or in lower LiOH concentration.

Such a structural difference in the $LiCoO_2$ film depending on reaction condition is also revealed in the cyclic voltammograms. The $LiCoO_2$ films fabricated at higher temperature than 100 °C show a good reversibility, indicating that these films are excellent electrodes in the electrochemical activity. When the cyclic voltammogram of the film (Figure 5) is compared with that of $LiCoO_2$ powder crystallized in hexagonal structure, one more anodic peak at lower potential of 3.7 V appears on the film. An anodic peak at around 4.1 V for both film and powder corresponds to the Li-deintercalation from 3a octahedral sites from layered $LiCoO_2$ hexagonal structure, but another peak at 3.7V for film is attributed to further Li-deintercalation from tetrahedral/octahedral sites from the cubic spinel one. It is, thus, obvious that such an electrochemical character is associated with the mixed phase of hexagonal and spinel $LiCoO_2$ on the film.

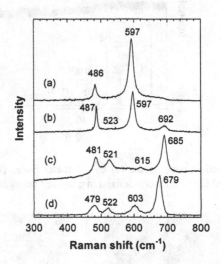

Figure 4. Micro-Raman spectra excited at 514.5 nm for (a) $LiCoO_2$ powder crystallized in a hexagonal structure, lithium cobalt oxide films fabricated at (b) 200 °C and (b) room temperature, and (d) the reference $LT-LiCoO_2$ powder prepared by solid state method at 400 °C.

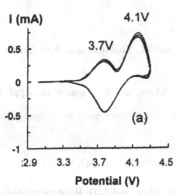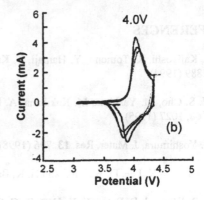

Figure 5. Cyclic voltammograms in 1M LiClO$_4$ (PC) electrolyte for (a) the LiCoO$_2$ film fabricated at 200 °C and (b) powder crystallized in a hexagonal structure; scan rate = 5 mV/s.

We have now reached the point where we consider the film formation mechanism. Film formation can be interpreted in terms of chemical dissolution of cobalt metal in LiOH solution and subsequent precipitation under our present hydrothermal condition. It is suggested that the dissolved cobalt species, cobaltite HCoO$_2^-$, is changed to CoO$_2^-$ at higher temperature and in higher LiOH concentration followed by heterogeneous nucleation by the reaction with Li$^+$ ions in the solution, then crystal growth to form LiCoO$_2$ crystallites. The dissolution and oxidation of cobalt metal to high valent dissolved cobalt species has been probed by an increase ($0 \rightarrow \sim 0.6$V) of potential of cobalt electrode with respect to Ag/AgCl reference electrode. The direct and spontaneous LiCoO$_2$ film formation by the dissolution-precipitation mechanism is the reason why no electrochemical anodic process is necessary for the oxidation process of cobalt in contrast to the nickel system; additional anodic oxidation step for the formation of trivalent nickel from divalent one was necessary for the transformation of Ni(OH)$_2$ to Li$_{1-x}$Ni$_{1+x}$O$_2$ as previously reported [8,9].

CONCLUSIONS

Films of LiCoO$_2$ could be easily fabricated by hydrothermal method, one of Soft Solution-Processing. With varying reaction conditions such as fabrication temperature and concentration of LiOH solution, we enabled to control surface morphology and film thickness, even the structure of LiCoO$_2$ phases like hexagonal or cubic spinel structure. It is, therefore, revealed that this Soft Solution-Processing can be useful to fabricate electrode films for application in rechargeable lithium microbatteries.

ACKNOLEDGMENTS

This work was supported by the "Research for the Future" program of the Japan Society for the Promotion of Science (JSPS-RFTF-96R06901). We wish to thank T. Watanabe, I. Sasagawa, N. Kumagai and H. Fujita for their helpful cooperation.

REFERENCES

1. K. Kajiyoshi, K. Tomono, Y. Hamaji, T. Kasanami, and M. Yoshimura, J. Am. Ceram. Soc. 77, 2889 (1994).

2. W. S. Cho, M. Yashima, M. Kakihana, A. Kudo, T. Sakata, and M. Yoshimura, Appl. Phys. Lett. 66, 1027 (1995).

3. M. Yoshimura, J. Mater. Res. 13, 796 (1998).

4. S. J. Lee, J. K. Lee, D. W. Kim, and H. K. Baik, J. Electrochem. Soc. 143, L268 (1996).

5. C. B. Wang, J. B. Bates, F. X. Hart, B. C. Sales, R. A. Zhur, and J. D. Robertson, ibid., 143, 3203 (1996).

6. K. A. Striebel, C. Z. Deng, S. J. Wen, and E. J. Cairns, ibid., 143, 1821 (1996).

7. C. B. Wang, J. B. Bates, F. X. Hart, B. C. Sales, R. A. Zhur, and J. D. Robertson, ibid., 143, 3203 (1996).

8. K. S. Han, P. Krtil, and M. Yoshimura, J. Mater. Chem. 8, 2043 (1998).

9. K. S. Han, S. W. Song, and M. Yoshimura, Chem. Mater. 10, 2183 (1998).

10. K. S. Han, S. W. Song, and M. Yoshimura, J. Am. Ceram. Soc. 81, 2465 (1998).

11. S. W. Song, K. S. Han, and M. Yoshimura, ibid., submitted (1999).

12. S. W. Song, K. S. Han, I. Sasagawa, T. Watanabe, and M. Yoshimura, Solid State Ionics, submitted (1999).

13. T. Ohzuku, and A. Ueda, J. Electrochem. Soc. 141, 2972 (1994).

14. R. J. Gummow, D. C. Liles, and M. M. Thackeray, Mat. Res. Bull. 28, 235 (1993).

15. W. G. Fately in *Infrared and Raman Selection Rules for Molecular and Lattice Vibrations: The Correlation Method*, Wiley-Interscience, New-York, 1972.

16. W. Huang, and R. Frech, Solid State Ionics, 86-88, 395 (1996)

EFFECT OF PRECURSOR SOL AGEING ON SOL-GEL DERIVED RUTHENIUM OXIDE THIN FILMS

S. Bhaskar, S. B. Majumder, P. S. Dobal, R. S. Katiyar, A. L. M. Cruz* And E. R. Fachini*
Department of Physics, University of Puerto Rico, San Juan, PR-00931-3343 USA
*Department of Chemistry, University of Puerto Rico, san Juan, PR-00931 USA

ABSTRACT

In the present work we have optimized the process parameters to yield homogeneous, smooth ruthenium oxide (RuO$_2$) thin films on silicon substrates by a solution deposition technique using RuCl$_3$.x.H$_2$0 as the precursor material. Films were annealed in a temperature range of 300°C to 700°C, and it was found that RuO$_2$ crystallizes at a temperature as low as 400°C. The crystallinity of the films improves with increased annealing temperature and the resistivity decreases from 4.86µΩ-m (films annealed at 400°C) to 2.94µΩ-m (films annealed at 700°C). Ageing of the precursor solution has a pronounced effect on the measured resistivities of RuO$_2$ thin films. It was found that the measured room temperature resistivities increases from 2.94µΩ-m to 45.7µΩ-m when the precursor sol is aged for aged 60 days. AFM analysis on the aged films shows that the grain size and the surface roughness of the annealed films increase with the ageing of the precursor solution. From XPS analysis we have detected the presence of non-transformed RuCl$_3$ in case of films prepared from aged solution. We propose, that solution ageing inhibits the transformation of RuCl$_3$ to RuO$_2$ during the annealing of the films. The deterioration of the conductivity with solution ageing is thought to be related with the chloride contamination in the annealed films.

INTRODUCTION

Oxide electrodes, such as ruthenium oxide (RuO$_2$), lanthanum strontium cobalt oxide (LSCO), strontium ruthenate (SrRuO$_3$), and yittrium barium copper oxide (YBCO) are attractive candidates to be used as bottom electrodes for ferroelectric memory devices [1]. Among these RuO$_2$ is a single component oxide and offers relatively lower resistivity and excellent diffusion barrier characteristics for lead based perovksite oxides [2]. The high thermal stability (~ 800°C) and ease of etching characteristics make it suitable for integrated devices. A variety of deposition techniques by rf-Sputtering [3], dc magnetron sputtering [4] and chemical vapor deposition [5] have been employed to synthesize RuO$_2$ thin films. Recently it has been demonstrated [6] that solution deposition technique offers an alternative approach to synthesize high quality RuO$_2$ films.

In case of the solution deposition technique, several process parameters needed to be optimized to obtain high quality RuO$_2$ thin films in a repetitive manner. Thus it has been demonstrated that increasing the annealing temperature has its effect to improve the crystallinity and grain growth of the deposited film. The grain size, porosity and surface roughness affects the measured resistivity. However, limited effort has been made to study the effect of other process variables on the properties of RuO$_2$ thin films. In case of the preparation of sol-gel derived ferroelectric films, it has been observed that ageing of the precursor has its effect in controlling the texture growth [7], microstructure and electrical properties of the deposited films [8].

In the present work, we have investigated the effect of solution ageing on the microstructure and electrical properties of the RuO$_2$ films.

EXPERIMENT

RuCl$_3$.x.H$_2$O was dissolved in absolute alcohol to prepare a solution of strength 0.05M/L. We have found that thicker sol affects the wetting of RuCl$_3$ solution on Si substrates and the concentration in the of 0.05 - 0.1 M/L was found to be optimum for the deposition of smoother films. Prior to deposition, Si substrates were cleaned after ultrasonification in acetone and absolute alcohol. In some cases these substrates were heated at 400°C/30s to grow oxide layers to improve the the film-substrate adhesion. The precursor solution was spun coated on the substrate at at 3000rpm for 20s. Just after deposition, films were directly inserted to a preheated furnace (kept at a temperature range 300°C - 700°C.) and annealed for 5mins followed by quench at room temperature. The coating and firing cycle was repeated for 10 times to yield films about 700nm thick. To study the ageing characteristics the solution was kept in a desiccator for a predetermined time and spin coated on the substrate. The deposited films were characterized in terms of their phase formation behavior, microstructure and electrical properties. Phase analysis was performed using X-ray diffractometer (Siemens D5000) with CuK$_\alpha$ radiation (1.5405Å), to obtain the diffraction data. The surface morphology of the films were observed using an Atomic Force Microscopy(Nanoscope IIIa Multimode AFM Digital Instruments) and the images were obtained with non contact AFM mode. Chemical composition and peak energy shifts in the samples were analyzed using X-ray photoelectron spectrometer (XPS) (Physical Electronics, PHI5600 ESCA system) using Al K$_\alpha$ radiation. Room temperature electrical resistivity of the RuO$_2$ films were measured using the conventional four probe Van der Paw method.

RESULTS AND DISCUSSIONS

X-ray diffraction and microstructure analysis

Figure 1 shows the X-ray diffractogram(XRD) of RuO$_2$ films deposited from the freshly

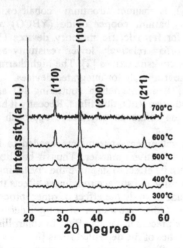

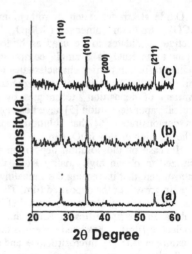

Figure 1 XRD patterns of RuO$_2$ films deposited from the freshly prepared solution and annealed at different temperatures.

Figure 2. XRD patterns of RuO$_2$ films from (a) freshly prepared (b) 7 days and (c) 60 days aged solution.

prepared solution and annealed in the temperature ranging 300°C - 700°C for 10 minutes.
As shown in the figure the film remains amorphous at 300°C and crystallizes to tetragonal structure at 400°C. The crystallite sizes for these films were calculated using Scherer formula [9] and Table I shows the variation of the crystallite size as a function of annealing temperature along with the measured resistivities for each films.

Table I Variation of the crystallite size and film resistivity with the annealing temperature.

Annealing temperature (°C)	Crystallite Size(nm)	Resistivity ($\mu\Omega$-m)
400°C	47	4.86
600°C	90	3.57
700°C	96	2.92

As shown in Table I, the crystallite size increases with the increase in annealing temperature, indicating better crystallinity of the films. The marginal decrease of film resistivity could be due to the improved crystallinity. The reduction of crystal imperfections (especially in the grain boundary region) and film porosity could increase the electron mean free path and thereby increase the conductivity [10]. Table I. shows the lowest value of resistivity was observed in the RuO_2 film annealed for 700°C for 10min.

Table II Summary of the crystallite size and film resistivity with precursor solution ageing.

RuCl₃ precursor solution	Crystallite Size(nm)	Resistivity ($\mu\Omega$-m)
Freshly Prepared	96	2.9
7 days aged	86	11.5
60 days aged	19	45.7

Therfore to study the effect of solution ageing on the film resistivity, we have heat treated the RuO_2 films at 700°C. Figure 2 shows the XRD of RuO_2 films prepared from (a) freshly prepared (b) 7 days aged and (c) 60 days aged solution. Table II summarizes the value of crystallite size and the film resistivities with precursor solution ageing. It is shown that the aged solution yields film with poor crystallinity and thereby increased resistivity. Also it may be noted that films from freshly prepared solution is textured along (101) direction and with ageing of the solution the preferred orientation changes along (110) direction.

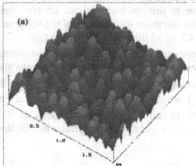
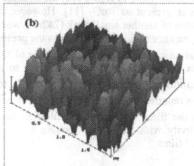

Figure3 AFM images of thin films from (a) freshly prepared and (b) 60 days aged solution.

AFM images of RuO$_2$ films deposited at 700°C from (a) freshly prepared and (b) 60 days are shown in Figure 3. The films from freshly prepared solution are smooth and have uniform grain size distribution with an average roughness value of 12nm. With solution ageing, the surface roughness increases to 22nm and the microstructure becomes non uniform. The observed increase of film resistivity in the aged film could be due to the non uniform grain size distribution and high surface roughness value.

XPS Analysis

To understand the role of solution ageing on the observed increase in film resistivity , we have performed XPS analysis on the films from freshly prepared, 7 days, and 60 days aged solutions. Figure 4 shows the XPS spectrum of RuO$_2$ film deposited from freshly prepared solution. The binding energy of various peaks were calibrated with respect to that of carbon (C1s ≈ 285 eV). Ru doublet peak energies 285eV (Ru3d3/2) and 281eV (Ru 3d5/2) correspond to RuO$_2$.

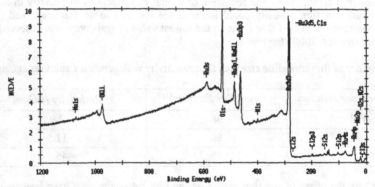

Figure 4 The XPS spectrum of RuO$_2$ film deposited from freshly prepared solution and annealed at700°C/10min.

In addition small concentration of chlorine and silicon are also detected in the spectrum. A slow scan XPS was performed to resolve Ru doublet in the binding energy range 277 to 297eV for the films deposited from freshly prepared, 7 days and 60 days aged solutions and annealed at 700°C/10mins. Note in figure 5(a) the doublet Ru 3d3/2 (285eV), Ru 3d5/2 (281eV) indicate that Ru is present as RuO$_2$ [11]. However, in addition to this doublet we can observe the appearance of another small peak (282eV) whose intensity increases with ageing of the solution. The appearance of this peak indicates the presence of RuCl$_3$ in addition to RuO$_2$ in the film [12].

Observation clearly indicates that in case of the films deposited from freshly prepared solution, RuCl$_3$ is almost transformed completely to RuO$_2$ whereas the ageing of the solution has its effect to inhibit the transformation of RuCl$_3$ to RuO$_2$ during the heat treatment of the films. It has been reported that in case of RuO$_2$ films prepared by ion beam sputtering incorporation of Fe ions in the films deteriorate the film conductivity [13]. In our case the deterioration of the conductivity with ageing is thought to be related with the chloride contamination in the annealed films.

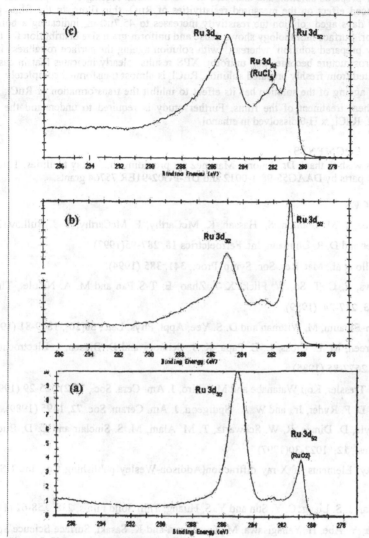

Figure5 Slow scan XPS spectra of RuO₂ films deposited from (a) freshly prepared (b) 7 days aged (c) 60 days aged solution.

CONCLUSIONS

We have studied the effect of precursor solution ageing on RuO₂ thin films synthesized by a solution based technique. Films annealed at 300°C were amorphous and RuO₂ crystallizes at a temperature as low as 400°C. X-Ray analysis indicates that with the increase in annealing temperature the crystallite size increases, indicating better crystallinity of the films. Room temperature resistivity measurements indicate lowest resistivity (≈ 2.94μΩ-m) for film deposited from freshly prepared solution and annealed at 700°C/10min. Ageing of the precursor solution has pronounced effect on the measured resistivities of RuO_2 thin films. In the film prepared from the 60 days aged solution the resistivity increases to 45.7μΩ-m, indicating a nonmetallic type behavior. Surface morphology shows dense and uniform grain size distribution for the films from freshly prepared solution whereas, with solution ageing the surface roughness increases and the microstructure becomes non uniform. XPS results clearly indicates that in case of the films deposited from freshly prepared solution, $RuCl_3$ is almost transformed completely to RuO_2 whereas the ageing of the solution has its effect to inhibit the transformation of $RuCl_3$ to RuO_2 during the heat treatment of the films. Further study is required to understand the solution chemsitry of $RuCl_3 .x.H_2O$ dissolved in ethanol.

ACKNOWLEDGMENTS

Authors wish to thank Dr. Antonio Martinez for providing the X-Ray facilities. This work is supported in parts by DAAG55-98-1-0012 and DE-FG02-91ER 75764 grants.

REFERENCES

1. G. Teowee, J. M. Boulton, S,. Hassan, K. McCarthy, F. McCarthy, T. J. Bulkowski, T. P. Alexander and D. R. Uhlmann, Int. Ferroelctrics **18**, 287-95(1997).

2. O. Auciello et al., Mat. Res. Soc. Symp. Proc., **341**, 385 (1994).

3. E. Kolawa, F. C. T. So, W. Flick, X.-A Zhao, E. T-S Pan and M. A. Nicolet, Thin Solid Films **173**, 217-24 (1989).

4. L. Krusin-Elbaum, M. Wittman and D. S. Yee, Appl. Phys. Lett., **50**[26], 1879-81 (1987).

5. M. L. Green, M. E. Cross, L. E. Papa, K. J. Schnoes and D. Brasen, J. Electrochem Soc., **132**(11), 2677-85 (1985).

6. James F. Tressler, Koji Watanabe and Masahiro, J. Am. Cera. Soc., **79**(2) 525-29 (1996).

7. C. Chen, D. F. Ryder, Jr., and W. A. Spurgeon, J. Am. Ceram. Soc. **72**, 1495 (1989).

8. T. J. Boyle, D. Dimos, R. W. Schwartz, T. M. Alam, M. S. Sinclair and C. D. Buchheit, J. Mater. Res., **12**, 1022-30(1997)

9. D. Cullity, Elements of X-ray diffraction(Addison-Wesley publishing Co., Inc USA 1967) p261.

10. S. Y. Mar, J. S. Liang, C. Y. Sun and Y. S. Huang, Thin Solid Films, 238, 158-62 (1994).

11. Y. Kaga, Y. Abe, H. Yanagisawa, M. Kawamura and K. Sasaki, Surface Science Spectra, **6**, 1 68-74(1999).

12. Handbook of X-Ray Photoelectron Spectroscopy (Perkin-Elmer Corporation, Physical Electronics Division).

13. H. Kezuka, R. Egerton, M. Masui, T. wada, T. Ikehata, H. Mase and M. Takeuchi, Appl. Surface Sci., **65/66**, 293-297 (1993).

PENTADIONATE: AN ALTERNATE SOL-GEL METHOD FOR THE SYNTHESIS OF FERROELECTRIC $Ba_{1-x}Sr_xTiO_3$

Pramod K. Sharma*, K. A. Jose, V. V. Varadan and V. K. Varadan
Center for the Engineering of Electronics and Acoustic Materials,
Department of Engineering Science and Mechanics, The Pennsylvania State University,
University Park, PA 16802, USA

ABSTRACT

Sol-gel synthesis of transitional metal oxides is one of the important technique to obtain the pure and homogenous materials at low temperature. In this work, Barium strontium titantate (BST) was synthesized by sol-gel processing using 2,4-pentadionate as the precursors of metal oxides. The $Ba_{1-x}Sr_xTiO_3$ was prepared for four values of x i.e. 0.2, 0.4, 0.5 and 0.6. The obtained powders were heated from 400 °C to 800 °C at the interval of 100 °C to determine the temperature of crystallization. The phase of the final products was investigated by x-ray diffraction (XRD). Two particle sizes (<100 nm and 200-400 nm) were observed under the scanning electron microscope (SEM) of the heat treated xerogel. The dielectric properties were determined by impedance analyzer at the frequency of 1MHz and explained in detail for the different values of x.

INTRODUCTION

There has been increasing interest in the synthesis of ceramic materials by low temperature (e.g. hydrothermal), sol-gel and growth controlled method) [1-3]. The materials synthesized by a low temperature route are highly pure, homogenous and with controlled stoichiometric. It is widely known that by sol-gel process glasses of silica, ceramics of titania and many other multi component metal oxides can be prepared [4,5]. They may be obtain in the form of fibers, thin films, coatings and monolith [6,7]. Further the quality of the ceramic produced by sol-gel method is determined by the state of suspension. The good dispersion of particles gives optimum packing state which influences the sinterability in the firing process and the physical and chemical properties of the final product. In the synthesis of multi component metal oxide, it is necessary to prepare gels of high homogeneity in which the cation of various kinds are uniformly distributed in the atomic scale. As reported in the literature that the most promising way to obtain gel of high homogeneity would be the employment of sol-gel technique of alkoxides or the organic precursor of metal [8,9].

There has been recent interest in the use of perovskite ceramics for various applications. Perovskite ceramic ABO_3 (e.g. $BaTiO_3$) have a phenomenon of

217

"ferroelectricity". These materials usually have high dielectric constant and applied for a series of application ranging from high dielectric constant capacitors to later developments in piezoelectric transducers, positive temperature coefficient devices and electro-optics. Materials with two or three metal oxides have dominated the field throughout their history.

The substitution on A site in $BaTiO_3$ by Sr gives $Ba_{1-x}Sr_xTiO_3$. For capacitors ceramics of the perovskite structure (ABO_3), doping in small amount at A site can greatly affects dielectric properties [10-12]. There has been recent interest in the use of $Ba_{1-x}Sr_xTiO_3$ phase shifting devices in antenna and radar [13,14]. A desirable material's properties include low temperature curie peak (Tc), low dielectric constant, low loss tangent and large change in dielectric constant with applied bias field which corresponding good tunability. BST ceramics meet all of these properties. However it is documented that the properties of BST depends on several factors e.g. microstructure, grain size, density, pore and particle size. Therefore, the new routes always in search to obtain the desirable properties of BST.

In this paper, we developed a new sol-gel method for the synthesis of BST. We have also investigated the conditions for the effect of Sr, calcination and characteristics of BST. This work also illustrates the microstructure, dielectric constant, loss tangent and tunability, which are important characteristics for use of the BST in phase shifters in antenna.

EXERIMENTAL

1. Synthesis

Titanium tetra iso-propoxide ($Ti(O-C_3H_7)_4$) and catalyst, were mixed in the appropriate molar ratio with methoxyethanol solvent and refluxed for 2 hrs at 80 °C. Separate solutions of Ba and Sr were prepared by dissolving 2,4-pentadionate salts of Ba and Sr, in methoxyethanol. Mild heating was required for a complete dissolution of salts. The metal salt solution was then transferred to the titania sol slowly. The solution was refluxed for another 6 hrs. The sol was then hydrolyzed by a particular concentration of water. It is important to note that direct addition of water leads to precipitation in the sol. Therefore a mixture of water/solvent has to be prepared, and then added to the sol drop by drop. The resultant sol was refluxed for 2 hrs to complete hydrolysis. This sol was kept in an oven at 90 °C to obtain the xerogel. (refer to Fig.1).

2. Characterization

The xerogel was studied by x-ray diffraction (XRD). XRD was used for the phase identification with a Scintag diffractometer (DMC 105) with copper Kα radiation for 2θ from 25° to 70°.

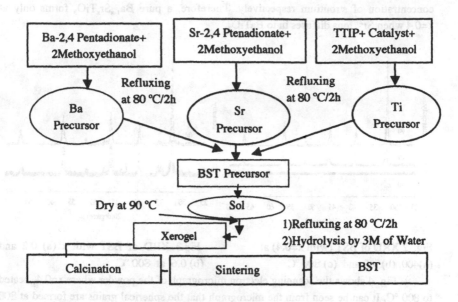

Fig. 1 Schematic diagram of preparation of BST.

For electrical measurements, each powder was pressed into a pellet under a load of 2.5 tons and was subsequently sintered at temperature of 1200 °C for 0.5 hr in air. Acrylic polymer (3 wt % of the powder) was used as a binder. Ag paste was painted on both sides of the pellets, and they were dried at 70 °C for 24 hrs, to form ohmic contact electrodes. Permitivity values at 1 MHz were measured at room temperature with an impedance analyzer (HP 4192A).

RESULTS AND DISCUSSION

The x-ray diffraction pattern of the powders are shown in Fig. 2 after heat treatment of xerogel at 400, 500, 800 °C. The BST powder studied initially at x=0.4 mole concentration. The result confirms that x-ray diffraction of the powder at room temperature is amorphous. This amorphous behavior of the powder exists up to the temperature of 400 °C. Crystalline peak started to appear when the powder was heated at 500 °C. A pure phase of $Ba_{1-x}Sr_xTiO_3$ was formed at 800 °C. Once optimize the heat treatment, the powders with x=0.2 and 0.6 were heated and investigated by x-ray diffraction technique. Fig. 3 shows the x-ray diffraction of the powders with x=0.2 and 0.6 mole, the Ba rich compound at initial stage and Sr rich compound at later stages of concentration of strontium respectively. Therefore, a pure $Ba_{1-x}Sr_xTiO_3$ forms only at x=0.4 when Sr^{2+} ions diffuses in to $BaTiO_3$.

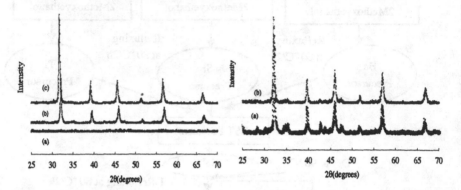

Fig. 2 XRD of BST (with x=0.4) at (a)400, (b)500 and (c) 800 °C.

Fig.3 XRD of BST with x (a) 0.2 and (b) 0.6 at 800 °C.

Fig. 4 shows the scanning electron micrograph of the powder where x=0.4, heated to 800 °C. It can be seen from the micrograph that the spherical grains are formed at 800 °C. Two particle sizes are observed (a) particles between 200 to 400 nm in diameter which formed due to the aggregation and (b) particles with a diameter less than 100 nm. Overall, the particles are in submicron order and have spherical morphology.

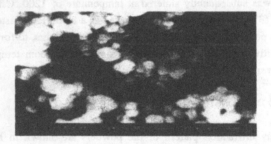

Fig. 4 Scanning electron micrograph of BST where x=0.4, heated to 800 °C.

Fig. 5 shows a variation in dielectric constant as a function of x in $Ba_{1-x}Sr_xTiO_3$. The dielectric constant was measured by impedance analyzer using the equation $(\varepsilon)=Cd/\varepsilon_0 A$ where (C) is capacitance, (d) is thickness of the sample, (ε_0) is air permittivity, (A) is area of electrode. The dielectric constant (ε) and loss of tangent were measured at 1 MHz. The dielectric constant of the sample was observed to be lowest for x=0.2 to a value of 948. It increases to the maximum value of 1175 at x=0.4 and then decreases for x ≥ 0.5. The higher dielectric constant at x=0.4 indicates that curie points of these compounds lie in the vicinity of room temperature. Study is underway to determine the Curie temperature of these materials Another important feature of the present study is the observation of extremely low tangent loss. The loss of tangent as measured for the powder with x=0.2 was the largest with a value of 0.098 and decreases with increasing x value to a lowest value *i.e.* 0.035 for x= 0.6 as shown in Fig. 6.

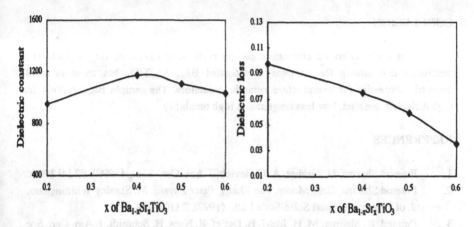

Fig. 5 Dielectric constant of BST with x   Fig. 6   Dielectric loss of BST with x

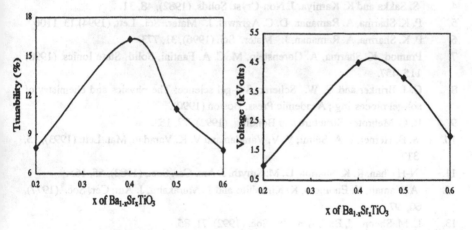

Fig. 7 Tunability constant of BST with x   Fig. 8   Bias voltage of BST with x

The dielectric tunability of the samples is determined by the equation tunability(%)= ((ε_{vo}-ε_{v})/ ε_{vo}))× 100. The tunability measurements were taken with an applied electric field that ranged from 0 to 4.0 kV at the frequency of 1 MHz. Fig. 7 shows the tunability as a function of x for the samples of $Ba_{1-x}Sr_xTiO_3$. The bias voltage as a function of x for the samples of $Ba_{1-x}Sr_xTiO_3$ is shown in Fig. 8. It is apparent from the Figs.7 and 8 that the tunability and applied voltage respectively, were increased with increasing value of x=0.2 to 0.4. They decrease with further increasing value of x. Tunability behaves similar to the dielectric constant.

CONCLUSION

This study offers an alternative sol-gel route to synthesis of $Ba_{1-x}Sr_xTiO_3$. It is concluded that among the samples investigated $Ba_{0.6}Sr_{0.4}TiO_3$ has most favorable dielectric properties for manufacture ceramic capacitors. The sample $Ba_{0.6}Sr_{0.4}TiO_3$ has high dielectric constant, low loss tangent and high tunability.

REFERENCES

1. Pramod Sharma, H. Fischer, A. Craievich, J. Am. Cer. Soc., (1999), 82 [4] 1020.
2. Pramod Sharma, Greg Moore, Fan Zhang, Peter Zavalij, M. Stanley Whittingham, J. of Electrochem.and Solid-State Lett, (1999), 2 [10], 494.
3. Pramod K. Sharma, M. H. Jilavi, B. Detlef, R. Nass, H. Schmidt, J. Am. Cer. Soc., (1998), 81 [10], 2732 .
4. S. Sakka and K. Kamiya, J. Non-Cryst. Solids, (1982), 48, 31.
5. P. K. Sharma, A. Ramanan, D. C. Agrawal, J. Mater. Sci. Lett, (1994),13, 1106
6. P. K. Sharma, A. Ramanan, J. Mater. Sci. (1996), 31, 773.
7. Pramod K. Sharma, A. Gorenstein, M. C. A. Fantini, Solid State Ionics, (1998) 115, 457.
8. C. J. Brinker and G. W. Scherer, Sol-gel science: The physics and chemistry of sol-gel processing ; Academic Press, Boston (1990).
9. R. C. Mehrotra. Structure and Bonding, (1992), 77, 153.
10. S. B. Herner, F. A. Selmi, V. V. Varadan and V. K. Varadan, Mat. Lett. (1993), 15, 317.
11. N-H Chan, R. K. Sharma, D. M. Smyth, J. Am. Cer. Soc., (1982), 65, 165.
12. A. Yamaji, Y. Enomoto, K. Kinoshito and T. Murakami, J. Am. Cer. Soc. (1977), 60, 97.
13. L. M. Sheppard, Bull. Am. Cer. Soc. (1992), 71, 85.
14. F. Selmi, D. K. Ghodgaonkar, R. Hughes, V. V.Varadan, V. K. Varadan, Proceedings SPIE," Structure, sensing and control," (1992), p. 97.

GROWTH AND STUDIES OF Li (Mn, Co) OXIDES FOR BATTERY ELECTRODES

S. Nieto-Ramos* and M.S. Tomar* and R.S. Katiyar[#]
*Physics Department, University of Puerto Rico, Mayaguez, PR 00681-9016, e.mail:
m_tomar@feynman.upr.clu.edu
[#]Physics Department, University of Puerto Rico, San Juan, PR 00931

ABSTRACT

There is interest in lithium intercalation oxide materials for cathodes in rechargeable batteries.
We have synthesized $LiMO_x$, (where M = Mn, Co) by a less expensive solution route. Reagent
grade acetates or hydroxides as precursors for lithium, manganese, and cobalt, respectively, with
methoxy ethenol and acetic acid as solvents were used. Powders with different compositions
were achieved at annealing temperature below 700^0 C. Thin films were deposited by spin
coating. X-ray diffraction, Raman spectroscopy, and impedance spectroscopic results are
presented. These studies indicate that this synthetic route is suitable to produce good quality
lithium-based oxides useful for cathode in lithium-ion rechargeable battery.

1. INTRODUCTION

Lithium intercalation materials are important as cathode in rechargeable lithium-ion battery.
Normally the cathode materials are selected from $LiCoO_2$, $LiNiO_4$, and $LiMn_2O_4$, depending on
the intercalation voltage required[1]. Each of these materials can react reversibly with a certain
amount of lithium between the cutoff voltages ~ 2.5 V to 4.2 V. $LiMn_2O_4$ has been extensively
studied in the past decade, because manganese-based cathodes are environmentally benign and
less expensive. However, it exhibits Jahn-Teller distortion at T~ 280 K, which may influence the
cyclability of $LiMn_2O_4$ cathode. But, Jahn-Teller effect can be suppressed[2] by increasing the
average oxidation state of manganese through aliovalent cationic substitutions for manganese,
e.g. by partial replacement of Li by Mn in Li $_{1+x}Mn_{2-x}O_4$. Recent computer simulation[3] indicates
the possibility of designing new cathode materials, where cobalt, manganese, and nickel could be
partially replaced by non-transition metals. Several new materials have been proposed and more
possibility exists. Therefore, there is an interest in economic synthetic routes of these interesting
intercalation materials both as powder and thin films.

Several synthesis routes have been used for perovskites and other oxides[4,5]. Most
versatile precursors are metal alkoxides, which are very reactive toward nucleophilic reagents
such as water. The hydrolysis and condensation of transition metal alkoxides produces high
quality oxide materials by sol-gel processing. However, alkoxides are very expensive, it is
desirable to find the economic routes for the synthesis of these materials. Lithium intercalation
oxides have also been synthesized by similar routes[6-8]. We report here the synthesis of $LiMn_2O_4$,
and $LiCoO_2$ using a solution-based route, where metal acetates or hydroxides were used as
precursors with suitable organic solvents. Both powders and thin films were prepared. X-ray
diffraction, Raman spectroscopy, and impedance spectroscopy were used for characterization.

2. EXPERIMENTAL

Precursor solutions were prepared using reagent grade chemicals of lithium acetate
$(LiOOCCH_3).2H_2O$, manganese acetate $(CH_3CO_2)_2Mn.4H_2O$, and cobalt acetate
$(CH_3CO_2)_2Co.4H_2O$, for lithium, manganese, and cobalt, respectively. Methoxy ethanol

($C_3H_8O_2$), 2-ethylhexanoic acid ($C_8H_{16}O_2$), and acetic acid solvents were purchased from Fisher Scientific.

For a particular ratio of the metals, the required amounts of the precursors were separately mixed in methoxy ethenol and 2-ethylhexanoic acid, and boiled. These solutions were mixed hot and refluxed to form a homogeneous solution. The final precursor solution was cooled and stored in sealed container. The part of this solution was slowly evaporated on a hot plate to dryness and then annealed at various temperatures to get the powder. The resulting powders were annealed in a box furnace at different temperatures to study the evolution of conversion for $LiMn_2O_4$ and $LiCoO_2$. Other portion of the solution was used for thin films by spin coating at 1500 rms. Films need much less time for the transformation to stoichiometric material.

3. RESULTS AND DISCUSSION

3.1 Xray Diffraction Studies

After the synthesis by the solution route, powder samples of $LiCoO_2$ and $LiMn_2O_4$ were analyzed by x-ray diffraction (CuKα-radiation). Figure 1 shows the x-ray diffraction pattern for the transformation of $LiCoO_2$ annealed for three hours in a box furnace at different temperatures.

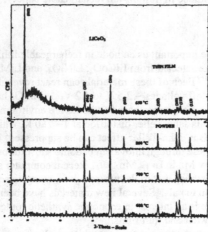

Figure 1. X-ray diffraction of $LiCoO_2$ for the samples annealed at different temperatures

The x-ray diffraction of thin film deposited by spin coating on quartz substrate, and annealed film is also shown in Figure 1. The evolution of x-ray diffraction patterns of $LiMn_2O_4$ powder and film are shown in Figure 2. All the diffraction peaks are labeled in the Figure 1 and Figure 2, which coincide with $LiCoO_2$ and $LiMn_2O_4$. As revealed by these x-ray diffraction patterns of Figures 1 and 2, a complete transformation takes place at the annealing temperature ~ 650^0 C. Evidently, the solid state reaction process needs much higher process temperature and a long annealing time.

3.2 Raman Spectroscopic Characterization

Raman spectroscopy is very sensitive tool to analyze the homogeneity, composition, and the required phases in the material through the symmetry allowed phonon modes. Raman spectra also indicates the structural phase transition, local order, disorder due to impurities, vacancies or doping. Two powder samples of LiMn₂O₄ and LiCoO₂, prepared by solution method and annealed at temperature of 600^0 C and 700^0 C, were chosen for Raman studies, and their Raman phase shift spectra are shown in Figure 3 and Figure 4, respectively. These samples were grinded in an agate crucible for each annealing step to maintain good mixing.

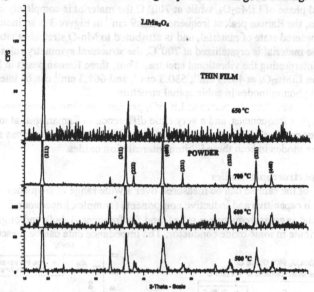

Figure 2. X-ray diffraction of LiMn₂O₄ powders by solution route, and annealed at different temperatures

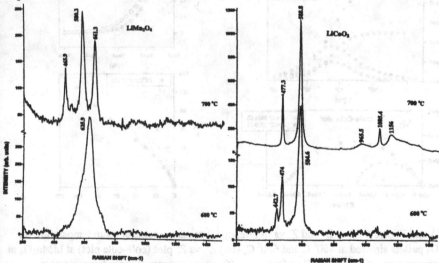

Figure 3 and Figure 4. Room temperature Raman spectra of LiMn₂O₄ and LiCoO₂ samples

The materials synthesized here are neither doped nor inhomogeneous in composition. Therefore, it is expected that structural phase disorder could dominate the Raman shift as indicated Figure 3 and 4.

As revealed by x-ray diffraction of Figure 1, sample annealed at 500^0 C, were partially converted to spinel phase of $LiMn_2O_4$, while at 700^0 C the material is completely converted to spinel phase. Thus, the Raman peak at frequency 625.9 cm^{-1} in Figure 3 is an indication of amorphous or disordered state of material, and is attributed to Mn-O stretching vibration. However, when the material is crystallized at 700^0C, the structural symmetry rules play important role in interpreting the vibrational spectra. Thus, three Raman peaks in Figure 3 of the ordered structure in $LiMn_2O_4$ at 465.9 cm^{-1}, 580.3 cm^{-1}, and 661.3 cm^{-1} can be interpreted due to symmetry allowed phonon modes in cubic spinel structure.

The room temperature Raman spectra of $LiCoO_2$ prepared in similar conditions show new Raman peaks at higher frequencies, and a very little difference in Raman shift at lower frequencies. Lower number of Raman peaks (< 21) could be due to weaker cross section of some Raman active modes in both these lithium intercalation oxides.

3.3 Impedance Spectroscopic Studies

The impedance[9,10] of the sample can be measured over a wide range of frequency. Since impedance has both capacitive and inductive components (complex impedance $Z^* = Z' - j Z''$, where $j = (-1)^{1/2}$, and angular frequency $\omega = 2\pi f$, and f = frequency), bulk and/or grain boundary regions may contribute in ionic oxide conductor. The impedance date can be presented in the

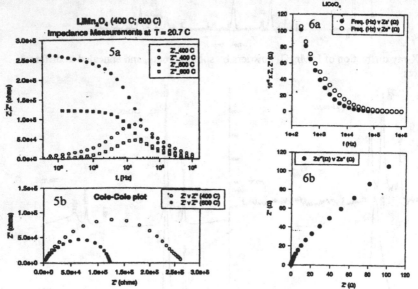

Figure 5 and Figure 6. (5a) Z″ and Z′ vs frequency (f) measurements at room temperature on $LiMn_2O_4$ pallets sintered at 400^0 C and 600^0C, (6b) Z″ vs Z′ plot (cole-cole plot) at $LiMn_2O_4$ at room temperature, (6a) Z″ and Z′ vs frequency (f) measurements at room temperature on $LiCoO_2$ pallets sintered at 600^0C, (6b) Z″ vs Z′ plot (cole-cole plot) at room temperature for $LiCoO_2$ sintered at 600^0 C

form of imaginary, Z'' (capacitive) against real, Z' (resistive) impedances. Complex impedance plane representation, Z'' vs Z' is appropriate for ionically conductive ceramics[7]. Each parallel RC (resistance-capacitance) gives rise to a semicircle from which the component R and C values can be extracted. Thus, single circle represents the bulk contribution, and a second circle at higher frequencies may represent additional grain boundary contribution. Resistance (R) values can be obtained from the intercepts on the Z' axis, and the capacitance (C) values can be obtained using the relation, $RC = 1/\omega_{max}$, to the frequency at the maximum of each semicircle.

LiMn$_2$O$_4$ and LiCoO$_2$ powders prepared at 700^0 C were pressed to form the pellets. Some pellets were sintered at 400^0 C and 600^0 C for two hours and studied by Novocool BDS impedance spectroscopy. The samples were cooled to room temperature and mounted for impedance measurements. Figure 5 and Figure 6, respectively, shows the Z', Z'' vs frequency (f) at room temperature of LiMn$_2$O$_4$ and LiCoO$_2$ pallets. The same data was used to plot Z'' vs Z' (cole-cole plot) in Figure 5 and Figure 6. Imaginary components of the impedance (Z'') of both samples of LiMn$_2$O$_4$ samples under investigation show their two respective peaks at the same frequency. They correspond to the semicircles on cole-cole plot of Figure 6 (b), which represents one parallel RC element, i.e. it is a bulk contribution of ionic conductor.

Since spinel LiMn$_2$O$_4$ is an ionic conductor, the observed one parallel RC behavior is valid. In LiCoO$_2$, full semicircle could not be achieved in the frequency range of the measurement, but it shows similar trend as for LiMn$_2$O$_4$ sample. Thus, the grain boundary contribution over a wide frequency range in LiMn$_2$O$_4$ and LiCoO$_2$ sintered samples seems to be is negligible. A decrease of the real component of impedance (Z') with increasing frequency in Figure 5 and 6, is also expected in ionic conductors. The absence of grain boundary contribution to the impedance suggests that these materials would behave as a stable structure during lithium intercalation in rechargeable battery.

4. CONCLUSIONS

Thus, analysis by x-ray diffraction, impedance spectroscopy, and Raman studies indicates that polycrystalline stoichiometric LiMn$_2$O$_4$ and LiCoO$_2$ thin films can be produced by the solution-based route at the substrate temperature $\sim 650^0$ C, but the film density is low. Currently, we are working to improve the film density in these spin coated thin films for cathode application in rechargeable thin film Li-ion/polymer battery.

ACKNOWLEDGEMENTS

This work was initiated with the support of Research and Development Center, University of Puerto Rico, Mayaguez, through the grant # FDP-11-97. Current effort on oxide materials is supported by DOD (Grant No. DAAG55-98-1-0012), and is gratefully acknowledged.

REFERENCES

1. J.P. Pereira-Ramos, N. Baffier, and G. Pistoia, in , *Lithium Batteries*, G. Pistoria, (Editor), Elsevier, New York, (1994), p. 281
2. G. Ceder, M.K. Aydinol, and A.E. Kohan, Comput. Mater. Sci., **8**, 161 (1996)
3. R.J. Gummow, A. De Kock, and M.M. Thackeray, Solid State Ionics, **69**, 59 (1994)
4. J. Livage, M. Henry, and C. Sanchez, Prog. Solid-State Chem. **18**, 259 (1988)
5. P. Barboux, J.M. Tarascon, and F. Shokoohi, J. Solid-State Chem. **94**, 209 (1991)
6. A. Kasbani, M.S. Tomar and E. Dayalan, J. Mater. Res. **10**, 2404 (1995)

7. K. West, B. Z-Christiansen, T. Jacobsen, and S. Skaarup, Electrochimica Acta, **38**, 1213 (1993)

8. Y-M Chiang, Y-I. Jang, H. Wong, B. Huang, D. R. Sadoway, and P.Ye, J. Electrochem. Soc. **145**, 887 (1998)

9. J. Ross Macdonald, *Impedance Spectroscopy*, Wiley, (1987)

10. T. Ohzuku, A. Ueda, M. Nagayama, Y. Iwakoshi, and H. Komori, Electrochimica Acta, **38**, 1159 (1993)

FABRICATION OF CRACK-FREE La$_{0.2}$Sr$_{0.8}$CoO$_{3-x}$ MEMBRANES ON ASYMMETRIC AND POROUS CERAMIC SUPPORTS—EFFECTS OF A METALLIC COVERING

Z-D. CAO *, L. HONG * X. CHEN **
*CHEE Department, National University of Singapore, 10 kent Ridge Crescent Singapore 119260
**Institute of Materials Research & Engineering of Singapore, 3 Research Link, Singapore 117260

ABSTRACT

Crack-free La$_{0.2}$Sr$_{0.8}$CoO$_{3-x}$ (LSCO) membranes have been fabricated on porous ceramic supports such as MgO and CeO$_2$ through dip coating and sintering. To avoid cracks in a LSCO membrane during sintering, a silver coating on top of the LSCO powder-packed layer that is formed via dip coating was found successful. The cracks would be otherwise unavoidable due to the shrinkage mismatch between the membrane and the support during sintering. The silver can be readily removed from the sintered LSCO membrane chemically without causing damages in the membrane.

INTRODUCTION

The ceramic membranes with the mixed conductive property have found applications in solid oxide fuel cells (SOFC) [1], oxygen separation [2-3], catalytic membrane reactors [4], and others. The mixed conducting membranes themselves are solid electrochemical cells, of which the bulk phase functions as the electrolyte of oxygen or hydrogen ions and the surface of the two sides as the electrodes. The electrochemical separation mechanism requires the membrane to be completely dense. As a result, an infinite permselectivity can be ensured on the basis of a gas-tight matrix. Besides the infinite permselectivity, to achieve a high flux, the membrane must be as thin as possible, and hence must be supported by a porous ceramic substrate, which is essential for cushioning the membrane as well as for guaranteeing mass transport in the feeding side. This paper uses perovskite oxide La$_{0.2}$Sr$_{0.8}$CoO$_{3-x}$ (LSCO) which has been known to have electronic-oxygen ionic mixed conductivity [5] as membrane material.

To fabricate a LSCO membrane on a porous support, the dip-coating method was used in the present work. A powder-packed layer is deposited on the porous support while it is dipped into a suspension consisting of LSCO powder and a volatile liquid. Particles of the layer are then consolidated via sintering at 1000°C to form a membrane. From the viewpoint of real applications where ceramic tubes are always the preferential support of ceramic membranes, the dip-coating way is of the simplest and most practical among the several known methods [6]. The main difficulties associated with the use of dip coating to make a dense ceramic film are a low powder-packed density in the deposit layer and the shrinkage mismatch between the membrane and the support while sintering. The first problem causes the difficulty to makes the membrane gas-tight. The second problem brings about defects and cracks in the membrane. This paper focuses on the second problem; a solution is proposed by means of the introduction of a silver-coating onto the powder-packed layer prior to sintering. The fabrication of a crack-free LSCO membrane on a porous MgO disk has been used as follows to demonstrate the idea.

Mat. Res. Soc. Symp. Proc. Vol. 606 © 2000 Materials Research Society

EXPERIMENT

Preparation of LSCO Powder

The Pechini method [6] was employed: EDTA (10.04 g, 34.4 mmol) was introduced with magnetic stirring into 25 ml H_2O placed in a 100-ml beaker. Aqueous ammonia solution was then added dropwise to covert EDTA into a water-soluble ammonium salt. To the resulting solution (pH ≈ 8 ~ 9), $La(NO_3)_3 \cdot 6H_2O$ (1.488 g, 3.436 mmol), $Sr(NO_3)_2$ (2.91 g, 13.75 mmol) and $Co(NO_3)_2 \cdot 6H_2O$ (5.00 g, 17.18 mmol) were added. A dark-brown solution was obtained, which changed gradually into a highly viscous gel under stirring on a hot plate at about 150 ~200°C. Following the formation of the highly viscous gel, the temperature of the hot plate was increased to about 400°C. After pyrolysis, a black powder was obtained. The powder was calcined under an airflow (5L/min) at 900 °C for 2 h in a Carbolite furnace to partially sinter the oxide powder. Then, the powder was ground and sieved by a sieve (Opening: 45 μm).

Preparation of the LSCO Suspension

A certain amount of LSCO powder was dispersed into a solution of polyvinylbutyral resin (Butvar-79, Monsanto) in a mixture of toluene and methylethyl ketone (MEK) to form a suspension. The deflocculating agent (Fish oil, 3-5 % by weight of the powder) was introduced into the suspension subsequently. The suspension was then ball-milled for three days to break up the agglomeration of the oxide particles. An ink-like colloidal suspension was therefore obtained.

Preparation of Porous Disks

A typical procedure is as follows: A MgO fine powder (Aldrich, -325 mesh) and a certain amount of carbon black powders were dry mixed and then introduced into a solution of Butvar-79 (2-3% by weight of the powder) in toluene / ethanol (v / v =1). The solvent was evaporated from the slurry with stirring. The dry lumps left were crushed by grinding and sieved until a fine powder was obtained. Green disks were prepared on a Carter's hydraulic press, which were then sintered at 1600°C for 1 h. The sintered MgO disks were polished firstly by grinding paper (Carbimet, grit 600), then further by polishing cloth (Buehler) to make a flat surface. Finally, the surface of the disks was carefully cleaned before dipping.

Fabrication of LSCO Membrane

A MgO disk was stuck onto an appliance with the polished surface up. Then the disk was horizontally dipped into a LSCO suspension until half of it is inside the suspension. After being held for a few seconds, the disk was gradually raised out of the suspension. The suspension was sucked into the substrate by the capillary force of the pores inside the disk. After the solvent was evaporated by a hair fan, a few repetitions of the dip coatings were carried out to obtain a powder-packed layer of a certain thickness. The obtained powder-coating layer was further covered by a silver layer via dipping in a high purity and well degassing silver paint (SPI). After the silver paint was dried, the disk was first slowly heated up to 400°C at the rate of 0.5°C/min, held for 1h, then to 1000°C at the rate of 2°C, held for another 1h. Finally a slow cooling rate (2°C/min) was set to cool down the item. A crack-free and mesoporous LSCO membrane was obtained.

RESULTS AND DISCUSSIONS

By using the dip-coating based ceramic membrane fabrication technique, it is imperative to carefully control the processing conditions that include the formulation of a colloidal suspension, the porous structure of a support and the sintering strategy. The experiments done were arranged to achieve the optimal conditions for each step of the processing.

Properties of the LSCO Suspension

In light of the coating suspension, it is known that the solid content of a suspension consisting of a powder and a liquid medium, the particle-size distribution of the powder, and the viscosity of the suspension affect the coating thickness. The coating thickness of the resultant powder-packed layer will eventually have an effect on the quality of the ceramic membrane. According to the Zeta potential measurement, the mean particle size of LSCO powder after a long ball-milling time is 1.4 μm and the related polydispersity is 0.038. Butvar-79, the binder of the LSCO powder-packed layer, exhibits a different viscosity with the change in solvent composition as shown in Figure 1. In contrast to the content of Butvar-79, the content of LSCO powder exerts a minor effect on the viscosity of suspension (Fig.2). It is found that the suitable range of the viscosity to attain an even casting layer is from 60 to 80 cp. For this reason, we selected a binary solution comprising of an equal volume of toluene and MEK as the solvent and 2 % by weight of the concentration of Butvar-79. The viscosity of the resulting solution is 71 cp. When LSCO powder (5 % by weight) is dispersed in the solution, the viscosity drops by 14 % due to the physical adsorption of the polymer chains on the particles. However, the viscosity increases with a rise in the solid content and reaches 76 cp at 20 % of solid content. It is likely that this dependency relates to the generation of an inter-particulate network with the particles as the crosslinking points and the adsorbed polymer chains as the connections. On the other hand, the higher the content of LSCO particles, the greater will be the packing density in the powder-packed layer. To fabricate a dense membrane, a high powder packing density is especially important. Although an increase in the solid content in the suspension can help to increase the packing density in the powder-packed layer, the appropriate solid content was found to be between 15 % and 20 % as far as the viscosity and the weight ratio of binder/powder are concerned. On the contrary, reduction in the particle size of LSCO powder has much larger room than the increase in the solid content to pursue a gas-tight LSCO membrane. We are inclined to believe that the mean particle size of LSCO powder that we prepared by following the Pechini method has already reach the limit of the method.

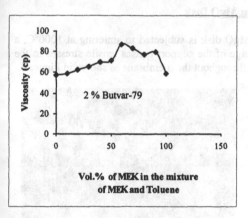

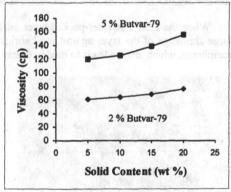

Fig. 1 *Solvent effect on the viscosity of the solution of Butvar-79 in toluene and MEK*

Fig. 2 *Variation of the viscosity with the content of the LSCO content in the suspension*

Characterizations of the Porous MgO Disks

The porosity of the MgO disks can be changed by either varying the content of pore former in the green body or the sintering temperature. Table I shows the relationship of the content of the pore former (carbon black) and the porosity of the sintered MgO disks. Figure 3 illustrates the change in the porosity of the MgO disk (based on 4% carbon blacks) in relation to the sintering temperature. Figure 4 shows the pore size distribution of the MgO disk (8% carbon and 1600°C), which has a porosity of about 37 %. This particular type of MgO disks was chosen as the support for LSCO membrane by compromising its porosity and the mechanical strength.

Table I. The porosity of MgO disks sintered at 1600°C

Content of Carbon black in the green disks (wt %)	4	8	12
Porosity (%)	28	37	44

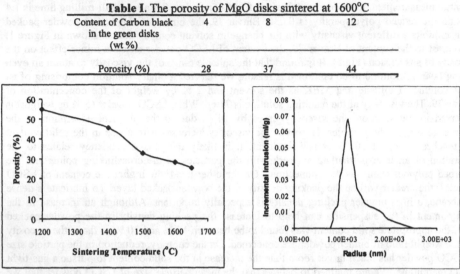

Fig.3 *Sintering curve of a porous MgO disk*   **Fig. 4** *Pore sizes distribution of MgO disk*

Formation of Crack-Free LSCO Membrane on MgO Disk

When the LSCO powder-packed layer on MgO disk is subjected to sintering at 1000°C, a large shrinkage of the layer against a nil shrinkage of the support causes tensile stresses in the membrane, which always leads to muddy cracks throughout the membrane as shown in Figure 4.

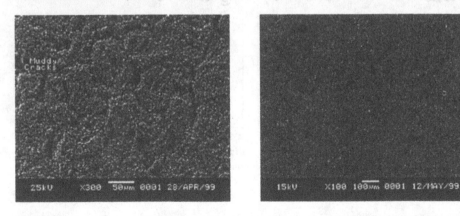

Fig. 4 *Muddy-cracks across a LSCO membrane formed on MgO disk*   **Fig. 5** *A muddy-crack free LSCO membrane formed on MgO disk*

To avoid the muddy cracks, a silver metal coating on the LSCO powder-packing layer was found effective. Figure 5 displays a crack-free LSCO membrane obtained after sintering. Silver has the melting point falling in the temperature range of sintering LSCO material and is inactive to LSCO material at high temperatures. At the initial stage of the sintering of LSCO, the silver topping is still in the form of a metal skin, which should shrink largely than the LSCO layer underneath. As a result, the compressive force acting on the LSCO layer can balance to a certain extent the tensile force imposed by the MgO disk. When Ag melts in a higher temperature stage, it converts to small liquid droplets. The droplets of Ag would automatically plug surface pores of the partially sintered LSCO layer. These pores are thought of the loci contributing mainly to the shrinkage. By virtue of the Ag-plugging in, the shrinking of the LSCO layer can be held up partly, which lessens the degrees of mismatch. In addition to this mechanical effect, the presence of silver, an excellent electronic conductor, can probably assist the entrance of oxygen molecules

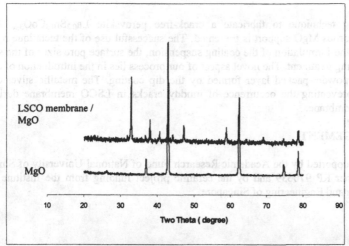

Fig. 6 *XRD of LSCO membrane on MgO disk*

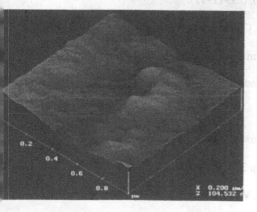

Fig. 7 *The Surface profile of the microdomain of LSCO membrane shown in Fig. 5*

Fig. 8 An *amplification of Fig. 5*

into the LSCO lattice via the reduction reaction, favouring a shift of the equilibrium between lattice oxygen and gaseous oxygen on the LSCO surface toward the lattice oxygen side. As a result, the composition stress caused by the loss of the surface lattice oxygen at high temperature can also be compensated. The composition stress is another possible origin leading to the cracks. It is likely that the mechanical effect and the cathodic effect of silver help LSCO membrane to avert muddy cracks. The silver beads left on the LSCO membrane after sintering can be removed chemically. The XRD of the resulting LSCO membrane / MgO-disk shows the perovskite structure of LSCO, which can be formed only when the annealing temperature is above 950°C. The AFM inspection displays a typical surface morphology resulted from the sintering of particles (Fig.7). A corresponding SEM picture (Fig. 8) is also provided to support this conclusion.

CONCLUSIONS

A processing technique to fabricate a crack-free perovskite $La_{0.2}Sr_{0.8}CoO_{3-x}$ (LSCO) membrane on a porous MgO support is presented. The successful use of the technique relies on the three factors: the formulation of the coating suspension, the surface pore sizes of the support and the pre-sintering treatment. The novel aspect of our process lies in the introduction of a silver coating onto the powder-packed layer formed by the dip coating. The metallic silver coating plays a role in preventing the occurrence of muddy cracks in LSCO membrane during the sintering of the membrane.

ACKNOWLEDGEMENT

This research is supported by the Academic Research Fund of National University of Singapore under grant number RP 970639 and by the ceramic project funding from the Institute of the Material Research and Engineering of Singapore.

REFERENCES

1. H. U. Anderson, Solid State Ionics, **52**(1-3), p 33 (1992).

2. J. Kilner, S. Benson, J. Lane and D. Waller, Chemistry & Industry, **Nov.** 17, p 907 (1997).

3. H. J. M. Bouwmeester and A. J. Burggraaf in *The CRC Handbook of Solid State Electrochemistry*, edited by P. J. Gellings and H. J. M. Bouwmeester, CRC Press, Boca Raton, 1997, pp 481-553.

4. E. Drioli and L. Giorno, Chemistry and Industry, **Jan.** 1, p19 (1996).

5. K. Keizer and H. Verweij, Chemtech, **Jan.** p 37 (1996).

6. *Ceramic Films and Coatings*, edited by J.B. Wachtman and R.A, Haber, Noyes Publications, Park Ridge, New Jersey, 1993.

7. M. P. Pechini, US Patent 3,330,697 (1967).

Alternative Chemical Processing Methods and Characterization of Electronic Ceramics

CHEMICAL SYNTHESIS OF PURE AND DOPED LaGaO₃ POWDERS OF OXIDE FUEL CELLS BY AMORPHOUS CITRATE/EG METHOD

A. C. TAS, H. SCHLUCKWERDER, P. MAJEWSKI, and F. ALDINGER
Pulvermetall. Lab., Max-Planck-Institut für Metallforschung, Stuttgart 70569, Germany

ABSTRACT

Powders of $LaGaO_3$, $La_{0.9}Sr_{0.1}GaO_{2.95}$ and $La_{0.8}Sr_{0.2}Ga_{0.83}Mg_{0.17}O_{2.815}$ were prepared by the amorphous citrate/EG method. The calcination behavior of the precursor powders of the above phases were studied in the temperature range of 200°-1400°C, in an air atmosphere. Characterization of the samples were performed by XRD, TG/DTA, FTIR, SEM, ICP-AES, and carbon and nitrogen analyses.

INTRODUCTION

Sr- and Mg-doped $LaGaO_3$ ceramics are known [1] to have superior oxygen ion conducting properties as compared, for instance, to yttria-stabilized zirconia electrolytes. In recent years, several researchers have synthesized the Sr- and Mg-doped $LaGaO_3$ ceramics by mainly using the conventional, 'solid-state reactive firing' method [2-4]. Stevenson, *et al.* [5] have tried the glycine-nitrate combustion synthesis for the preparation of Sr- and Mg-doped $LaGaO_3$ powders. The route of co-precipitation (w/NH₄OH addition) from an aqueous mixture of the acetates of La, Sr, and Mg, and of Ga-nitrate have also been attempted by Huang, *et al.* [6]. In this paper, a chemical preparation route is outlined for the synthesis of pure and doped $LaGaO_3$ ceramics. The product powders are fully characterized (phase distribution, particle size and morphology, IR behavior, thermogravimetric and chemical analyses).

EXPERIMENT

The powders were synthesized from $La(NO_3)_3.9H_2O$ (Merck, Germany), $Ga(NO_3)_3.xH_2O$ (Sigma, USA), $Sr(NO_3)_2$ (Merck) and $Mg(NO_3)_2.6H_2O$ (Merck). A Pechini-type process [7-8] was employed to prepare the polymeric precursors. For each precursor, nitrate salt solutions corresponding to a 5 g yield of $LaGaO_3$ ceramics were mixed in 50 mL of boiled, de-ionized water. Citric acid monohydrate (60 wt%)-ethylene glycol (40 wt%) mixture was added (mole ratio of citric acid-to-total cations=1.88) to the cations solution. The resulting solution was evaporated until first a yellow gel, then a dark brown resin, formed. The resins were finely ground by using an agate mortar/pestle, and then calcined in air at various temperatures (200°-1400°C). Each calcination batch was heated to the specified temperature at a rate of 5°C/min, annealed at this temperature for 6 h, and furnace cooled to room temperature. The phase distribution in the powders was analyzed as a function of calcination temperature using a Siemens D-5000 X-ray diffractometer and CuKα radiation (30kV, 20 mA). Pyrolysis and decomposition of 150 mg of the ground, amorphous resin samples were monitored by simultaneous differential thermal and thermogravimetric analysis (STA501, Bahr GmbH) in air at a rate of 5°C/min. FTIR analyses of the samples were performed (IFS 66, Bruker GmbH) by mixing them (1 wt%) with dry KBr to form the pellets. The residual C and N contents of the uncalcined and calcined powders were studied by the combustion-IR absorption method (CS-800, Eltra GmbH). ICP-AES analyses were performed to get the quantitative elemental information (JY-70Plus, Instruments SA). The morphology of the powders was studied by scanning electron microscopy (DSM 982 Gemini, Zeiss GmbH).

RESULTS AND DISCUSSION

The phase evolution and distribution behaviors of pure (i.e., LG: $LaGaO_3$) and doped (i.e., LSG: $La_{0.9}Sr_{0.1}GaO_{2.95}$ and LSGM: $La_{0.8}Sr_{0.2}Ga_{0.83}Mg_{0.17}O_{2.815}$) lanthanum gallate samples are depicted by the XRD spectra of Fig.1. LGsamples (after calcination in air at 1200°C for 6 h) were found to be orthorhombic with the lattice parameters of a=5.489, b=5.519, c=7.751 Å, which are in close agreement

237

A. OCZAŚ, ..., .. ŁUKOWSKA, ..., .. MAJEWSKI, and P. VLEUGELS
Published by ... Max Planck Institut für Silberwang, but not 20500, Germany

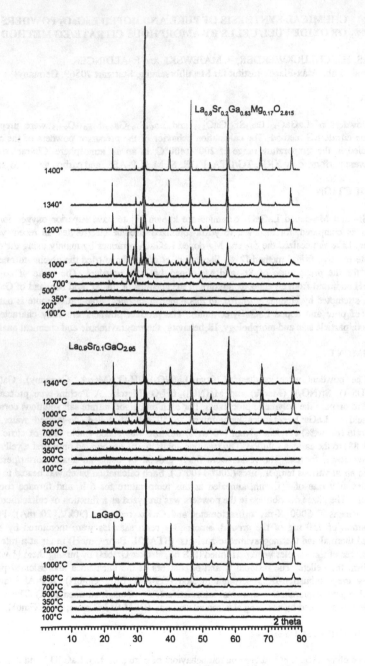

Fig. 1 XRD spectra of precursor powders of LG, LSG and LSGM samples

238

with the previously reported values [3]. LSG samples (after calcination at 1340°C for 6 h) were also orthorhombic with the lattice parameters of a=5.491, b=5.523, c=7.764 Å. Although the single-phase LG was formed from the starting X-ray amorphous resins at about 850°C, single-phase LSG could only be produced after calcination at 1340°C for 6 h. The resins of the LSG stoichiometry, after calcination at 1200°C, still contained the secondary phases of $La_4Ga_2O_9$ and $SrLaGa_3O_7$. The resins of the LSGM stoichiometry, on the other hand, contained about 3-4 wt% of the above-mentioned secondary phases even after calcination at 1400°C for 8 h. The crystal structure of the LSGM samples of this study were found to be non-cubic (i.e., orthorhombic), in sharp contrast to many earlier reports [2-4, 6], and this finding about the crystal structure of LSGM is in accordance with the recent study of Ishihara, *et al.* [9].

The results of the simultaneous TG/DTA analyses are given in Fig. 2. The DTA traces of LSG and LSGM resins showed exotherms at ~400°, 490°, and 605°C. The first exotherm was associated with charring of the polymer, the second with the pyrolysis of the organics, and the last one resulted from char burnout. TG analysis showed that most of the weight loss occurred at temperatures between 250° and 500°C. This corresponded to the range where polymer burnout occurred. The results of residual carbon analyses (as a function of calcination temperature) are shown in Table 1. The nitrogen content of the 100°C-calcined precursor samples were found to be in the range of 1.05 to 1.2 wt%, but with increasing calcination temperatures (starting from 200°C) it decreased to levels below the reliable detection level (i.e., 100 ppm) of the equipment used.

ICP-AES analysis results (in terms of mole ratios) of the 1340°C-calcined samples of LSG (i.e., La/Sr=8.984, La/Ga=0.912) and LSGM (La/Sr=3.990, La/Ga=0.958, La/Mg=4.693) compositions confirmed the theoretical mole ratios of elements present. Amorphous citrate/EG method, therefore, is shown to be able to yield LSG and LSGM ceramics of high elemental uniformity.

Table I. Results of residual carbon analyses (wt%)

Temp. (°C)	$LaGaO_3$	$La_{0.9}Sr_{0.1}GaO_{2.95}$	$La_{0.8}Sr_{0.2}Ga_{0.83}Mg_{0.17}O_{2.815}$
100	31.7 *(3)*	33.3 *(2)*	32.7 *(6)*
350	10.3 *(3)*	10.4 *(1)*	13.8 *(2)*
700	0.530 *(3)*	0.550 *(5)*	0.280 *(4)*
850	0.059 *(1)*	0.143 *(9)*	0.168 *(3)*
1000	0.042 *(2)*	0.050 *(3)*	0.060 *(2)*
1340	0.010	0.0124 *(2)*	0.0143 *(4)*

FTIR plots (as a function of calcination temperature) of the LG, LSG, and LSGM samples are given in Fig. 3. The broad band at 3500-2500 cm^{-1} is due to O-H stretching. The presence of the citrate ion was detected by the band at 2990-2874 cm^{-1} at low calcination temperatures. Dissolved or atmospheric CO_2 was indicated by the band at 2350 cm^{-1}. The carboxylate anion (COO$^-$) stretching was seen by two bands in the range of 1740-1380 cm^{-1}, and the structural CO_3^{-2} was observed by the broad band at 1500-1300, also at 1080-1030, and 800 cm^{-1}. The existence of a covalent carbonyl bond (C=O stretching vibration) was found by the bands at 1730-1700 and 1190-1075 cm^{-1}, and the bands at 1440-1300 and 1070-1030 cm^{-1} indicated the trace presence of nitrate ions in the low temperature (100°C) samples. Because of the appearance of bands due to CO_2 adsorbed on the metal cations, the decrease in the intensity of the bands due to citrate/carboxylate groups and the lower intensity of the water stretching band, one may conclude that the metal-carbonyl links begin to break after heating to >350°C. Samples calcined at 700°C basically show the CO_3^{-2} and OH$^-$ ions in their IR spectra. After calcination at 1200°C, all IR bands attributed to anion vibrations disappeared in all samples.

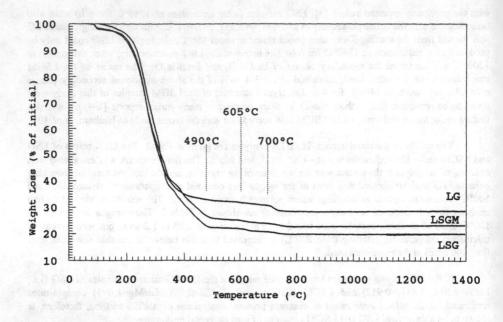

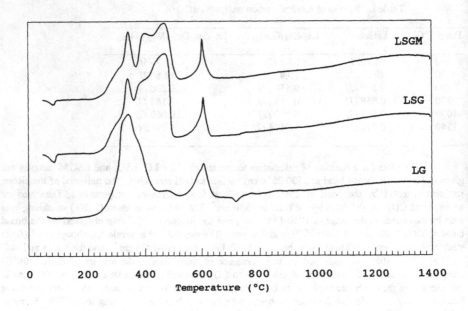

Fig. 2   TG (*top*) and DTA (*bottom*) spectra of precursors of LG, LSG and LSGM samples

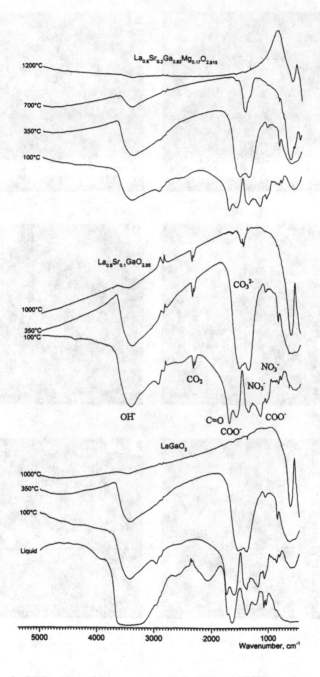

Fig. 3 FTIR spectra of the precursors of LG, LSG and LSGM samples

241

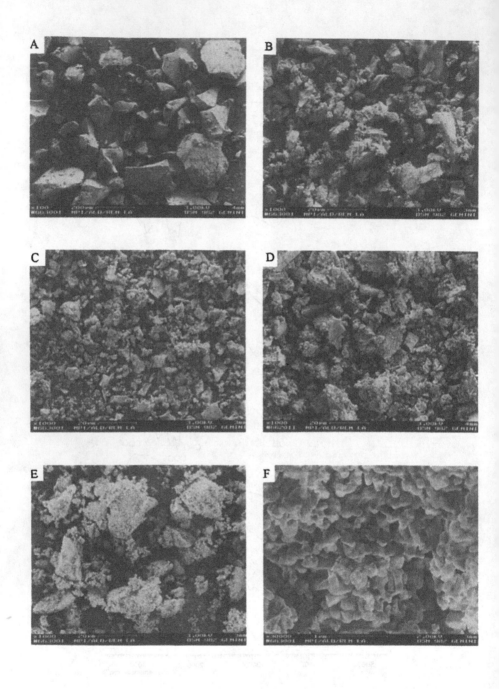

Fig. 4   SEM micrographs of LSGM samples (a) 100°, (b) 350°, (c) 500°, (d) 700°, (e) & (f) 1000°C

242

The powder morphology of LSGM powders, as a function of calcination temperature, was given by the SEM pictures of Fig. 4. The initially X-ray amorphous particles (of irregular morphology) began decomposing above 500°C (Figs. 4(c) and 4(d)), and after calcination at 1000°C, the bigger chunks consisted of smaller particles (~0.2 μm), fused together (Figs. 4(e) and 4(f)). The fracture surface of a pellet of LSGM powders (CIPped at 650 Mpa and then calcined at 1400°C for 8 h) indicates (Figs. 4(g) and (h)) that after sintering, the final grain size was still ≤ 2 μm.

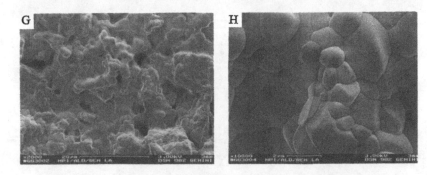

Fig. 4   (*cont.*) SEM micrographs of LSGM pellets; (g) and (h) 1400°C, 8 h

CONCLUSIONS

A Pechini-type, polymeric precursor route has been studied for the synthesis of powders of LG, LSG and LSGM. Although single-phase LG and LSG powders were successfully synthesized after calcination at 800° and 1340°C, respectively; the LSGM powders still contained 3-4 wt% of secondary phases in them after calcination at 1400°C for 8 h. According to the results of FTIR analyses, LSG samples have shown, at all temperatures, an increased affinity for the atmospheric CO_2, as compared to LG powders. Calcination of LG, LSG and LSGM precursors at temperatures in excess of 1000°C was found to be necessary to reduce the C-content to the ppm levels.

REFERENCES

1.   T. Ishihara, H. Matsuda, and Y. Takita, J. Am. Chem. Soc., **116**, 3801 (1994).
2.   M. Feng and J. B. Goodenough, Eur. J. Sol. State Inorg. Chem., **31**, 663 (1994).
3.   P. Huang and A. Petric, J. Electrochem. Soc., **143**, 1644 (1996).
4.   K. Huang, R. S. Tichy, and J. B. Goodenough, J. Am. Ceram. Soc., **81**, 2565 (1998).
5.   J. W. Stevenson, T. R. Armstrong, W. J. Weber, J. Electrochem. Soc., **144**, 3613 (1997).
6.   K. Huang, M. Feng, and J. B. Goodenough, J. Am. Ceram. Soc., **79**, 1100 (1196).
7.   M. Pechini, U.S. Patent No. 3 330 697 (11 July 1967).
8.   P. A. Lessing, Am. Ceram. Soc. Bull., **68**, 1002 (1989).
9.   P. R. Slater, J. T. S. Irvine, T. Ishihara, and Y. Takita, J. Sol. State Chem., **139**, 135 (1998).

SILICON CLEANING METHODS COMPARED at METAL CONCENTRATIONS BELOW 1E10 atoms/cm²

Joseph Ilardi, Rajananda Saraswati, George Schwartzkopf
Mallinckrodt Baker, Inc., Research and Technology Development Department, Phillipsburg, NJ

ABSTRACT

An alkaline aqueous silicon wafer cleaner has been developed which reduces trace metal contamination levels on "p-type" unpatterned silicon wafers to below 1E10 atoms/cm² without acidic cleaning. This patented technology uses a specially formulated buffering system consisting of an oxidation resistant chelating agent and salt of a weak acid. The aqueous cleaner maintains a stable pH of about 9.5 over a wide range of dilutions, temperatures, hydrogen peroxide concentrations and bath aging times. A single-step bath, megasonic bath or spray clean with this formulation leaves the chemical oxide on the silicon wafer surface free of particles, atomically smooth, free of organics and lower in trace metal contamination levels than similar surfaces cleaned with conventional formulations.

An analytical method was developed which allows the reliable detection of trace metals on silicon wafer surfaces down to 1E8 atoms/cm² for aluminum, calcium, copper, iron, nickel, sodium and zinc. The procedure uses an ICP-MS with a concentric nebulizer and a desolvator. The acids used were ultrapure to keep the blanks to a very minimum and analyses were run in a class 10 clean room. The trace metals on the wafer were extracted using known amounts of ultrapure acids and were directly aspirated using a special arrangement with the concentric nebulizer. The J.T. Baker cleaner was compared to and outperformed the conventional RCA-1&2 clean and dilute RCA-1&2 chemistries using ultrapure ammonium hydroxide, hydrochloric acid and hydrogen peroxide.

INTRODUCTION

Trace metal contamination has long been known to affect gate oxide integrity. The reduction of trace metal concentrations in process chemicals is becoming more important as integrated circuit feature sizes and gate oxide thickness decrease. Pre-gate silicon wafer cleaning to remove trace metals usually employs the RCA clean [1] or one of its variants. This two-step process uses an ammonia containing hydrogen peroxide solution (RCA-1) followed by aqueous hydrogen peroxide and hydrochloric acid mixtures (RCA-2). The effects of various RCA chemistries on gate oxide integrity have been studied [2,3] with the usual emphasis on metals such as calcium, copper, iron, sodium and zinc. Aluminum is an important metal contaminant that is frequently omitted perhaps because of its low sensitivity to TXRF detection.

For a silicon cleaning process to be successful it must leave the resulting thin chemical oxide in a condition that will insure the integrity of the gate. The cleaned chemical oxide should have a low surface microroughness, low particle count, low trace metal concentration and be free of organic contamination. Recent thermodynamic investigations [4] have shown that Al_2O_3 can exist as a stable phase within a silicon dioxide matrix especially if the SiO_2 has come into contact with an aqueous ammonium hydroxide/hydrogen peroxide solution, i.e. RCA-1. Once alumina becomes embedded in the oxide it cannot be easily removed even by the action of hydrochloric acid, for example with RCA-2. The presence of alumina in the cleaned chemical oxide can generate flaws in the subsequent gate oxide.

Mat. Res. Soc. Symp. Proc. Vol. 606 © 2000 Materials Research Society

This study describes a unique aqueous alkaline cleaning process [5,6] which effectively replaces the separate basic and acidic steps of the RCA process with a highly effective single step. Trace metals were determined using a novel ICP-MS technique. The cleaned wafers were scanned with a mixture of ultrapure acids and the resulting droplet was directly aspirated from the wafer surface into the ICP instrument avoiding any environmental contamination normally associated with the droplet handling procedure. This resulted in analytical sensitivities approaching 1E8 atoms/cm^2.

EXPERIMENTAL

All cleaning experiments were carried out in a Class 100 Cleanroom using 75 mm diameter p-type prime silicon wafers. Cleaning solutions were brought to 70°C in an agitated bath which was continuously filtered through a 0.1 μm filter. The five cleaning sequences compared were: 1. JTB-111; 2. RCA-1 followed by RCA-2; 3. RCA-1 only; 4. very dilute RCA-1 using ultrapure ammonium hydroxide and hydrogen peroxide; 5. very dilute ultrapure hydrochloric acid. Table I further describes the cleaning mixtures studied.

Table I. Cleaning Mixtures Tested (grade indicates the maximum concentration of any one metal)

Cleaning Mixture	Ratio	Chemical Grade	Components
JTB-111	1:0.22:5	10 ppb	JTB-111:hydrogen peroxide:water
(a) RCA-1	1:1:5	10 ppb	ammonium hydroxide:hydrogen peroxide:water
(b) RCA-2	1:1:5	10 ppb	hydrochloric acid:hydrogen peroxide:water
RCA-1	1:1:5	10 ppb	ammonium hydroxide:hydrogen peroxide:water
Dilute RCA-1	1:1:50	100 ppt	ammonium hydroxide:hydrogen peroxide:water
Dilute HCl	1:51	100 ppt	hydrochloric acid:water

The wafers were placed in the hot cleaning solution for ten minutes, removed, and rinsed for two minutes at room temperature in a cascade DI water rinse vessel followed by an ultrapure nitrogen drying step. For the experiments in which an RCA-1 treatment was followed by an RCA-2 treatment, the wafers were rinsed in DI water between treatments and directly placed into the RCA-2 solution without drying. After cleaning, the wafers were placed in specially cleaned single wafer holders and moved to our Class 10 Ultraclean Analytical Laboratory for analysis.

The surface of each treated wafer was scanned with a small droplet of an ultrapure aqueous HF/HCl mixture. The presence of HCl in the droplet assured that any copper present in the scanning droplet remained dissolved without the deposition of elemental copper onto the silicon surface. This scanning process consists of the systematic movement of the droplet across the entire wafer surface. Extra care must be employed to achieve the complete scanning of the entire wafer surface. After scanning the wafer surface for several minutes, a portion of the droplet was directly aspirated into the ICP-MS thereby eliminating sources of contamination.

A Hewlett Packard HP4500 ICP-MS equipped with a concentric nebulizer and desolvator was used. This unit was also equipped with a gold/platinum shield on the torch for cold plasma operations eliminating the formation of argon oxide which normally interferes with the detection of

^{56}Fe and ^{40}Ca isotopes. Whenever possible, two isotopes were used for each analyte. Obtaining identical results for different isotopes of an element indicated that isobaric interferences were absent. Three replicates were performed for each instrumental measurement, and results were calculated from the mean and standard deviation of the mean of the analyte and internal standard signals. A calibration check of the instrument was performed before and after each sample measurement. All standard solutions were spiked with the same internal standard mix and diluted in the same manner as the sample matrix. The acids and deionized water used for diluting the standards and samples were J.T. Baker Ultrex II grade. The described method improvements in the form of lower blanks and higher instrumental sensitivity were achieved through the use of cold plasma, and facilitated the reporting of results approaching 1E8 atoms/cm^2.

RESULTS

Experimental results are shown in Tables II and III. Except for the RCA-1&2 cleaning process, all other treatments evaluated consisted of a single cleaning step followed by a single rinse step. A value of 1E10 atoms/cm^2 was chosen, at or above which the cleaner was considered to be ineffective. Rather than subtracting the analyzed values for the blank (scanning) acid, these values were converted to surface concentrations permitting direct comparisons of the results without error propagation.

Table II . ICP/MS Results for Scanning Droplets (in ppt - average of four wafers for treatments; average of three acids for the blanks)

	Blank Acids	JTB-111	RCA-1&2	RCA-1 only	Dilute RCA-1	Dilute HCl
Na	14	10	400	1300	6	2
Al	22	10	25	3000	40000	4000
Ca	14	680	1160	990	400	400
Fe	14	64	41	120	4400	2300
Ni	10	11	10	70	6	6
Cu	4	8	0	20	100	37
Zn	14	40	120	1200	40000	1

1. Aluminum

The highest performance cleaner for removing aluminum contamination was JTB-111. The order of effectiveness was JTB-111>RCA-1&2>RCA-1>dilute HCl>dilute RCA-1. We attribute the effectiveness of JTB-111 for removing and/or preventing aluminum incorporation into the oxide surface to its alkalinity and sequestering capability. The dilute RCA-1 left aluminum levels on the wafers almost two orders of magnitude higher than the incoming wafers. The dilute HCl left approximately three times the starting aluminum contamination.

247

Table III. Results Calculated as Surface Concentrations (E8 atoms/cm^2, average of four wafers or three blanks, ± one standard deviation shown for the JTB-111 treatments)

	Blank Acids	JTB-111	RCA-1&2	RCA-1 only	Dilute RCA-1	Dilute HCl
Na	39	29±1	1200	3700	18	4
Al	53	24±2	60	8000	100000	10000
Ca	23	1100±100	1900	1600	600	600
Fe	16	75±6	48	140	5100	2700
Ni	11	12.0±0.6	11	80	6	7
Cu	4	8±4	0	20	110	38
Zn	14	40±10	120	1200	40000	1.3

2. Copper

All treatments studied gave acceptable copper values. The full RCA clean gave slightly better copper removal than JTB-111. Both the full RCA and JTB-111 treatments outperformed RCA-1, dilute RCA-1 and dilute HCl.

3. Iron

Three of the treatments studied left the wafer surfaces with acceptable iron levels. The performance order of these cleans was RCA-1&2>JTB-111>RCA-1. Surprisingly, dilute HCl removed iron poorly giving values greater than 1E11 atoms/cm^2.

4. Sodium

Dilute HCL and dilute RCA-1 as well as JTB-111 all showed acceptable sodium values of less than 1E10 atoms/cm^2.

5. Zinc

Of the five cleaning formulations studied only the dilute ultrapure HCl and JTB-111 reduced zinc levels to below 1E10 atoms/cm^2.

6. Nickel

All five cleaners were capable of leaving less than 1E10 atoms/cm^2 of this metal on the surface. The poorest performer for nickel was RCA-1.

7. Calcium

All five treatments left more than 1E10 atoms/cm^2 of calcium on the wafer surface. The calcium concentration of the untreated wafers was about 1E12 atoms/cm^2 which may be an unrealistic calcium challenge for these cleaners.

These data were evaluated to determine an order of effectiveness for the five cleaners studied. Disregarding calcium, of the five cleaners studied, only JTB-111 reduced all six metals to below 1E10 atoms/cm^2. The two acid-last treatments, RCA-1&2 and dilute HCl lowered four out of six metals to acceptable levels. The other alkaline treatments tested, 1:1:5 RCA-1 and 1:1:50 RCA-1 only reduced two out of six metals to less than 1E10 atoms/cm². Figure 1 illustrates that overall, JTB-111 gave the cleanest surfaces.

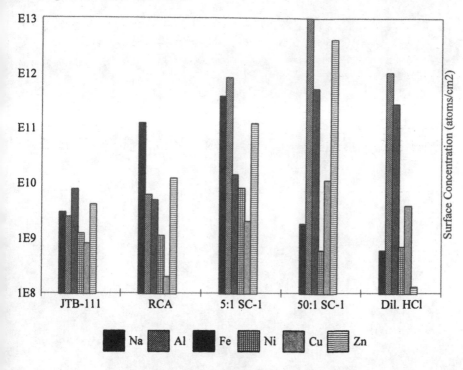

Figure 1. Comparison of Treatments

CONCLUSIONS

An analytical methodology has been developed which provides reliable trace metal data approaching 1E8 atoms/cm². This technique was used to prove that a properly formulated alkaline cleaner will outperform conventional two step or dilute RCA formulations made with ultrapure chemicals as well as ultrapure, dilute HCl.

REFERENCES

1. W. Kern, D.A. Puotinen, RCA Review 31 (2), 187 (1970).
2. C. P. D'Emic, S. Cohen, M. A. Zaitz in Science and Technology of Semiconductor Surface Preparation, edited by G. S. Higashi, M. Hirose, S. Raghavan, S. Verhaverbeke,
(Mater. Res. Soc. Proc. 477, Pittsburgh, PA, 1997) pp. 233-239.
3. T. Ohmi, M. Miyashita, M. Itano, T. Imaoka, I. Kawanabe, IEEE Trans. on Elec. Dev. 39 (3), 537 (1992).
4. J. N.Gerd, L. C. Kimerling, J. Electronic Materials, Vol. 24, No. 4, 1995 pp. 397-404.
5. W. A. Cady, M. Varadarajan, J. Electrochem. Soc., Vol. 143, No. 6, p. 2,064 (June 1996).

6. G. Chen, T. Gilton, B. Cartensen, W. Lee, G. Waldo, G. Mitchell, W. A. Cady, W. C. Greiner, J. M. Ilardi, (to be published in Mater. Res. Soc. Proc., Pittsburgh, PA, 1999).

β-ACETOXYETHYL SILSESQUIOXANES: CHLORIDE-FREE PRECURSORS FOR SIO₂ FILMS VIA STAGED HYDROLYSIS

Karin A. Ezbiansky*, Barry Arkles**, Russell J. Composto*, Donald H. Berry*,
*Department of Chemistry and Department of Materials Science and Engineering, University of Pennsylvania, Philadelphia, Pennsylvania 19104-6323, **Gelest, Inc. 612 William Leigh Drive, Tullytown, PA 19007-6308.

ABSTRACT

Silsesquioxanes containing β-acetoxyethyl (BAE) groups are processible resins that can be employed as spin-on-glass precursors to dielectric silica films. Thermal treatment >250 °C results in extrusion of ethylene from the $CH_2CH_2OCOCH_3$ moiety with formation of Si-$OCOCH_3$ groups, which undergo facile hydrolysis to a silica network. A minor pathway involving extrusion of acetic acid leaves some silicon vinyl groups, leading to residual organic carbon in the material. However, addition of a fluoride ion catalyst greatly accelerates the major reaction, resulting in lower conversion temperatures (<200 °C), quantitative extrusion of ethylene, and essentially pure silica. Alternatively, BAESSQs can be processed photochemically (λ<200 nm) to cleanly yield silica at ambient temperature.

INTRODUCTION

The spin-on-glass (SOG) approach to ceramic films involves coating a surface with a soluble precursor solution and subsequent thermal conversion to the desired material. The inherent processibility of the precursors makes this approach attractive for gap-filling and planarization applications. Typical SOG precursors are based on sol-gel chemistry, in which hydrolysis of silicon alkoxides produces meta-stable hydroxy-substituted species, which subsequently undergo condensation to a three-dimensional network - i.e. silica or related materials. A significant advantage of sol-gel processing is the relatively low reaction temperatures required compared with other methods.[1,2] On the other hand, SOG precursors based on sol-gel chemistry usually require highly unstable solutions generated by the hydrolysis of alkoxysilanes, which have limited shelf-life and necessitate removal of large amounts of alcohol and water during curing.

We previously reported a new approach to SOG materials based on "hydrolysis in stages" using β-chloroethylsilsesquioxanes (BCESSQ) as the silica precursor.[3] This contribution describes the preparation and processing of a new chloride-free analogue, β-acetoxyethylsilsesquioxane (BAESSQ).

EXPERIMENT

All solvents and reagents were used as received without further purification. Thermogravimetric analysis (TGA/MS) analyses were performed on a Texas Instrument SDT 2960 simultaneous DTA-TGA with a Fison Thermalab mass spectral analyzer. Ceramic residue yields are reported as the percentage of the sample remaining after completion of the heating cycle. Film compositions were determined by RBS using 2.0 MeV He⁺ ions for silicon and oxygen, and 3.4 MeV He²⁺ ions for carbon as described previously.[4] X-ray powder diffraction spectra of the ormosil products were obtained on a Rigaku Geigerflex powder X-ray diffractometer using Cu K α-radiation with a graphite monochromator. AFM images were obtained in tapping mode.

Preparative scale tube furnace pyrolyses were performed in boron nitride or silica boats in a quartz tube wrapped with heating tape (190 °C) or in Lindberg 55035, 54233 and 54434 tube furnaces equipped with Eurotherm temperature controllers (350 and 1200 °C). Thermolyses were performed under a stream of air saturated with water. Samples were dried at 200 °C under flowing dry air or vacuum prior to analysis. Elemental analyses were performed at the Nesmeyanov Institute of Organoelement Compounds (INEOS), Moscow, Russia.

RESULTS

β-Acetoxyethylsilsesquioxanes were prepared from the triethoxide via conventional hydrolysis routes.[1,2] The material is a pale-yellow viscous resin. The thermal behavior of the resin was examined by TGA/MS. The first major onset of weight loss occurs at ca. 250 °C, and is accompanied by evolution of ethylene and acetic acid (Figure 1). This is similar to the chloride analogue,[3] except the onset of weight loss is ca. 100 ° higher for the BAESSQ.

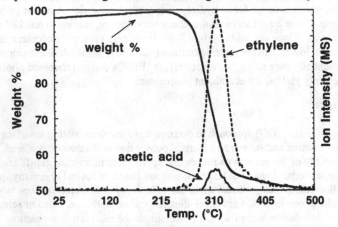

Figure 1: TGA/MS of BAE resin (10 °/min)

The products of the bulk thermolysis of β-acetoxyethylsilsesquioxane in a tube furnace at various temperatures were examined by elemental analysis (Table I) and XRD. The products treated at 350 and 450 °C are orange-brown solids with calculated empirical formulas $Si_{1.0}O_{2.2}C_{0.43}H_{1.63}$ and $Si_{1.0}O_{2.1}C_{0.34}H_{1.0}$, respectively.

Table I: Composition of BAESSQ processed without catalyst.

Processing Temp. (°C)	% C	% H	% Si	% O[a]
350	7.34	2.36	40.20	50.12
450	6.14	1.56	42.56	49.75
1200	4.87	-	46.08	49.05

[a] % oxygen calculated by difference.

It was previously shown by CP-MAS studies of the BCESSQ derived materials that residual carbon and hydrogen are present as vinyl and methyl groups.[5,6] Assuming similar behavior for the BAE resin suggests the analogous mechanism shown in Figure 2.

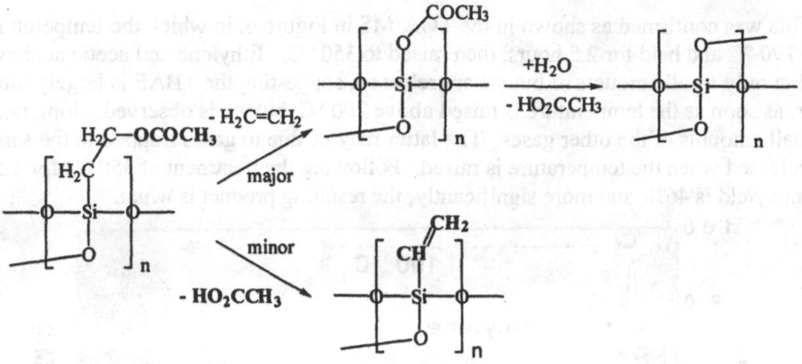

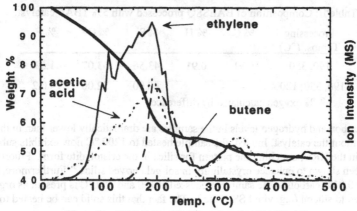

Figure 2: Proposed mechanism for uncatalyzed thermal conversion of BAESSQ.

A sample of BAE resin heated to 1200 °C for 2 hours resulted in a black residue that is amorphous (XRD). This is fully consistent with conversion of a silica containing some organic groups to silicon oxycarbide or "black glass", the well-known product of organosilsesquioxane pyrolysis.[7,8]

Although residual organic carbon is not inherently detrimental for all potential applications, greater control over the thermal conversion process was desired. Thus, catalysts for the BAE decomposition were explored. Addition of 5 mol % tetrabutylammonium fluoride (TBAF) to the BAE resin results in a striking change in the thermal behavior observed in the TGA/MS (Figure 3). Some initial weight loss is associated with residual water and solvent introduced with the TBAF. However, ethylene and acetic acid evolution is rapid starting ca. 100 °C. As the temperature approaches 200 °C, weight loss accelerates, and a new species, butene, is observed in the MS. Butene is the most volatile product of the known decomposition of TBAF. The mass (54%) is fairly constant up to 300 °C, at which point another burst of ethylene and acetic acid is observed. These results suggest the following: (1) TBAF is an effective catalyst for the conversion of BAE groups <200 °C; (2) TBAF decomposes rapidly at 200 °C; and (3) any BAE groups remaining after catalyst decomposition undergo uncatalyzed conversion >250 °C. The obvious implications are that processing temperatures below the TBAF decomposition threshold will be most effective, and TBAF can be subsequently removed by heating >200 °C.

Figure 3: TGA/MS of BAE resin with 5% TBAF.

This was confirmed as shown in the TGA/MS in Figure 4, in which the temperature was raised to 190 °C and held for 2.5 hours, then raised to 350 °C. Ethylene and acetic acid evolve rapidly, but only small amounts of butene are released, suggesting the TBAF is largely intact. However, as soon as the temperature is raised above 200 °C, butene is observed, along with rather small amounts of the other gases. The latter may be due to gases trapped in the sample that are released when the temperature is raised. Following the treatment at 350 °C for 1.5 hours, the ceramic yield is 46%, and more significantly, the resulting product is *white*.

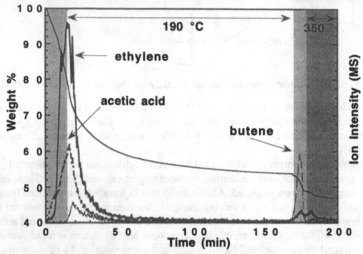

Figure 4: TGA/MS of BAESSQ with 5% TBAF held at 190 °C for 2.5 hours, then heated to 350 °C.

Bulk pyrolysis of BAE resin with 5% TBAF was performed following similar parameters as the TGA/MS (2.5 hours at 190 °C, 1.5 hours at 350 °C). A second sample was treated identically, but then heated at 1200 °C for 2 hours. The resultant solids from both experiments were white in color, and the elemental analyses are summarized in Table II.

Table II: Composition of BAESSQ processed with 5% TBAF catalyst.

Processing Temps. (°C)	% C	% H	% Si	% O[a]	% F
190, 350	0.91	0.91	43.56	53.07	1.53
190, 350, 1200	-	-	48.00	52.00	< 0.2

[a] % oxygen calculated by difference.

The carbon and hydrogen levels in the samples are dramatically lower than in those processed without the catalyst. Indeed, the sample heated to 1200 °C now exhibits substantial crystallinity in the XRD, matching the pattern for silica in the cristobalite form. Fluoride has previously been shown to enhance crystallinity in sol-gel derived silica.[9] Furthermore, it is not clear that the 0.91% carbon in the sample processed at 190 and 350 °C is present as organic groups bonded to silicon (e.g. vinyl SSQ, etc.) The fact that this solid can be heated to 1200 °C without significant conversion to the intensely black oxycarbide strongly suggests any residual carbon is in a more labile form.

Preliminary experiments confirm that BAESSQ can be used as a SOG to prepare films. Solutions of BAESSQ in diglyme (1 - 40 wt%) were deposited on silicon substrates by spin-coating to yield as-cast films ranging in thickness from 200 - 6000 Å. Processing at 350 or 450 °C for 4 hours resulted in ca. 50% decrease in thickness, although there was little change after 20 min. Analysis of the films by AFM and SEM reveals extremely smooth, featureless surfaces (RMS roughness ~ 0.35 nm). Film compositions (RBS) correlate closely to the elemental analysis measured for the bulk samples. A similar correlation between bulk and film composition was previously observed in the case of BCESSQ. Although dielectric measurements on these films have yet not been obtained, it is worth noting that films prepared from BCESSQ exhibited dielectric constants ~3.5 (1 MHz).[10] This value is intermediate between that of thermally grown oxide (~4.0) and methyl SSQ (~2.7), and is consistent with the composition - i.e. mainly SiO_2 with some organic SSQ sites. We are hopeful that the films prepared from the chloride-free BAE analog will exhibit comparable dielectric behavior. Studies of films prepared from BAESSQ with TBAF catalyst are also underway.

Finally, we have previously demonstrated that BCESSQ films can be converted to essentially carbon-free films at ambient temperatures using photochemical irradiation (λ<200 nm).[10] BAESSQ films behave similarly, and exposure to 185 nm UV radiation produces smooth films with carbon levels below the detection limits of RBS (~1 atom%). However, BAESSQ films require ca. 2x longer exposure times than BCESSQ films to obtain complete conversion.

CONCLUSIONS

β-Acetoxyethylsilsesquioxane (BAESSQ) is a soluble resin that undergoes conversion to silica-rich materials in bulk or thin films under relatively mild thermal or photochemical conditions. Onset of thermal conversion occurs above 250 °C, but results in ~10% residual carbon as organic SSQ sites. Photochemical treatment (193 nm) yields substantially cleaner conversion. The thermal conversion of BAESSQ is dramatically accelerated in the presence of TBAF, allowing processing <200 °C and yielding a silica with <1% carbon, which does not appear to be covalently bound to silicon. Studies employing this TBAF catalyzed BAESSQ SOG to prepare dielectric films are currently in progress.

ACKNOWLEDGMENTS
We would like to thank the National Science Foundation for support of this work under the MRSEC program (DMR 96-32598).

REFERENCES

1.   U. Schubert, N. Hüsing, and A. Lorentz, Chem. Mater. **7**, 2010 (1995).

2.   C. Sanchez and F. Ribot, New J. Chem. **18**, 1007-1047 (1994).

3.   B. Arkles, D. Berry, L. Figge, R. Composto, T. Chiou, H. Colazzo, J. Sol-Gel Sci. Tech. **8**, 465-469 (1997).

4.   Q. Pan, G. Gonzalez, R. Composto, W. Wallace, B. Arkles, L. Figge, D. Berry, Thin Solid Films. **345**, 244-254 (1999).

5. L. Figge, PhD thesis, University of Pennsylvania, 1996.

6. K. Ezbiansky, L. Figge, B. Arkles, R. Composto, D. Berry, manuscript in preparation.

7. F. I. Hurwitz, P. Heimann, S.C. Farmer, D. M. Hembree, J. Mater. Sci. **28**, 6622 (1993).

8. L. Bois, J. Maquet, F. Babonneau, Chem Mater. **7**, 975-981 (1995).

9. L. Villaescusa, P. Barrett, M. Camblor, Chem. Mater. **10**, 3966-3973 (1998).

10. J. Sharma, D. Berry, R. Composto, H. Dai, J. Mater. Res. **14** (3), 990-994 (1999).

HIGH DENSITY PLASMA ETCHING OF Ta$_2$O$_5$-SELECTIVITY TO Si AND EFFECT OF UV LIGHT ENHANCEMENT

K.P Lee*, H.Cho*, R. K. Singh*, S. J. Pearton*, C.Hobbs** and P.Tobin**
* Department of Materials Science and Engineering, University of Florida, Gainesville FL 32611
** APDRL, Motorola, Austin, TX 78721

ABSTRACT

Etch rates up to 1200 Å·min^{-1} for Ta$_2$O$_5$ were achieved in both SF$_6$/Ar and Cl$_2$/Ar discharges under Inductively Coupled Plasma conditions. The etch rates with N$_2$/Ar or CH$_4$/H$_2$/Ar chemistries were an order of magnitude lower. There was no effect of post deposition annealing on the Ta$_2$O$_5$ etch rates, at least up to 800 °C. Selectivities to Si of ~1 were achieved at low source powers, but at higher powers the Si typically etched 4-7 times faster than Ta$_2$O$_5$. UV illumination during ICP etching in both SF$_6$/Ar and Cl$_2$/Ar produced significant enhancements (up to a factor of 2) in etch rates due to photo-assisted desorption of the TaF$_x$ products. The UV illumination is an alternative to employing elevated sample temperatures during etching to increase the volatility of the etch products and may find application where the thermal budget should be minimized during processing.

INTRODUCTION

Ta$_2$O$_5$ is an important material for storage capacitors in dynamic random access memories [1-5], as well as being a potential replacement for SiO$_2$ in metal-oxide-semiconductor transistors [1-4,6-9]. There has been a lot of work on the correlation between deposition conditions and electrical properties and on use of post-growth annealing in O$_2$ ambients to reduce the leakage currents in capacitor structures [10-12]. Ta$_2$O$_5$ has also found application as an etch mask during surface or bulk micromachining of Si [13], as an insulating in thin film electroluminescent display devices [14,15] and as a detection layer in sensors [16]. There has been relatively little work on dry etch patterning of Ta$_2$O$_5$ [17,18], which is the preferred method of forming small structures due to the difficulty in wet etching [19,20]. Previous reports have found relatively slow etch rates for Ta$_2$O$_5$ in fluorocarbon-based [17,18] plasma chemistries such as CF$_4$, C$_2$F$_6$, CHF$_3$ and CF$_3$Cl.

In this work we report on the use of ultraviolet (UV) light irradiation during Inductively Coupled Plasma (ICP) etching to produce enhancements in etch rate in both Cl$_2$- and F$_2$- based chemistries. Maximum etch rates of ~1200Å·min^{-1} in SF$_6$/Ar were achieved with UV irradiation to enhance etch product desorption.

EXPERIMENTAL

Amorphous Ta$_2$O$_5$ films ~1000Å thick were deposited on (100), p-type (B-doped, 1 Ω-cm) Si wafers. The precursors used were Ta(C$_2$H$_5$O)$_5$ and O$_2$. The deposition temperature was ~350°C. Post-growth annealing at 600-800 °C for 1 min in an O$_2$ ambient was performed to examine whether there was any effect of film densification on the etching properties, as previously reported for wet etching experiments [20]. The 800 °C annealing is close to the recrystallization condition.

Samples ~5x5 mm^2 were masked with Apiezon wax for etch rate experiments. The etch depth was established by stylus profilometry to an accuracy of ±10% after removal of the mask in acetone. All of the etching was performed in a Plasma Therm 790 series reactor. The samples were thermally bonded to a Si carrier wafer on a He backside cooled, rf powered (13.56 MHz) chuck. This power was used to control the incident ion energy. The ion flux and plasma dissociation was controlled by the power into the ICP source (2 MHz, 100-1000W). Gas injection into this source was metered through electronic mass flow controllers at a typical load of 15 standard cubic centimeters per minute (sccm).

In some cases a Hg arc lamp (400 W) was installed on 1 inch diameter quartz window on the top of the ICP source, ~20 cm from the sample position. This was used to provide UV illumination of the sample surface during plasma etching. Sample heating due to the lamp is minimal (<10 °C), as determined by a fluoro-optic probe mounted on the Si carrier wafer. The absorption depth of the UV light is expected to be ~100 Å when TaClx or TaFx covers the surface [21], so that virtually all the incident light makes it to the etching surface.

RESULT AND DISCUSSION

Table I summarizes the maximum etch rates achieved for Ta_2O_5 in the different plasma chemistries investigated, along with the selectivity over Si. Both SF_6- and Cl_2-based mixtures produce etch rates at least an order of magnitude higher than $CH_4/H_2/Ar$ or N_2/Ar, and there is always reverse selectivity with respect to Si i.e. the latter etches 2-6 times faster than Ta_2O_5 in all of the plasma chemistries. There was little difference in the results obtained with O_2 or Ar addition to SF_6, even though O_2 addition is often found to enhance the atomic neutral fluorine density in

Table I. Comparison of maximum etch rates achieved for ICP etching of Ta_2O_5 in different plasma chemistries, and selectivity over Si under same conditions.

Plasma Chemistry	Maximum Ta $_2O_5$ Etch Rate (Å·min^{-1})	Ta $_2O_5$ Selectivity over Si
SF_6/Ar	1200	0.16
SF_6/O_2	1200	0.20
SF_6/CH_4	1000	0.25
SF_6/H_2	950	0.16
Cl_2/Ar	1050	0.32
$CH_4/H_2/Ar$	50	0.45
N_2/Ar	100	0.65

F_2-based discharges. In an attempt to slow the Si etch rate and possibly achieve selective Ta_2O_5-to Si etching, we tried both CH_4 and H_2 addition to SF_6 to enhance polymer deposition on the Si.

Figure 1 shows a comparison of Ta_2O_5 and Si etch rates, with the resultant selectivities, as a function of discharge composition in SF_6/Ar. The etch rates for Ta_2O_5 increase more slowly with increasing SF_6 concentration than do the rates for Si, with the consequence that the maximum selectivity for Si over Ta_2O_5 is achieved in pure SF_6 discharges. The dc self-bias increases as the SF_6 concentration increases, indicating that the positive ion density is decreasing. Note that Ta_2O_5 etch rates of ~1200 Å·min^{-1} are achieved in the SF_6-based mixtures, at self-biases in the range −215 to −290 V. We could not detect the etch products for Ta_2O_5 with optical emission spectroscopy, but assume they are probably TaF_x and O_2. In the case of Si, we readily observed the SiF_x etch products, with emission lines in the range 400-430 nm.

The effect of ICP source power on the etch rates is shown in Figure 2 for fixed plasma composition. The Si etch rate increases either modestly or not at all over the range 300-1000 W, while the Ta_2O_5 etch rate tends to decrease at the higher powers. This decrease is at least partially caused by the fall-off in dc self-bias, which is suppressed as the positive ion density in the discharge increases at high powers. It is clear that because Si etches in atomic fluorine even without ion bombardment, whereas Ta_2O_5 does not, that there will always be a faster etch rate for the former in non-polymer-forming plasma chemistries.

The results for Cl_2/Ar etching at fixed source power (750 W) and rf chuck power (250 W) are shown in Figure 3 as a function of discharge composition. The rates increase as Cl_2 is added, which is often an indicator of some chemical contribution to the etch mechanism. Indeed the enhancement factor is 10-20 for etch rates obtained in pure Cl_2 relative to pure Ar, much larger than expected based on a purely physical sputtering enhancement due to the heavier mass of Cl_2^+ (74) versus Ar^+ (40), which are the most abundant ions present. The ratio of molecular to atomic chlorine is in the range 5-10 under our conditions. The selectivity for Ta_2O_5 over Si is in the range 0.3-1. Similar maximum etch rates (~1200 Å·min⁻¹) for Ta_2O_5 in Cl_2 were obtained as with SF_6 discharges.

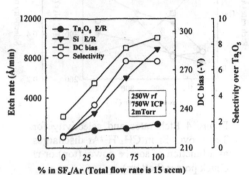

Figure 1. Etch rates and selectivities for Si over in SF_6/Ar discharges (250 W rf chuck power, 750 W source power, 2 mTorr), as a function of SF_6 percentage.

The $CH_4/H_2/Ar$ chemistry has proven useful for dry etching materials such as InSnOx and III-V compound semiconductors (GaN, GaAs, InP) through the formation of metal organic and hydride etch products. In the case of Ta_2O_5, possible products include $(CH)_xTa_y$, H_2O and O_2. We examined the effect of varying source and chuck power, at fixed discharge composition (chosen from past experiments in other materials). Figure 4 shows that under all of our conditions, Ta_2O_5 etches at a slower rate than Si.

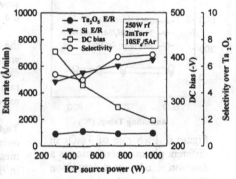

Figure 2. Etch rates and selectivity for Si over Ta_2O_5 in SF_6/Ar discharges (250 W rf chuck power, 2 mTorr) as a function of ICP source power.

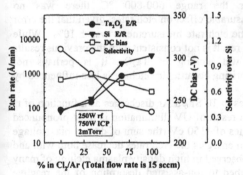

Figure 3. Etch rates and selectivity for Ta_2O_5 over Si as a function of Cl_2 percentage in Cl_2/Ar discharges (250 W rf chuck power, 750 source power, 2 mTorr).

Another possible etch chemistry is based on N_2/Ar, in the hope that reactive nitrogen neutrals would extract oxygen as N_2O or NO_2, with the Ta being removed by Ar^+ sputtering. the etch rates for Ta_2O_5 are again very low, and slower with N_2/Ar mixtures than with simple Ar^+ sputtering. We also examined the rate of source power and chuck power, but again always found that the Ta_2O_5 etched

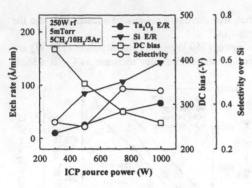

Figure 4 Etch rates and selectivity for Ta₂O₅ over Si in CH₄/H₂/Ar discharges as a function of source power (top) or rf chuck power (bottom).

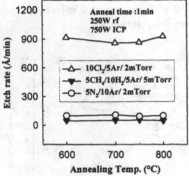

Figure 6. Etch rate of Ta₂O₅ in different plasma chemistries as a function of post-deposition annealing temperature.

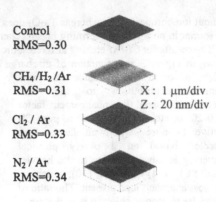

Figure 5. AFM scans of Ta₂O₅ surfaces before and after etching in CH₄/H₂/Ar, Cl₂/Ar or N₂/Ar discharges (350 W rf chuck power, 750 W source power, 2 mTorr).

slower than Si, with selectivities for Ta₂O₅/Si of 0.4-0.8.

The etched Ta₂O₅ surface morphologies with the CH₄/H₂/Ar, N₂/Ar and Cl₂/Ar chemistries were similar to those of control samples, as shown in the example of Figure 5. This is consistent with equirate removal of the Ta and O, or else one would expect significant surface roughening.

Figure 6 shows the effect of annealing of the Ta₂O₅ on the dry etch rates. Figure 6 shows that over the range 600-800 °C there was no measurable effect on etch rates, given that the error in the etch rate measurements was ± 10%. While this result is not consistent with the previous results on wet etching of Ta₂O₅, it is perhaps not surprising because dry etching is less influenced by material quality.

A final point of interest is the Ta₂O₅ etch rates in 10 SF₆/5 Ar discharges as a function of rf chuck power. The enhancement in etch rates as a result of UV illumination is most pronounced at low chuck powers, corresponding to ion energies of ~150 eV (the sum of dc self-bias voltage and the plasma potential of ~22 eV). At higher ion energies the etch rates decrease both with and without the UV illumination. This is commonly observed in high density plasma etching of many different materials systems, and is usually ascribed to ion-assisted desorption of the reactive neutrals before they can react with the atoms in the substrate [22,23].

To investigate the effect of ion flux on the Ta₂O₅ etch rate, we varied ICP source power for two different chuck powers, as shown in Figure 8. At the low chuck power condition (100 W, top), the etch rate enhancement with UV illumination increases with source power and reaches a factor of approximately two in the range 500-750W. By sharp contrast, at the higher rf chuck power condition (Figure 8, bottom) there is essentially no increase in etch rate as a result of UV illumination. Several groups have reported that UV irradiation dramatically enhances the etch rate of Cu in Cl_2-based high density plasmas, through absorption of UV photons by involatile $CuCl_x$ etch products, transformation to more volatile species (e.g. Cu_2Cl_3) and subsequent non-thermal desorption of these species [24,25]. In the case of Ta_2O_5, an analogous situation would involve photodesorption of TaF_x species since the oxygen should be removed as O_2 or oxyfluorides. The etch rate of the Ta_2O_5 was too low for us to detect the etch products by

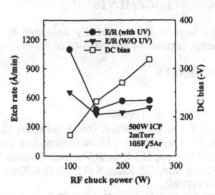

Figure 7. Ta_2O_5 etch rate with and without UV illumination in ICP 10 SF_6/Ar discharges (500 W source power, 2 mTorr) as a function of rf chuck power.

optical emission spectroscopy, although we could determine that the atomic fluorine concentration was not increased by UV illumination, ruling this out as a possible mechanism for the higher rates.

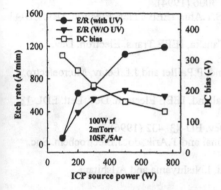

SUMMARY AND CONCLUSIONS

A number of different plasma chemistries were examined for dry etching of Ta_2O_5 under ICP conditions. The maximum etch rates (~1200 $Å \cdot min^{-1}$) were achieved with SF_6 or Cl_2 based chemistries, while the rates with $CH_4/H_2/Ar$ or N_2/Ar were at least an order of magnitude lower. Under practical etching conditions, the etch rates for Si were always lower than for Ta_2O_5 (up to a factor of 6), and we were unable to achieve selectivity for Ta_2O_5 over Si. The etched surface morphologies were smooth over a wide range of plasma conditions and chemistries.

UV illumination during ICP etching of Ta_2O_5 in both SF_6/Ar and Cl_2/Ar plasma chemistries produces significant enhancements in etch rates. The increased rates are likely due to photo-assisted desorption of the $TaCl_x$ and TaF_x etch products. The use of UV illumination is an alternative to

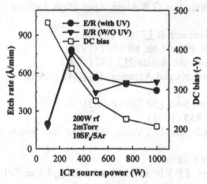

Figure 8. Ta_2O_5 etch rate with and without UV illumination in ICP 10 SF_6/5Ar discharges as a function of source power for two different rf chuck powers, 100 W (top) and 200 W (bottom).

employing elevated sample temperatures during etching to increase the volatility of the etch products and may find application where the thermal budget should be minimized during processing.

ACKNOWELDGEMENTS

The work at UF is partially supported by a DOD MURI monitored by AFOSR (H. C. DeLong), contract F49620-96-1-0026.

REFERENCES

1. See for example, C.Chaneliere, J.L.Autran, R.A.B.Devine and B.Balland, Mat. Sci. Eng. Rep. R22 269 (1998).
2. J.H.Yun and S.W.Rhee, Thin Solid Films 292 324 (1997).
3. B.C.Lai and J.Y.Lee, J.Electrochem. Soc. 146 226 (1999).
4. S.O.Kim and H.J.Kim, J. Vac. Sci. Technol. B12 3006 (1994).
5. K.W.Kwon, C.S.Kang, S.O.Park, H.K.Kang and S.T.Ahn, IEEE Trans. Electron. Dev. ED-43 919 (1996).
6. T.Mizuno, T.Kobori, Y.Saitoh, S.Sawada and T.Tanaka, IEEE Trans. Electron. Dev. ED 39 4 (1992).
7. R.A.B.Devine, C.Chaneliere, J.L.Autran, B.Balland, P.Paillet and J.L.Letay, Microelectro. Eng. 36 61 (1997).
8. J.L.Autran, R.A.B.Devine, C.Chaneliere and B.Balland, IEEE Electron. Dev. Lett. EDL-18 447 (1997).
9. H.Shimada and T.Ohmi, IEEE Trans. Electron. Dev. ED-43 432 (1996).
10. T.Aoyama, S.Saida, Y.Okayama, M.Fujisuki, K.Imai and T.Arikado, J.Electrochem. Soc. 143 977 (1996).
11. S.Kamiyama, P.Y.Lesaicherre, H.Suzuki, A.Sakai, I.Nishiyama and A.Ishitani, J.Electrochem. Soc. 140 1617 (1993).
12. J.P.Chang, M.L.Steigerwald, R.M.Fleming, R.L.Opila and G.B.Alers, Appl. Phys. Lett. 74 3705 (1999).
13. A.K.Chu, Y.S.Huang and S.H.Tang, J. Vac. Sci. Technol. B 17 455 (1999).
14. K.Kukli, J. Ihanees, M.Ritala and M.Leskela, Appl. Phys. Lett. 68 3737 (1996).
15. J.Sun, G.Zhong, X.Fan, G.Fu and C.Zhong, J.Non-Cryst. Solids 212 192 (1997).
16. D.H. Kwon, B.W.Cho, C.S.Kim and B.K.Sohn, Sensors and Actuators B34 441 (1996).
17. Y.Kuo, J.Electrochem. Sco. 139 579 (1992).
18. S.Seki, T.Unagami and B.Tsujiyama, J.Electrochem. Soc. 130 2505 (1983).
19. C.H.An and K.Sugimoto, J.Electrochem. Soc. 139 853 (1992).
20. K.W.Kwon, C.S.Kang, T.S.Park, Y.B.Sun, N.Sandler and D.Tribula, Mat. Res. Soc. Symp. Proc. 284 505 (1993).
21. Handbook of Optics, ed. M.Bass (McGrow-Hill, NY 1995).
21. R.J.Shul, M.Lovejoy, D.L.Hetherington, D.J.Rieger, J.F.Klem and M.R.Melloch, J.Vac. Sci. Technol. B 13 27 (1995).
23. O.A.Popov (ed), High Density Plasma Sources (Noyes Publishing, Park Ridge, NY (1994).
24. K-S.Choi and C-H.Han, J. Electrochem. Soc. 145 L37 (1998).
25. M.S.Kwon and J.Y.Lee, J.Electrochem. Soc. (in press).

STUDY ON ZrO$_2$ DEPOSITED DIRECTLY ON Si AS AN ALTERNATIVE GATE DIELECTRIC MATERIAL

WEN-JIE QI, RENEE NIEH, BYOUNG HUN LEE, YOUNGJOO JEON, LAEGU KANG, KATSUNORI ONISHI, AND JACK C. LEE
Microelectronics Research Center, The University of Texas at Austin, Austin, TX 78758

ABSTRACT

Reactive-magnetron-sputtered ZrO$_2$ thin film has been deposited on Si directly for gate dielectric application. Both structural and electrical properties of the ZrO$_2$ film have been investigated. An amorphous structure for 30Å ZrO$_2$ and a semi-amorphous structure for 200Å ZrO$_2$ have been revealed. The sputtered film shows a good stoichiometry and a good structural stability of ZrO$_2$ based on the X-ray photoelectron spectroscopy and Rutherford backscattering spectroscopy data. Thin equivalent oxide thickness of about 11.5Å was obtained without the consideration of quantum mechanical effects. A low leakage of less than 10^{-2} A/cm^2 at ±1V relative to the flat band voltage was obtained for this 11.5Å equivalent oxide thickness Pt/ZrO$_2$/Si structure. High effective dielectric breakdown and superior reliability properties have been demonstrated for ZrO$_2$ gate dielectric.

INTRODUCTION

High dielectric constant (high-k) materials have received a lot of attention recently for application as alternative gate dielectrics. ZrO$_2$ seems to be a promising candidate for gate dielectric application. Based on a thermodynamic study [1], ZrO$_2$ is stable in contact with Si. This indicates that there is no reaction between ZrO$_2$ and Si, and therefore, ZrO$_2$ can be deposited directly on Si without a barrier layer. This also suggests that polysilicon may still be usable for ZrO$_2$ gate dielectric without using an upper barrier layer. Both can significantly simplify the process and are major advantages over some other high-k materials such as TiO$_2$ [2,3], Ta$_2$O$_5$ [4,5], and strontium titanate [6]. ZrO$_2$ has a dielectric constant of about 25, a wide bandgap of around 5.16eV to 7.09eV [7-9], and good chemical inertness. In fact, ZrO$_2$ also has many excellent mechanical and optical properties such as hardness and high refractive index.

There have been a wide variety of studies on ZrO$_2$ in the past. For years, researchers have been investigating ZrO$_2$ for mechanical and optical applications. There have been many reports on its structural properties [10-14]. For electronics applications, reports on ZrO$_2$ as an insulator can be found in the literature [15-18]. Unfortunately, all these earlier works focused on relatively thick films (>300Å) and not much attention was paid to the interface properties which are most critical for gate dielectric applications. Therefore, ZrO$_2$ is still a "new" material as a gate dielectric, and it is primarily the thermodynamic stability that makes ZrO$_2$ an attractive candidate. In this paper, we investigated ZrO$_2$ as an alternative gate dielectric material. Material and electrical properties will be presented. By using ZrO$_2$, equivalent oxide thickness of as thin as 11.5Å with a low leakage was obtained. ZrO$_2$ also exhibits excellent reliability properties.

EXPERIMENTS

P-Si (100) with a resistivity of 5~25 Ω·cm was used as the substrate. Field oxide of 3500~4000Å was grown and patterned by photolithography to form the active area. Before ZrO$_2$ deposition, the wafers were piranha cleaned, dipped in HF solution ′ .40), rinsed in DI water, and spin-dried with N_2. ZrO_2 was reactive-magnetron-sputtered from a 4-inch diameter Zr target (99.7% pure). The base vacuum of the sputtering chamber is about 2×10^{-7} Torr. The target-to-substrate distance is about 30cm. The substrate can be heated by lamps mounted above the wafer, and sputtering temperature ranged from room temperature to 300°C. The pressure was 40mTorr, and the power ranged from 200W to 400W. The ZrO_2 films were deposited in the following manner. Initially, a thin layer of Zr was sputtered in an Ar ambient. Next, 2sccm of O_2 was introduced to the chamber during sputtering in order to oxidize the Zr. The total pressure during the Ar + O_2 sputtering remained at 40mTorr, and there was some ZrO_2 deposition occurring at this time. The initially sputtered Zr layer serves as a barrier layer for oxidation. This process also prevents the target from being oxidized. Each sputtering cycle consists of one Ar-only sputtering and one Ar + O_2 sputtering and oxidization process. After ZrO_2 sputtering, the film was furnace annealed or rapid thermal annealed in a N_2 ambient. Pt was sputtered as the electrode. It was patterned by lithography and etched with aqua regia solution (H_2O:HCl:HNO_3 = 5:7:1) at 80°C. Al was evaporated on the backside of the wafer to ensure a low contact resistance. ZrO_2 material properties were characterized by ellipsometer, X-ray diffraction (XRD), Transmission Electron Microscopy (TEM), Rutherford Backscattering Spectroscopy (RBS), and X-ray Photoelectron Spectroscopy (XPS). Capacitance-Voltage (C-V), Leakage-Voltage (I-V), and time-dependent dielectric breakdown (TDDB) were measured using HP4156 and 4194.

RESULTS AND DISCUSSIONS

Structural and material properties of sputtered ZrO_2 thin film

Fig. 1 shows the refractive index of sputtered ZrO_2 after furnace annealing for 5minutes in N_2 ambient. For films sputtered at 300W and 300°C, the refractive index is close to the ideal (2.2); while for films sputtered at 200W and room temperature, the refractive index is lower. For both films, the refractive index remains fairly stable even after 800°C annealing. Russak et al [10] reported that for 2000Å thick ZrO_2, higher sputter rates resulted in higher refractive index and denser film. The sputter rate at 200W is about 17.5Å/min while at 300W is about 85.0Å/min. The higher sputter rate contributes to the higher refractive index. Since annealing of the film did not change the refractive index significantly, the higher refractive index likely comes from the higher sputter rate at 300W, rather than the higher sputtering temperature.

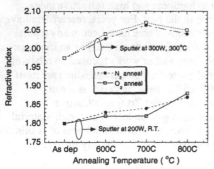

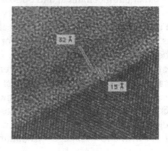

Fig. 1. The refractive index of sputtered ZrO_2 films, higher sputter rate results in higher refractive index.

Fig. 2. TEM pictures of ZrO_2/Si after 550°C, 5', N_2 anneal. An amorphous ZrO_2 structure can be seen with a 15Å interfacial layer.

The crystallinity of ZrO_2 was also studied. For gate dielectric application, an amorphous structure is desirable for low leakage and high impermeability to impurities. For thicker films (>300Å), polycrystalline structure was reported [7, 10-12, 15-17]. Here we investigated thin films down to 30Å. Fig.2 shows a TEM picture of ~30Å ZrO_2/Si structure. The ZrO_2 film was annealed at 550°C for 5 minutes. An amorphous structure can be seen. In fact, this layer remains amorphous even after at least 700°C anneal. We believe that the major difference between previous studies and our study is the film thickness. For thinner films, the nucleation barrier height may be higher because of the surface and interface energy. At the same time, when the film thickness is comparable to the dimension of a nucleus, it would be more difficult for the nucleus to form and hence grow. In fact, the grain size from the literature [7,15,16] was reported to be from 50Å to 600Å; even the smallest grain size reported is larger than the film thickness we are dealing with here. Similar phenomena occurred in $TiSi_2$ for the C49 to C54 phase transformations [19]. Fig. 3 shows the XRD spectra for a 200Å ZrO_2 film on Si after furnace annealing from 600°C to 900°C. It shows cubic crystal structure, although monoclinic structure is typical for bulk ZrO_2 below 1400K. This is consistent with previously reported sputtered ZrO_2 film [10]. From the XRD patterns, the intensities of the diffraction peaks remain almost the same, which indicates that there is no significant grain growth upon high temperature annealing. Similar results were reported by M. Balog et al [7]. Fig. 4 is the TEM picture of 200Å ZrO_2 on Si after 900°C, 5 minutes annealing. It shows a semiamorphous structure with partially crystallized ZrO_2 film. This is consistent with the relatively weak diffraction shown in the XRD spectra. It should be noted that for ZrO_2 film on Si substrate there is an interfacial layer, which is grown due to the excess oxygen in the deposition chamber or during the annealing, and NOT the reaction between ZrO_2 and Si. With process optimization, this interfacial layer could be minimized, and this interfacial layer should be helpful to reduce the interface states.

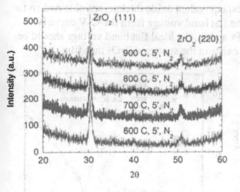

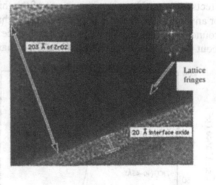

Fig. 3. XRD patterns of 200Å ZrO_2 on Si show a cubic crystal structure with no significant grain growth after 900°C, 5 minutes annealing.

Fig. 4. TEM picture of 200Å ZrO_2 on Si shows a semi-amorphous structure and an interfacial layer. Inset is the selected electron diffraction showing spots and diffuse rings.(900°C, 5', N_2 annealing)

Fig. 5 gives the XPS spectra of ZrO_2. Typical ZrO_2 chemical bonding can be observed. The stoichiometry was further confirmed by RBS. Fig.6 gives the RBS spectra of ZrO_2 deposited on Si that shows $Zr_{0.33}O_{0.67}$, and the 1:2 stoichiometry within the resolution of RBS measurement.

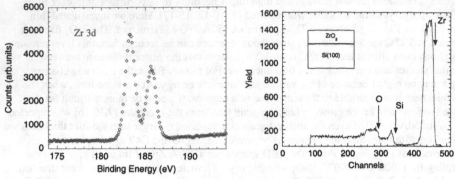

Fig. 5. XPS spectra indicate a typical ZrO_2 chemical bonding.

Fig. 6. RBS (Channeling) spectra show the ZrO_2 is stoichiometry.

Electrical properties of Pt/ZrO₂/p-Si structure

Well-behaved C-V characteristics were obtained for the Pt/ZrO₂/p-Si MOS structure as shown in Fig.7. The physical thickness of ZrO_2 is around 40Å, and the film was rapid thermal annealed at 600°C for 30 seconds in N_2 ambient. From the accumulation capacitance at –3V, an equivalent oxide thickness of 11.5Å can be obtained without taking the quantum mechanical effects into consideration. This is one of the thinnest equivalent oxide thicknesses obtained so far for any high-k materials reported. Also note that the flat band voltage from the C-V curve is around 0V. From the work function difference of Pt and Si, the ideal flat band voltage should be about 0.4V. This indicates that some fixed charge exists in the sputtered ZrO_2 thin film.

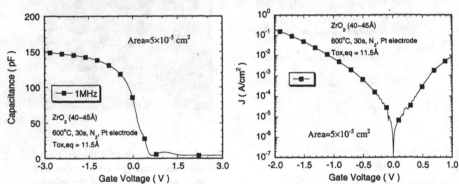

Fig. 7. High frequency C-V characteristics of Pt/ZrO₂/Si structure.

Fig. 8. The leakage property of the Pt/ZrO₂ /Si with 11.5Å equivalent oxide thickness.

Fig. 8 demonstrates the J-V curve of the 11.5Å Pt/ZrO₂/Si MOS structure. A leakage current of less than 10^{-2} A/cm² at ±1V can be seen. Considering the flat band voltage of 0V, the leakage values are taken at 1V above and below the flat band voltage. This low leakage is a result of the rather thick physical thickness of the ZrO₂, the wide bandgap, and the amorphous state of the film.

The long-term reliability properties of ZrO₂ were characterized. Fig. 9 shows the time-to-breakdown distributions under constant voltage stress measured at room temperature. The equivalent oxide thickness for this film is 15.8Å without the quantum mechanical consideration. The stress field was calculated based on the breakdown voltage and equivalent oxide thickness. Fairly tight distribution was observed for different stress fields. It can be seen that the effective stress fields were very high (>25MV/cm), which indicates that by using high-k films, the breakdown field is significantly improved. Fig. 10 shows the mean-time-to-breakdown (MTTF) as a function of the stress voltage (V_{DD}). Projected operating voltage for 10-year lifetime is – 2.49V, which is much higher than the 1997 SIA roadmap predictions (operating voltage of 0.9-1.2V for 1.5-2.0nm equivalent oxide thickness). These results suggest that ZrO₂ exhibits good reliability properties.

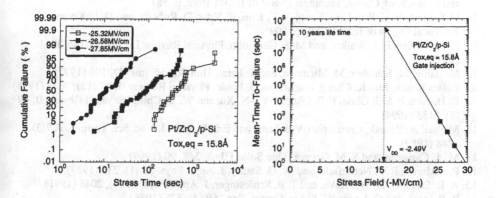

Fig. 9. Time-to-breakdown distribution for Pt/ZrO₂/Si with 15.8Å equivalent oxide thickness.

Fig. 10. Mean-time-to-failure as a function of stress field. A high V_{DD} of −2.49V can be extrapolated for 10-year life time.

CONCLUSION

ZrO₂ as an alternative gate dielectric material has been demonstrated. Experimental results indicate that higher sputter rates result in a higher refractive index and denser film. XPS and RBS data suggest that the sputtered ZrO₂ has a stoichiometric 1:2 Zr to O ratio. TEM and XRD results revealed that for ultra-thin ZrO₂ film (30Å), an amorphous structure was obtained. For film thickness of about 200Å, a semi-amorphous structure was observed. The film remains stable with no significant grain growth and phase transformation up to at least 900°C. C-V results of Pt/ZrO₂/Si show that a 11.5Å equivalent oxide thickness has been obtained with a low leakage of less than 10^{-2}A/cm² at ±1V relative to the flat band voltage. ZrO₂ exhibits superior reliability such as high effective dielectric breakdown and excellent TDDB properties. These results suggest that ZrO_2 holds good promise as the alternative gate dielectric for future ULSI devices.

ACKNOWLEDGEMENTS

The authors appreciate Dr. Vidya Kaushik from Motorola at Austin for many useful discussions and suggestions, as well as the help in TEM analysis. This work is partially supported by SRC/SEMATECH through the FEP research center.

REFERENCES

1. K. J. Hubbard, and D. G. Schlom, J. Mater. Res., **11**(11), 2757 (1996)
2. Xin Guo, T. P. Ma, T. Tamagawa, and B. L. Halpern, Technical Digest of IEDM 1998, p.377
3. B. He, T. Ma, S. A. Campbell, and W. L. Gladfelter, Technical Digest of IEDM 1998, p. 1038
4. Donggun Park, Qiang Lu, Tsu-Jae King, Chenming Hu, Alexander Kalnitsky, Sing-Pin Tay, and Chia-Cheng Cheng, Technical Digest of IEDM 1998, p. 381
5. H. F. Luan, B. Z. Wu, L. G. Kang, B. Y. Kim, R. Vrtis, D. Roberts, and D. L. Kwong, Technical Digest of IEDM 1998, p. 609
6. R. A. McKee, F. J. Walker, and M. F. Chisholm, Physical Review Letters, **81**(14), 3014 (1998)
7. M. Balog, M. Schieber, M. Michman, and S. Patai, Thin Solid Films, **47**, 109 (1977)
8. Balazs Kralik, Eric K. Chang, and Steven G. Louie, Physical Review B, **57** (12), 7027 (1998)
9. R. H. French, S. J. Glass, F. S. Ohuchi, Y. –N. Xu, and W. Y. Ching, Physical Review B, **49** (8), 5133 (1994)
10. Michael A. Russak, Christopher V. Jahnes, and Eric P. Katz, J. Vac. Sci. Technol. A **7**(3), 1248 (1989)
11. M. A. Cameron, and S. M. George, Thin Solid Films, **348**, 90 (1999)
12. P. J. Martin, R. P. Netterfield, and W. G. Sainty, J. Appl. Phys. **55** (1), 235 (1983)
13. A. K. Stemper, D. W. Greve, and T. E. Schlesinger, J. Appl. Phys. **70** (4), 2046 (1991)
14. R. P. Ingel, and D. Lewis III, J. Am. Ceram. Soc., **69** (4), 325 (1986)
15. Joseph Shappir, Ayal Anis, and Ida Pinsky, IEEE Tran. Electron Devices, ED-**33**, 442 (1986)
16. Cheol Seong Hwang, and Hyeong Joon Kim, J. Mater. Res., **8** (6), 1361 (1993)
17. M. Balog, M. Schieber, S. Patai, and M. Michman, J. Crystal Growth, **17**, 298 (1972)
18. G. A. Samara, J. Appl. Phys., **68** (8), 4214 (1990)
19. Z. Ma, L. H. Allen, Physical Review B, **49**, 13501 (1994)

A STUDY ON HYSTERESIS EFFECT OF BARIUM STRONTIUM TITANATE THIN FILMS FOR ALTERNATIVE GATE DILECTRIC APPLICATION

WEN-JIE QI, KEITH ZAWADZKI, RENEE NIEH, YONGJOO JEON, BYOUNG HUN LEE, AARON LUCAS, LAEGU KANG, JIAN-HUNG LEE, AND JACK C. LEE
Microelectronics Research Center, The University of Texas at Austin, 10100 Burnet Road, Building 160, Austin, TX 78758

ABSTRACT

Hysteresis effect of barium strontium titanate (BST) thin films for gate dielectric application has been studied. It is found that the "counterclockwise" hysteresis has strong sweep voltage and operating temperature dependence. It can be reduced or eliminated by proper thermal annealing or by using a barrier layer. A charge trapping and detrapping mechanism has been proposed.

INTRODUCTION

As metal-oxide-semiconductor (MOS) device dimensions are scaled down to $0.1\mu m$ and beyond, conventional SiO_2 will phase out due to its excessive leakage current. High-k dielectric materials with low leakage and good interface on Si will be needed. Ferroelectric materials were discovered in 1921, and their bulk material properties have been well-characterized [3]. However, it is only recently that thin-film ferroelectric materials have been studied and compared to their well-known bulk properties. Although there has been extensive research on preventing the interfacial layer formation in order to achieve a thinner equivalent oxide thickness [1,2], the high-k materials impose other challenges, e.g. hysteresis.

Ferroelectric materials have a phase transformation from ferroelectric phase to paraelectric phase at Curie temperature. Material in the paraelectric phase does not have a spontaneous polarization and hence should not have a hysteresis loop as possessed by the ferroelectric material. It is obvious that for gate dielectric application, ferroelectric materials are not acceptable because of the hysteresis. Hysteresis causes the threshold voltage to switch between two distinct values when the applied field is swept. For the BST material studied in this paper, the composition is $Ba_{0.5}Sr_{0.5}TiO_3$, the Curie temperature is below room temperature. Therefore, at room temperature, it should be in paraelectric phase which will not show the hysteresis loop. Consequently, the hysteresis observed in BST film should result from other mechanisms than ferroelectricity. The literature indicates that the ultra high-k materials such as barium strontium titanate (BST) exhibit higher hysteresis [4] than Ta_2O_5 [5] or TiO_2 [6,7]. The exact mechanism of hysteresis is not well understood. Interface charge trapping has been proposed for layered structure [2], while ferroelectricity [8,9] and mobile ion drift [10] have also been proposed.

In this paper, hysteresis behavior of RF sputtered BST films as a function of temperature, sweep voltage and stack structure has been studied. It was found that the hysteresis is due to the charge trapping and detrapping, and that it can be reduced by proper thermal annealing or use of a barrier layer. The interaction at the interface between BST and Si could be the source of interface states, traps, and defects that may cause hysteresis.

EXPERIMENTS

P-Si (100) wafers with a resistivity of 5~25Ω·cm were used as the substrate. Field oxide of 3500~4000Å was grown, patterned using photolithography, and etched in buffered HF solution to form the active area. Prior to BST deposition, the Si wafer was piranha cleaned and HF (1:40) dipped. BST film was then RF sputtered from a stoichiometric target with a composition of $Ba_{0.5}Sr_{0.5}TiO_3$. The base pressure of the sputter chamber was about 5×10^{-7} Torr, and the deposition temperature was 460°C. The sputtering procedure consisted of two steps to improve the BST/Si interface: a low power sputter at 20W followed by a higher power sputter at 50W. In some structures, Ta_2O_5 (35Å~75Å) were DC reactive-magnetron sputtered in an Ar+O_2 ambient to serve as a barrier layer. Pt (1500Å) was used as the top electrode. Pt was sputtered, patterned using photolithography, and etched using aqua regia solution (H_2O:HCl:HNO_3 = 5:7:1) at 80°C. The area of the capacitor is 5×10^{-5} cm^2. Annealing in N_2 or O_2 ambient was then performed in a conventional furnace. The Capacitance-Voltage (C-V) characteristics and hysteresis were measured using HP (Hewlett Packard) 4194, and the charge trapping property was measured using HP 4156.

RESULTS AND DISCUSSIONS

Fig. 1 demonstrates a typical C-V hysteresis loop of the Pt/BST/Si structure. A counterclockwise hysteresis loop can be seen. It is known that at room temperature the BST film is in a paraelectric phase [11] and indeed P-E (Polarization vs electric field) measurement of the Pt/BST/Ir (MIM) structure indicates no polarization hysteresis (not shown). To further understand the difference between Ir and Si substrate, the crystallinity of BST films on Ir and Si was examined. Glancing angle (5°) X-ray diffraction results are shown in Fig. 2. It can be seen that BST films show the same polycrystalline structure. Fig. 1 shows that the C-V hysteresis can be significantly reduced by thermal annealing, although the capacitance also drops, indicating an increase of the equivalent oxide thickness. This increase has been reported to be due to the interfacial layer growth [12]. The "counterclockwise" behavior and thermal annealing effect observations rule out ferroelectricity as the cause of hysteresis for these films. They suggest an improvement of the BST/Si interface during the annealing. Also, note the C-V curves shift almost parallel to each other, which seems to indicate that the hysteresis is not due to the interface states.

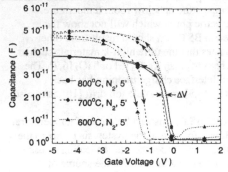
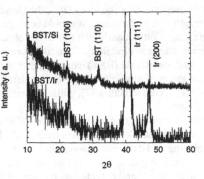

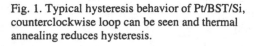

Fig. 1. Typical hysteresis behavior of Pt/BST/Si, counterclockwise loop can be seen and thermal annealing reduces hysteresis.

Fig. 2. Glancing angle (5°) XRD patterns show similar crystallinity for BST on Ir and Si.

The hysteresis observed here also has sweep voltage dependence for all BST structures. It increases almost linearly with the sweep voltage range as shown in Fig.3. This suggests some kind of charge response to the applied voltage. Interestingly, the hysteresis of BST films after O_2 anneal is less than N_2 anneal. It has been shown that there are oxygen vacancies in the BST films, and an O_2 anneal should serve to fill the oxygen vacancies. This result suggests that the oxygen vacancies may play a role in the hysteresis behavior.

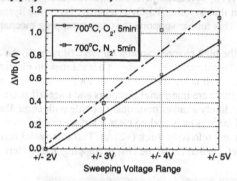

Fig. 3. The sweeping voltage dependence of the hysteresis. O_2 annealed samples show smaller hysteresis than N_2 annealed ones.

The hysteresis behavior was also measured as a function of operating temperature. Fig. 4 shows that for both BST and BST/Ta_2O_5, hysteresis is significantly reduced as the operating temperature is increased. This rules out the mechanism of mobile ion drift since it would cause an increase in hysteresis at high temperatures. Instead, the mechanism of charge trapping/detrapping at the high-k/Si substrate interface can be used to explain all of the above observations. Charge trapping leads to a parallel shift and "counterclockwise" hysteresis in C-V curves. The trapping/detrapping rates are sweep voltage dependent. At higher voltages, the leakage current is higher, which can lead to more charge trapping. And when the C-V is measured at the opposite voltage, the charge will be detrapped and then cause the flat band voltage change. At high operating temperatures, the charge trapping/detrapping is faster, and the hysteresis is then significantly lower. Furthermore, traps can usually be annealed out at high temperatures, which results in a lower hysteresis. Also note that the hysteresis is significantly reduced by the use of Ta_2O_5 as a barrier layer.

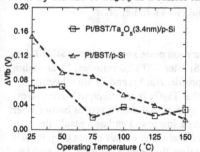

Fig. 4. The operating temperature dependence of hysteresis. Ta_2O_5 is effective to reduce hysteresis.

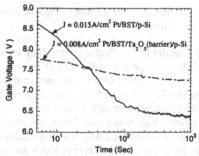

Fig. 5. The charge trapping properties of BST. Ta_2O_5 is useful to reduce the charge trapping.

In order to characterize the charge trapping properties of the BST/Si interface, the change in the gate voltage (ΔV_g vs time) was monitored under constant current stress. As shown in Fig. 5, these films show a very long time-to-breakdown at high stress current. It can be seen that the charge trapping rate for the BST/Si structure is worse than pure SiO_2. Interdiffusion at BST/Si interface has been reported and may be the cause of higher trapping rate [1]. However, the charge trapping rate of the BST/Ta_2O_5 (barrier layer)/Si structure is significantly reduced. This result is consistent with the fact that by using Ta_2O_5 as a barrier layer, the hysteresis is substantially reduced (Fig. 4). This further supports the charge trapping mechanism for hysteresis behavior.

In order to understand the high charge trapping for BST film on Si, the structural properties of BST on Si were characterized. Fig. 6 shows a transmission electron microscopy (TEM) picture of BST/Si structure. It is known that BST is not thermodynamically stable in contact with Si [13], therefore there are interfacial reactions and interdiffusion between BST and Si. The interfacial layer is most likely a rather poor quality oxide with some Ba, Sr, and Ti. As shown in Fig. 6, the interfacial layer is hazy and no clear interface can be seen; also this interfacial layer is not uniform and relatively thick (~25Å). This interfacial layer is probably the source of interface traps and defects that cause the charge trapping and hysteresis.

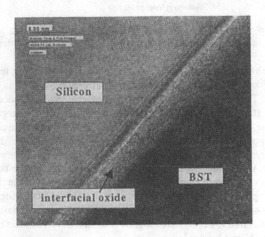

Fig. 6. TEM picture of BST/Si interface indicates a rather poor interface.

The barrier layer effect was examined by secondary ion mass spectroscopy (SIMS). Fig. 7 is the SIMS depth profile of BST/Ta_2O_5(barrier layer)/Si after 700°C, 5 minutes, N_2 annealing. Although some Si outdiffusion can be seen in the depth profile, Ba, Sr, and Ti are all being blocked by the Ta_2O_5 barrier layer. This result indicates that Ta_2O_5 is an effective barrier layer for the interdiffusion between BST and Si. This further supports that with a proper barrier layer, the interdiffusion can be reduced, hence, the interface can be improved and as a result, the hysteresis can be minimized. Moreover, it should be noted that some metals create deep level traps in Si. By using a barrier layer to block the interdiffusion, the deep level traps can also be reduced.

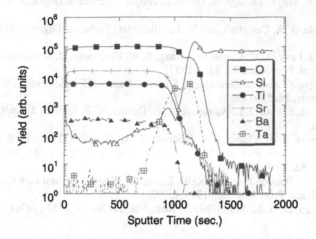

Fig. 7. SIMS depth profiles of BST/Ta$_2$O$_5$(barrier layer)/Si show that Ta$_2$O$_5$ is an effective barrier layer to block the BST/Si interdiffusion.

CONCLUSION

The hysteresis effect of BST/Si and BST/Ta$_2$O$_5$/Si stack structures has been studied. The results indicate that the hysteresis is due to the charge trapping and detrapping at the interface. The interface quality is directly related to the hysteresis. A better interface with less interdiffusion will exhibit less hysteresis. This hysteresis has strong sweep voltage dependence and operating temperature dependence. By using Ta$_2$O$_5$ as a barrier layer, the hysteresis can be reduced significantly. A thermal annealing which may reduce some of the defects.

ACKNOWLEDGEMENT

This work is partially supported by SRC/SEMATECH through the Front End Process Research Center.

REFERENCES

1. Yongjoo Jeon, Byoung Hun Lee, Keith Zawadzki, Wen-Jie Qi, Aaron Lucas, Renee Nieh, Jack C. Lee, Technical Digest of IEDM 1998, p.797
2. P. K. Roy, and I. C. Kizilyalli, Appl. Phys. Lett, **72**, 2835(1998)
3. Y. Xu, "Ferroelectric Materials and Their Applications", North-Holland, p.7, 1991
4. D. Roy, and S. B. Krupanidhi, Appl. Phys. Lett, **62**, 1056 (1993)
5. H. F. Luan, B. Z. Wu, L. G. Kang, B. Y. Kim, R. Vrtis, D. Roberts, and D. L. Kwong, Technical Digest of IEDM 1998, p.609

6. X. Guo, T. P. Ma, T. Tamagawa, and B. L. Halpern, Technical Digest of IEDM 1998, p.377

7. B. He, T. Ma, S. A. Campbell, and W. L. Gladfelter, Technical Digest of IEDM 1998, p.1038

8. C. Hubert, J. Levy, A. C. Carter, W. Chang, S. W. Kiechoefer, J. S. Horwitz, D. B. Chrisey, Appl. Phys. Lett, **71**, 3353(1997)

9. T. Kawakubo, K. Abe, S. Komatsu, K. Sano, N. Yanase, H. Mochizuli, IEEE Electron Device Letters, **18**, 529 (1997)

10. C. Chaneliere, S. Four, J. L. Autran, R. A. B. Devine, N. P. Sandler, J. Appl. Phys, **83**, 4823 (1998)

11. B. A. Baumert, L. -H. Chang, A. T. Matsuda, T. -L. Tsai, C. J. Tracy, R. B. Gregory, P. L. Fejes, N. G. Cave, W. Chen, D. J. Taylor, T. Otsuki, E. Fujii, S. Hayashi, K. Suu, J. Appl. Phys, **82**, 2558(1997)

12. Yongjoo Jeon, Byoung Hun Lee, Keith Zawadzki, Wen-Jie Qi, and Jack C. Lee, Mat. Res. Soc. Symp. Proc. Vol. **525**, 1998, p.193

13. K. J. Hubbard, D. G. Schlom, J. Mater. Res., Vol. **11**, No.11, 2757 (1996)

A STUDY OF THE OXYGEN SURFACE EXCHANGE COEFFICIENT ON La$_{0.5}$Sr$_{0.5}$CoO$_{3-\delta}$ THIN FILMS

X. CHEN*, S. WANG**, Y.L. YANG**, L. SMITH*, N.J. WU*, A.J. JACOBSON**, AND A. IGNATIEV*
* SVEC and MRSEC, University of Houston, Houston, TX 77204-5507
** Department of Chemistry and MRSEC, University of Houston, Houston, TX 77204-5500

ABSTRACT

La$_{0.5}$Sr$_{0.5}$CoO$_{3-\delta}$ (LSCO) can be used as a cathode material for low temperature solid oxide fuel cell applications. LSCO epitaxial thin films have been deposited on LaAlO$_3$ substrates by pulsed laser deposition (PLD). The transient behavior of the thin film conductivity with the pressure changes was recorded as a function of temperature and partial oxygen pressure. The surface exchange coefficient k of the LSCO thin film was obtained from analysis of the electrical conductivity relaxation data. The measured surface exchange coefficient increases with temperature and with final pressure but is not sensitive to the initial pressure. After prolonged annealing at 900°C, the k value was found to have greatly increased. The mechanism of the dependence of the measured k on pressure is discussed.

INTRODUCTION

La$_{0.5}$Sr$_{0.5}$CoO$_{3-\delta}$ (LSCO) is a perovskite type material that shows mixed ionic and electronic conductivity. Pervoskite oxides of this special type are widely used in high temperature devices such as solid oxide fuel cells (SOFC), oxygen sensors, and gas separation membranes [1,2].

Thin film structures of these devices can offer high performance, for example, a thin film structure can decrease the operating temperature of a SOFC, therefore reducing long-term degradation problems and material costs. High oxygen transport rates are important for the mixed conductors working in these devices. However, the oxygen transport in many perovskite oxide materials becomes surface exchange limited when the film thickness is smaller than 100μm [2,3]. Studies on dense ordered crystal LSCO thin films can provide fundamental information on gas exchange at the film surface.

The epitaxial LSCO films were deposited by pulsed laser deposition (PLD) in this work. The oxygen surface exchange coefficient for these ordered LSCO thin films were measured at high temperature by using electrical conductivity relaxation. The dependences of the measured surface exchange coefficient k on both the temperature and oxygen partial pressures were studied. The effect of high temperature annealing on k was also analyzed.

THEORETICAL BACKGROUND

The oxygen transport through a mixed conductor is determined by both the ambipolar diffusion coefficient D_a, and the oxygen surface exchange coefficient k [2]. The value of D_a/k is L_c. For samples with thickness L much greater than L_c, the transport is limited by the diffusion inside the film; for thin films with L much smaller than L_c, the transport is limited by the surface exchange rate. Electrical conductivity relaxation analysis is widely used to study the oxygen diffusion and surface exchange in mixed ion and electron conductors (MIEC). When an abrupt change in oxygen partial pressure occurs, oxygen begins to diffuse into or out of the MIEC. As a result, the oxygen concentration changes in the material. Through measuring the related electrical conductivity change in the MIEC, the oxygen concentration change can be monitored. Information on the oxygen chemical diffusion coefficient and the surface exchange coefficient can be extracted by analyzing the electrical conductivity relaxation data.

Consider a thin film on an oxygen diffusion isolating substrate, and suppose the ambient oxygen pressure changed from P_1 to P_2 at time $t=0$. The oxygen flux into the film after the pressure change can be described by [4]:

$$j(t) = K_{ex}(C_g - C_0(t)) \qquad (1)$$

Where K_{ex} (cm/s) is the apparent surface exchange coefficient, C is the oxygen concentration in the film. The subscription $_g$ denotes the equilibrium value under oxygen pressure P_2, and $_0$ denotes the value near film surface.

For a thin film of $L << L_c$, it is generally assumed that the concentration inside the film equilibrates as soon as oxygen flows in or out through the surface, i.e. $C_0(t)$ equals the oxygen concentration inside the film at time t. We have:

$$L\Delta C_0(t) = j(t)\Delta t \qquad (2)$$

Solving equation (1) and (2) by integration, and suppose the change in the film conductivity σ is proportional to the change in oxygen concentration [5], then:

$$(\sigma_g - \sigma_0(t)) / (\sigma_g - \sigma_0(0)) = e^{-K_{ex}t/L} \qquad (3)$$

By fitting the measured electrical conductivity relaxation data with Eq. (3), K_{ex} can be obtained. To account for changes in composition, a surface exchange coefficient k is defined by [4]:

$$k = K_{ex} / (\frac{1}{2} \, \partial \ln P / \partial \ln C) \qquad (4)$$

where P is the ambient oxygen pressure. The value of $\frac{1}{2} \, \partial \ln P / \partial \ln C$ is about 100 in the temperature and P_{O2} range of the experiment [6].

EXPERIMENT

The LSCO thin film was deposited on a LaAlO₃ (LAO) (100) single crystal substrate with the PLD method. A KrF excimer laser was used for the deposition. The deposition was performed at 150mtorr oxygen pressure, with a substrate temperature of ~550°C, and a pulse frequency of 7Hz. A deposition of 30min yields a film of thickness ~500nm.

The samples was mounted on a ceramic sample holder and placed inside a quartz tube in a furnace. Four gold electrodes were attached to the sample with gold paste and used as electrodes for four-point resistance measurements. A thermocouple is put beside the sample through another ceramic tube from the other side of the quartz tube. A 4-way valve is used to switch the gas flow from different gas sources through the quartz tube. Different oxygen sources with 100%, 50%, 30%, 10%, 5%, and 1% oxygen, balanced with nitrogen, were used. The ambient oxygen partial pressure was changed abruptly between these partial oxygen pressures. The gas flow rate was 80ml/min.

The relaxation measurements were first performed at temperatures from 700 to 760°C. The temperature range was selected to give relaxation times within the range of the equipment.

The sample was then annealed at 900°C for 50hrs, the temperature was decreased and the relaxation measurement was repeated again with the annealed sample.

Scanning electron microscopy (SEM), x-ray diffraction (XRD), and atomic force microscopy (AFM) measurements were performed the sample before and after the relaxation experiment.

RESULTS AND DISCUSSION

The measurements of the surface exchange coefficient were first performed at temperatures from 700 to 760°C. The dependence of the LSCO film resistance with time was fitted with Eqn.(3). Fig. 1 shows k values extracted from the fitting results of the experiments. It shows a linear relationship between lg (k) and the reciprocal of the absolute temperature. In Fig. 1, we can see the k values measured when the oxygen partial pressure was switched from 1.00 to 0.50atm are a little lower than the corresponding values for switches from 0.50 to 1.00atm. The relationship between k and the switching pressure was further studied (Fig. 2). Fig. 2a is the plot of the relationship between k and the initial partial pressure. It shows that, for a final pressure P_2, k is independent of the initial oxygen partial pressure P_1. Fig. 2b is the plot of the relationship between k and P_2. It shows that k increases with the final pressure for a specific P_1. The plot also shows a linear trend of lg (k) vs. lg (P_2), indicating k is proportional to P_2^n for the experimental pressure range. From the slope of the fitted line, n is ~0.3.

As indicated in Eqn.(1), the oxygen flux is related to P_2, but not related to P_1, the result that the measured k is related to P_2 but not P_1 is expected. The relationship between lg (k) and lg (P_2) can be used to study the reaction mechanism of the surface exchange. The oxygen interaction with the film surface can be roughly written as [3]:

$$\tfrac{1}{2}O_2 + V_O^{\cdot\cdot} \Leftrightarrow O_O^x + 2h^{\cdot} \qquad (5)$$

where O_2 is molecular oxygen in the ambient, $V_O^{\cdot\cdot}$ is an oxygen vacancy, O_O^x is an oxygen atom in the sample lattice, and $h^{\cdot}$ is an electron hole.

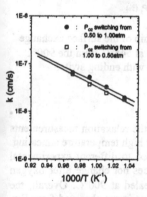

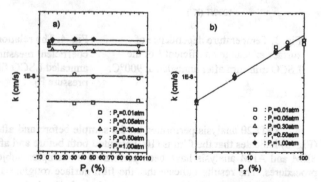

ig. 1. Temperature dependence of urface exchange coefficient k of a SCO thin film deposited on LAO.

Fig. 2. The relationship of switching pressure and the 700°C measured LSCO film surface exchange coefficient. (a) with beginning switch pressure P_1; (b) with ending switch pressure P_2.

If the concentration of $V_O^{\cdot\cdot}$ does not change with pressure, the oxygen surface flux rate is expected to be proportional to $P_2^{1/2}$ from Eqn.(5), and n of ½ is expected. In reality, the concentration of $V_O^{\cdot\cdot}$ decreases with the increase of pressure, and n is usually smaller than 0.5 as a result. This result is in agreement with the observation on similar oxide materials [7].

The k values measured for the pulsed laser deposited epitaxial thin film appears to be very low comparing with those from bulk polycrystal materials [8], in agreement with the observation in literature that the surface exchange coefficient of single crystal LSCO is lower than the polycrystal bulk material [9]. To investigate this phenomenon, the sample was annealed at 900°C for 50 hours. The k value of the film after annealing is shown in Fig. 3. We have obtained the k values here at a temperature range lower than the earlier measurement because the relaxation time has become too short for an accurate measurement at the previous temperature. By extrapolating the fitted lines to the same temperature where measurements were performed before annealing, we found that k increased by about two orders of magnitude.

The relationship between k and the ending switching pressure P_2 after annealing is measured, and is shown in Fig. 4. Similar to before annealing of the sample, the relationship between lg (k) and lg (P_2) is linear, indicating k is proportional to P_2^n in the experimental pressure range. n is found to be ~0.4 from the slope of the fitted line, which is smaller than 0.5, and is in agreement with the earlier discussion.

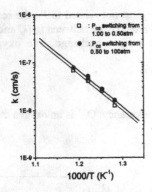

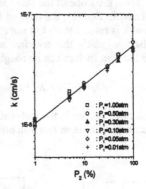

Fig. 3. Temperature dependence of surface exchange coefficient k of the LSCO thin film after annealed at 900°C.

Fig. 4. The relationship of the surface exchange coefficient measured at 545°C from the 900°C annealed LSCO film with ending switch pressure P_2.

XRD 2θ analysis performed on the sample before and after the relaxation measurements (Fig. 5) indicates that the film is (100) oriented both before and after high temperature annealing. SEM and AFM analysis have been performed on samples subjected to different experimental procedures. The results indicate that the film surface roughness does not change after oxygen pressure switch at 700°C, but became roughened after being annealed at 900°C. Overall, the crystal quality change of the film after the high temperature annealing is not observed from Fig. 5. The surface morphology change may be the reason for the large increase in k. More discussions will be discussed elsewhere [10].

CONCLUSIONS

Surface exchange coefficients of $La_{0.5}Sr_{0.5}CoO_{3-\delta}$ thin films deposited on $LaAlO_3$ substrate were measured. The conductivity relaxation curve shows exponential decay behavior, and are well fitted with the surface exchange limited model. k increases with temperature, and $lg\,k$ vs $1/T$ shows a linear relationship. k is dependant on the ending oxygen pressure in the measurement, but is not affected by the beginning pressure. Acting like a single crystal material, the surface exchange coefficient of epitaxial LSCO film is much smaller than bulk polycrystal material. However, high temperature annealing can largely increase k of the film. k is proportional to P_2^n in the experimental pressure range with n of 0.3~0.4, which agrees with defect chemistry model. The increase in

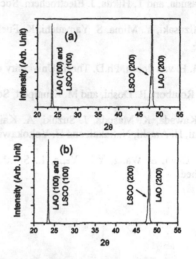

Fig. 5. the XRD analysis of LSCO thin films developed with PLD. (a) the virginal sample; (b) the 900°C annealed LSCO thin film.

roughness, which comes with high temperature annealing, results in a great increase in the k value, and is important in SOFC performance. More studies about the effects of surface morphology change on the conductivity relaxation measurement are needed.

ACKNOWLEDGEMENTS

The assistance of A. Zomorrodian, S. Perry and P. Van der Heide is greatly acknowledged. This work was partially supported by the MRSEC Program of the National Science Foundation under award number DMR–9632667, NASA; the Texas Center for Superconductivity; and the R. A. Welch Foundation.

REFERENCES

1. J.A. Kilner, R.A. de Souza, and I.C. Fullarton, Solid State Ionics, **86-88**, 703(1996).

2. B.C.H. Steele, Solid State Ionics, **75**, 157 (1995).

3. P.J. Gellings, and H.J.M. Bouwmeester, *The CRC Handbook of Solid State Electrochemistry*, (CRC Press, Inc., 1997), p. 481.

4. J.E. ten Elshof, M.H.R. Lankhorst, and H.J.M. Bouwmeester, J. Electrochem. Soc., **144 (3)**, 1060 (1997).

5. I. Yasuda, and T. Hikita, J. Electrochem. Soc., **141 (5)**, 1268 (1994).

6. J. Mizusaki, Y. Mima, S. Yamauchi, K. Fueki, and H. Tagawa, J. Solid State Chem., **80**, 102 (1989).

7. R. H. E. van doorn, Ph.D. Thesis,University of Twente, Enschede, The Netherlands (1996).

8. J.L. Routbort, R. Doshi, and M. Krumpelt, Solid State Ionics, **90**, 21 (1996).

9. T. Kawada, K. Masuda, J. Suzuki, A. Kaimai, K. Kawamura, Y. Nigara, J. Mizusaki, H. Yugami, H. Arashi, N. Sakai, and H. Yokokawa, Solid State Ionics, **121**, 271 (1999).

10. X. Chen, S. Wang, Y.L. Yang, L. Smith, N.J. Wu, A. Jacobson, and A. Ignatiev, to be published.

CRYSTALLOGRAPHIC ORIENTATION IN BULK POLYCRYSTALLINE SILICON CARBIDE PRODUCED BY A CHEMICAL VAPOR DEPOSITION (CVD) PROCESS

James V. Marzik *, William J. Croft **
* Performance Materials, Inc., Hudson, NH
**Mineralogical Museum, Harvard University, Cambridge, MA.

ABSTRACT

Polycrystalline, theoretically dense silicon carbide was deposited onto graphite substrates via the reductive pyrolysis of methyltrichlorosilane in a hot-walled chemical vapor deposition (CVD) chamber. The resulting product can be considered a bulk material with deposit thicknesses in the range of 4 to 8 millimeters. The material was characterized using powder x-ray diffraction and Laue back-reflection techniques. Under the deposition conditions investigated in this study, the crystallographic orientation varied as a function of distance from the substrate. The material exhibited a high degree of randomness in proximity to the substrate, and progressively showed a higher degree of preferred crystallographic orientation as the deposit progressed. This phenomenon is correlated with the microstructure of the material as well as such mechanical properties as hardness and fracture toughness.

INTRODUCTION

Polycrystalline silicon carbide possesses a unique combination of mechanical, thermal, and chemical properties [1], which makes it attractive for a number of applications. Its desirable properties include high thermal conductivity and thermal shock resistance, low thermal expansion, high strength at temperatures in excess of 1000°C, oxidation resistance at elevated temperatures, superior hardness and erosion resistance, and high elastic modulus. Silicon carbide produced by chemical vapor deposition (CVD) is theoretically dense and very pure compared to SiC produced by other ceramic processing methods. The high purity and thermal conductivity of CVD SiC make it a material of interest to the semiconductor community for such applications as wafer carriers during silicon wafer processing, as well as for structural components of wafer processing chambers such as chamber liners. There has also been recent interest in polycrystalline SiC as a substrate material for thin film growth of electronic grade single crystal films of gallium arsenide and silicon carbide. It has been suggested [2] that preferred crystallographic orientation of polycrystalline SiC substrate materials has a significant effect on the electrical properties of deposited thin films.

EXPERIMENTAL

The silicon carbide material used in this study was produced by a commercial chemical vapor deposition process [3] in which methyltrichlorosilane is reductively pyrolyzed in a graphite chamber in the presence of hydrogen (see Figure 1). Process temperatures ranged from 1300 to 1375°C and the chamber pressure was controlled at 0.25 atm. Temperatures were measured independently using thermocouples and optical pyrometry. All temperatures, pressures and gas flows were microprocessor controlled.

Mat. Res. Soc. Symp. Proc. Vol. 606 © 2000 Materials Research Society

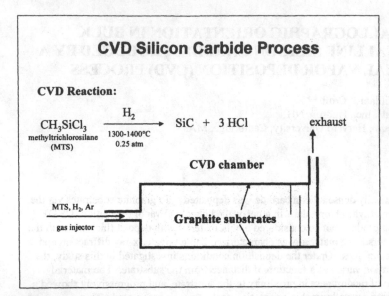

CVD Silicon Carbide Process

CVD Reaction:

$$CH_3SiCl_3 \xrightarrow[\substack{1300\text{-}1400°C \\ 0.25 \text{ atm}}]{H_2} SiC + 3 HCl$$

methyltrichlorosilane (MTS)

exhaust

CVD chamber

MTS, H_2, Ar
gas injector

Graphite substrates

Figure 1. Diagram of the CVD SiC process.

X-ray diffraction patterns were obtained using a Scintag Model XDS-2000 x-ray diffractometer and copper Kα radiation detected with a germanium crystal cooled to liquid nitrogen temperature. The diffraction patterns were taken in the range of $2° \leq 2\theta \leq 70°$ with a scan rate of $1° \, 2\theta$ per minute. The diffractometer was calibrated with a NIST silicon standard. Intensity ratios were calculated from normalized integrated intensity values. Samples were also examined using a Laue back reflection camera with a 0.1mm collimated beam. Both white x-ray radiation and copper Kα radiation were used for the Laue photographs. Silicon carbide samples for x-ray diffraction were typically sliced parallel to the substrate plane at thicknesses of approximately 0.7mm at various distances from the graphite substrate.

Vickers hardness measurements were made according to an ASTM standard method [4]. Fracture toughness was determined by a microindentation technique [5]. Hardness and fracture toughness measurements were performed using a Buehler Micromet 2004.

The microstructure was examined using polished silicon carbide samples. Samples were etched using a plasma etching technique. The plasma gas etchant was a mixture of tetrafluoromethane and oxygen. Samples were heated and cooled in argon.

RESULTS AND DISCUSSION

As stated earlier, the CVD silicon carbide used for this study was relatively thick, standalone material. Typical material used was in the 6-8 mm thickness range. X-ray diffraction patterns of the CVD silicon carbide used in this study were identified as and indexed according to the cubic structure of β-SiC. The intensity ratio of the diffraction peaks, however, varied depending on the distance of the sample from the CVD substrate as well as the orientation of the sample relative to the CVD growth direction. In samples sliced and measured parallel to the substrate plane, it was observed that there was a variation in the intensity ratio of the 111 and 220 reflections, which are the two strongest peaks of β-SiC. The value of this ratio systematically changed from the start of the deposit to the end of the deposit. Figure 2 shows a typical plot of the 220/111 peak intensity ratio at the start, mid-point and end of the deposit. The CVD SiC has a slight 111 orientation

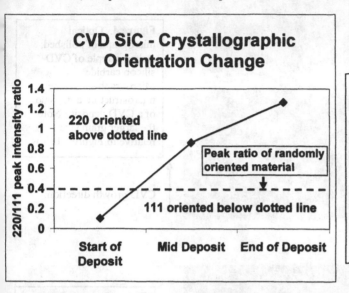

Figure 2. Plot of the 220/111 peak intensity ratio as a function of distance from the substrate. The horizontal dotted line shows the value for a randomly oriented material. As the deposit progresses, material becomes more 220 oriented.

at the start of the deposit and progressively changes to a 220 preferred orientation as the deposit proceeds. Diffraction patterns of CVD SiC samples, which were sliced perpendicular to the substrate plane, showed 220/111 peak intensity ratios much closer to that expected for a randomly oriented polycrystalline material. Thus orientation changes in the material are more apparent when the material is measured parallel to the substrate plane. Laue back reflection photographs of this material with the collimated beam placed at the start of the CVD deposit show complete rings indicative of a small-grained, randomly oriented material. Laue

End of deposit
- Larger grains
- Columnar growth pattern well defined
- Highly 220 oriented parallel to the substrate plane

Figure 3. Summary of crystallographic and microstructural features of CVD SiC used in this study.

- Small grains
- Columnar growth pattern begins
- Mostly random orientation of the grains with some 111 orientation observed parallel to the substrate plane

Start of deposit

photographs of the material at the end of deposit showed spotted rings, which is indicative of larger grained material. This is consistent with the observed microstructure of the material (see Figures 4 and 5). The Laue back reflection photographs did not give clear evidence of preferred orientation in the material. The preferred orientation observed in the x-ray diffraction patterns, however, suggest that the orientation effect is associated with the columnar microstructure

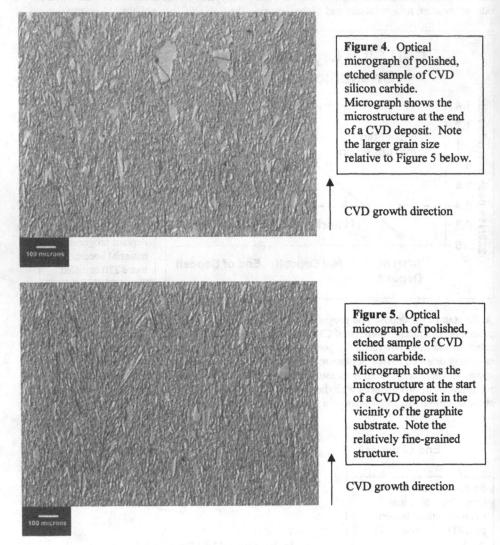

Figure 4. Optical micrograph of polished, etched sample of CVD silicon carbide. Micrograph shows the microstructure at the end of a CVD deposit. Note the larger grain size relative to Figure 5 below.

CVD growth direction

Figure 5. Optical micrograph of polished, etched sample of CVD silicon carbide. Micrograph shows the microstructure at the start of a CVD deposit in the vicinity of the graphite substrate. Note the relatively fine-grained structure.

CVD growth direction

typical of polycrystalline silicon carbide. The progressively increasing 220 orientation appears to be associated with the columnar growth habit of this CVD material. As the deposited material grows in thickness and the columnar growth becomes more defined, a preferred crystallographic stacking sequence develops on each "column" of silicon carbide. The 220 orientation was somewhat unexpected and seems to be inconsistent with the Law of Bravais as extended by Donnay and Harker [6] which states that a crystal prefers to grow perpendicular to the plane with the highest density of lattice points. Thus grain growth β-SiC, which has a face-centered cubic lattice, would be predicted to favor 111 orientation. There have been previous reports of preferred 111 orientation in polycrystalline silicon carbide [7,8]. This reported 111 orientation was more pronounced at low deposition rates (10-20 μm/hr). The rate at which our material was deposited was higher (50-75 μm/hr). An examination of the sphalerite structure of β-SiC may provide a clue to the crystallographic orientation reported herein. 111 orientation in this structure features stacking of alternating layers of silicon and carbon. Silicon carbide with 220 orientation contains equal numbers of silicon and carbon in its two dimensional stacking layers. It is suggested that higher chemical vapor deposition rates tend to kinetically favor 220 orientation in which equal numbers of silicon and carbon may be found at any active deposit surface during the growth of the material. Figure 3 summarizes features of the material both at the start and towards the end of the deposit. Microstructural examination of the polycrystalline CVD SiC studied here indicated that the grain size of the material increased as the deposit proceeded. This is shown in Figures 4 and 5. On a larger scale, the grains of the material were grouped in such a way as to give a columnar growth habit. The microstructure of the material can be correlated with such mechanical properties as hardness and fracture toughness. Results are listed in Table 1. The material exhibited higher values of hardness and fracture toughness towards the end of the deposit. In Batch A the difference in hardness and fracture toughness at the start and at the end of the deposit was statistically significant above the 90% confidence level. The data from Batch B followed the same trend but at lower confidence levels. More data would need to be compiled to determine whether the grain size or crystallographic orientation is the critical factor or if there are other critical variables.

Table 1. Summary of Fracture toughness and Hardness measurements of CVD SiC

		Fracture toughness[1] ($MPa\ m^{1/2}$)		Vickers Hardness[2] ($kg\ mm^{-2}$)	
		Mean[3]	Std Dev	Mean[4]	Std Dev
Batch A	Substrate side	2.75	0.24	2710	95
	Growth Side	2.95	0.30	2860	126
Batch B	Substrate side	2.98	0.16	2930	150
	Growth Side	3.08	0.27	2980	103

[1] 1000 g indent load   [3] sample size ranged from 5 to 10
[2] 500 g indent load   [4] sample size ranged from 10 to 30

CONCLUSIONS

Polycrystalline, theoretically dense silicon carbide was deposited onto graphite substrates via the reductive pyrolysis of methyltrichlorosilane in a hot-walled chemical vapor deposition (CVD) chamber. The resulting product can be considered a bulk material with deposit thicknesses in the range of 4 to 8 millimeters. The material was characterized using powder x-ray diffraction and Laue back-reflection techniques. Under the deposition conditions investigated in this study, the crystallographic orientation varied as a function of distance from the substrate. When measured parallel to the substrate plane, the material exhibited a high degree of randomness in proximity to the substrate, and progressively showed a higher degree of preferred 220 crystallographic orientation as the deposit thickness increased. The average grain size of the material also increased as the material deposit progressed. This phenomenon is attributed to the columnar microstructure of the material and the rate of material growth.

REFERENCES

1. URL: www.performancematerial.com/cvd_sic/cvd_sic_prop.htm for a summary of properties of the CVD silicon carbide used in this study

2. T.M. Sullivan, III-Vs Review, Vol. 12 (5), 34 (1999)

3. PerformanceSiC™ manufactured by Performance Materials, Inc., Hudson, NH.

4. 1998 Annual Book of ASTM Standards, Vol 15.01, Standard C1327-96a

5. A.G. Evans and E.A. Charles, J. Amer. Ceram. Soc., **59**, 371 (1976).

6. F. Donald Bloss, *Crystallography and Crystal Chemistry* (Mineralogical Society of America, Washington D.C., 1994), p. 338.

7. J.S. Goela and R.L. Taylor, Proceedings of the SPIE Conference on Window and Dome Technologies and Materials IV, Vol. 2286, 46 (1994).

8. J.S. Goela, L.E. Burns, and R.L. Taylor, Appl. Phys. Lett. **64**, 131 (1994).

STRUCTURE-PROPERTY RELATIONS IN SOL-COATED PMN CERAMICS: MICROSCOPY, DIELECTRIC AND ELECTROMECHANICAL RESPONSE

A. Sehirlioglu *, and S.M. Pilgrim, New York College of Ceramics at Alfred University, Alfred, NY, 14802

ABSTRACT:

Some of the most promising materials for electrostrictive response are $Pb(Mg_{1/3} Nb_{2/3})O_3$ (PMN) - ceramics; however, the properties of a given composition are only optimum in a limited range of temperatures. In a previous study, it was found that sol coating of PMN particles modified and improved the electromechanical and/or dielectric properties of the resulting product--doubling induced strain in some cases. Understanding the origin of these changes will help to produce an optimized PMN ceramic for a given application from a single source powder. This work concentrates on the Ti and Zn coatings which gave superior properties within the concentration matrix. The relation between the structure and the enhanced electrostrictive behavior is studied. Structural characterization is done by XRD, TEM and SEM. Electrical measurements of dielectric constant, loss, polarization and strain, the property determinations needed to complete the structure-property suite.

INTRODUCTION:

PMN ceramics are well known for their electrostrictive properties. Despite the low values of the electrostrictive coefficients, the high induced polarization of these ceramics results in high strain values (strain is proportional to the square of polarization).[1-3]

In this specific work PMN-PT-BT (B400090) particles are coated with Zn and Ti both separately and together using sol-gel method.. The samples will be denoted with the first letter of the coating materials and their mol percents (i.e. T4Z2 is the name for the sample with 4mol% of Ti and 2mol% of Zn coating). Among all these compositions, T2Z2 is the most superior composition.

The aim of using the sol-gel technique is to obtain a homogeneous coating of a liquid phase precursor material onto the surface of the PMN-PT-BT powders at temperatures below 100°C [4]. The hydrolysis/polycondensation of the organometallic precursors is carried out in the presence of the base material. As discussed this process and these additions result in improved properties.

EXPERIMENTAL PROCEDURE:

The base material is ball milled in ethyl alcohol medium for 15-20 hours. Then the organometallic precursors are prepared by adding the necessary amount to isopropanol in a small beaker. The precursors for coating materials are Zn-acetate, Ba-acetate and Ti-isopropoxide for Zn, Ba and Ti coatings respectively. After addition of organometallic precursors, the system is stirred continuously until the solution becomes clear. When the clear solution is produced, a few drops of HCl that works as a catalyst for the reaction and the base material are added concurrently. After the gel is formed, it is dried and crushed to obtain the particulate-coated powder.

The powder obtained is then dry pressed at 10000 psi after addition of 5wt% of a 3wt% PVA solution and sieving to 65 mesh. The pellets are then heated to 600°C at a heating rate of

Mat. Res. Soc. Symp. Proc. Vol. 606 © 2000 Materials Research Society

5°C/min to burnout the organics. The samples are kept for 2 hours at 600°C and then cooled by furnace cooling.

The sintering is done by using a double crucible method with a PbZrO₃ lead source powder. The sintering of samples is done at 1200°C. The system is heated at 5°C/min and kept at 1200°C for 4 hours. It is then cooled with furnace cooling. The samples are then polished as necessary for the characterization methods.

The sintered samples are electroded by Ag/Pt sputtering and Ag electroding paste for electronic and electromechanical measurements. The dielectric measurements are done in a system that includes a Delta 9039 furnace, a HP 3488A Switch/Control Unit and a HP 4284A Precision LCR Meter. A program written using Lab View by National Instruments controls the whole system. The samples are heated to 160°C with a ramp of 10°C/min and held at that temperature for 5 minutes until the system equilibrates. Then they are cooled with a rate of 2°C/min by using liquid nitrogen. The data is taken during cooling until -40°C at 4 different frequencies; 0.1,1,10 and 100 kHz.

The electromechanical measurements are done as discussed in Cho et al. (1Hz, ±1MV/m)[5].

The X-ray Diffraction measurements are done by using Bragg Brentano Parafocusing Powder Diffractometer. The Cu-Kα x-ray is generated with a voltage of 40kv and current of 20mA. The effect of other emissions is filtered using the program Jade.

For the grain size measurements an AMRAY 1810 Scanning Electron Microscope is used.

The TEM study is done by using JOEL-Jem 2000 FX Transmission Electron Microscopy. The EDS detector used is a PGT-Prism Light Element X-ray detector.

RESULTS:

i) Diffraction

The measurements are done by using the X-ray diffraction machine described before. All measurements are done between 20 and 120 degrees 2θ with 0.02 step size, 7 sec count time. Si is used as an internal standard. The following graph (Figure 1) shows the effect of composition change on lattice parameters. The standard deviation bars are for the machine error and calculated by the program Jade.

In samples T4 and T8 a second phase of Pb₂Nb₂TiO₉, which is accepted as a pyrochlore with Z=4, forms This phase is not observed in T2.The main peak for this phase can be seen in figure 2

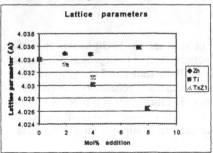

Figure1: The lattice parameters for the sol-coated PMN-ceramics

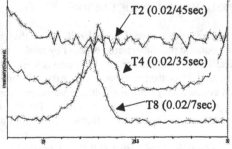

Figure2: Main peak of pyrochlore formed in Ti added samples

ii)Electrical/Electromechanical Response:

The 1kHz values of room temperature measurements of maximum dielectric constant (Kmax), temperature (T@Kmax) and loss (tanδ) at maximum dielectric constant are tabulated and then graphed. 5 samples for each composition are measured. To see the compositional effect on the properties the Student T-test is used. The following graphs (3, 4 and 5) are the final values for Kmax, T@Kmax and tanδ. The confidence level between each data step is shown in tables (Tables I and II).

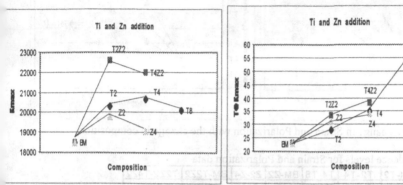

Figures 3&4: Change in Kmax and T@Kmax with change in composition

Table I: Confidence levels for Kmax and T@Kmax data

	BM-T2	T2-T4	T4-T8	BM-Z2	Z2-Z4	BM-T2Z2	T2Z2-T4Z2
Kmax	65%	27%	45%	43%	31%	93%	28%
T@Kmax	>99%	>99%	>99%	>99%	>99%	>99%	95%

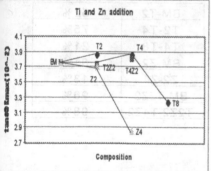

Figure5: Change in tanδ @ Kmax with change
in composition

Table II: Confidence levels for tanδ@Kmax data

	tanδ
BM-T2	24%
T2-T4	1.20%
T4-T8	>99%
BM-Z2	14%
Z2-Z4	98%
BM-T2Z2	7%
T2Z2-T4Z2	56%

The apparatus described in work of Cho et al is used to measure the strain and the polarization. The same statistical method and indication system that was preferred for dielectric measurements is used. The following graphs (6,7 and 8) are the final values for strain, polarization and Q coefficient. The confidence level between each data step is shown Tables III and IV.

The coefficients Q and g can be calculated from the graphs of strain versus the absolute value of polarization by fitting a second order polynomial. The second order coefficient of this polynomial is the Q value while the first order coefficient is the g value. In all cases, g is negligible (< 0.8 mV.m/N).

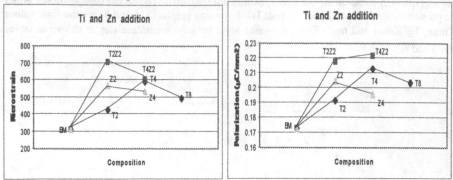

Figures 6&7: The change in Strain and Polarization with the change in composition

Table III: Confidence levels for Strain and Polarization data

	BM-T2	T2-T4	T4-T8	BM-Z2	Z2-Z4	BM-T2Z2	T2Z2-T4Z2
Strain	88%	98.30%	95%	99%	39%	>99%	76%
Polarization	90%	97%	86%	99%	70%	>99%	39%

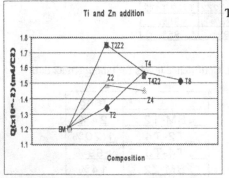

Figure 8: The change in Q coefficient with the change in composition

Table IV: Confidence levels for Q coefficient data

	Q (confidence)
BM-T2	52%
T2-T4	75%
T4-T8	21%
BM-Z2	81%
Z2-Z4	13%
BM-T2Z2	96%
T2Z2-T4Z2	98%

iii) Microscopy:

TEM/EDS and SEM is used to understand structural changes resulting from composition change. The TEM study showed that there is a second phase forming at the triple points which consists of Mg, Zn and O for the Zn and Ti/Zn added samples. The picture, diffraction pattern and EDS of this phase can be seen in figure 9. No second phase is observed at grain boundaries for single Ti additions.

Careful works on the base material showed that this phase is wholly absent without doping. This second phase is found to be crystalline and diffraction provides d-spacings of 2.3653 Å, 1.48 Å, 1.261 Å and 0.9328 Å.

The percentage differences between the calculated values and the periclase are 2.7%, 0.65%, 0.69% and 0.94% respectively for the points mentioned above. Despite this discrepancy the values arewell within the commonly accepted accuracy of TEM diffraction (10%). The second phase is thus tentatively identified as a Zn-doped periclase.

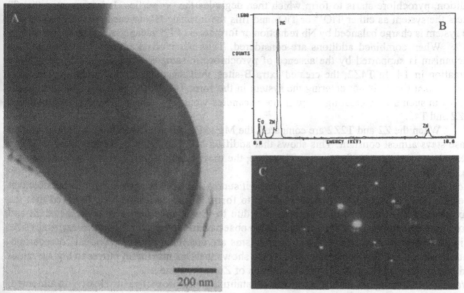

Figure 9: (A) Bright field image of the second phase (B) TEM/EDS of the second phase (C) Diffraction pattern of the second phase, zone (1-1-2).

The grain size measurements are done by randomly choosing a total of 60 lines on 6 different micrographs for each composition using SEM. The results graphed are shown in figure 10:

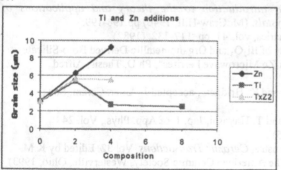

Figure10: Change of grain size with change in composition

CONCLUSION:

Zn addition to the system forms a second phase that is periclase with ZnO in it as a solid solution. The leaching of Mg from the grains causes the grains to grow as expected[6]. The change in valence of Pb from +2 to +4 to charge balance the system is not supported by the data as a decrease in Kmax and T@Kmax is expected in such a case[7]. This is not experimentally observed. It is suspected that the charge balance system involves oxygen vacancies.

Ti addition causes no second phase on the grain boundaries. At high values of Ti addition, pyrochlore starts to form which then degrades the properties. It is suggested that Ti enters the system as either TiO^{+2} or TiO_2 and fills the intrinsic B-site vacancies. In both cases, the system is charge balanced by Nb reduction or formation of Pb vacancies or Mg vacancies.

When combined additons are considered, Ti is believed to occupy the B-sites. This mechanism is supported by the absence of pyrochlore in sample T4Z2 despite of pyrochlore formation in T4. In T4Z2, the created extra B-sites shift the pyrochlore formation. This also suggests that the Ti is not entering the system in the form TiO_3^{-2}, forming a A-site vacant unit cell, as in such a case creating more B-site vacancies would not cause any difference between T4Z2 and T4.

When the Z2 and T2Z2 are compared, the Mg to Zn ratio in the EDS data of their second phase stays almost constant. This shows that addition of Ti does not cause any additional Mg vacancies to compensate for charge. Therefore, the mechanisms possible for charge balance are Nb reduction or formation of Pb vacancies.

Vacancies of Pb favor pyrochlore formation so if such a mechanism is working, it is possible that after a critical limit of V_{Pb} pyrochlore starts to form. For Ti additions, it is believed that the maximum in electromechanical properties is due to the balance between Ti replacement and pyrochlore formation. Such a maximum is also observed in the structural features e.g. grain size. This data suggests that two different mechanisms are dominant around a critical composition. Both structural and the electromechanical data shows that this maximum moves to higher values and higher doping levels of Ti with the addition of Zn to the system.

Overall, the coating process does affect substantially the properties in clearly documented, poorly understood manners.

REFERENCES:

1. B. Jaffe, W.R. Cook and H.L. Jaffe, *Piezoelectric ceramics*, (Academic Press, New York, 1971) Chp. 2.
2. W.G Cady, *Piezoelectricity: an introduction to the theory and applications of electromechanical phenomena in crystals*, (McGraw-Hill, 1946) pp.198-199.
3. S. Nomura and K. Uchino, Ferroelectrics, vol. 41, pp. 117-132, (1982)
4. K.G. Brooks, " A Comparative Study of Bi_2O_3 and Organometallic Derived Boro-Silicate Sintering Additives as Applied to Li-Zn Microwave Ferrites"; Ph.D. Thesis, Alfred University, NY, USA, 1990
5. Y.S. Cho, S.M. Pilgrim, H. Giesche and K. Bridger, , accepted J. Amer. Ceram. Soc., October 1999.
6. K. Uchino, M. Tatsumi, I. Hayashi and T. Hayashi, Jap. J. of App. Phys., Vol. 24 Supplement 24-2, pp.733-735, (1985)
7. R.E. Newnham and S. Trolier-McKinstry, *Ceramic Transactions*, Vol 32. Edited by K.M. Nair, J.P. Guha and A. Okamoto, (The American Ceramic Society, Westerville, Ohio, 1993) pp. 1-18.

THE CONTROL OF ZN FOR ZST MICROWAVE CERAMICS WITH LOW SINTERING TEMPERATURE

Yong H. Park *, Moo Y. Shin, Ji M. Ryu, Kyung H. Ko
Dept of MS&E, Ajou Univ., Suwon 442-749, Korea, *p021@chollian.net

ABSTRACT

ZnO addition for low-T sintering of ZST has serious side effects such as decrease of Q × f value. In this work, the curing of these side effects without any further chemical additives and causing degradation of other dielectric properties have been presented. After sintered at 1350℃ for 2h, samples were annealed in oxygen at 900～1100℃ for 5hr. It was observed that Q × f value of post-annealed sample at 900℃ could recovered up to 46000 from the as-sintered value of 40000. Because there were no formations of second phases or significant changes in lattice constants, which could affect microwave properties of ZST. However, it was found that only for the specimens annealed at 900℃, Zn was almost depleted from grin inside and diffused toward grain boundary. So, it is suggested that the out-diffusion of Zn is responsible for the recovery of Q × f value. Moreover, when the amount of Zn incorporation increased via successive calcination and sintering of pre-mixed powder of ZST and ZnO, Q × f value of 33000 also could enhanced up to 39000 due to the redistribution of Zn near grain boundary by out-diffusion.

INTRODUCTION

In the past decades, the $(Zr_{0.8}Sn_{0.2})TiO_4$(ZST) dielectric ceramics has been well known as materials for microwave devices, such as dielectric resonator, duplexer, etc[1-3]. However ZST ceramics is known to be hardly sintered.($\leq$ 1600℃) In order to overcome this problem various sintering aids has been developed and their effects were analyzed in many reported[4-5]. Among another sintering aids, ZnO have been a potential additives for sintering of ZST at low T($\leq$ 1400℃). But, it also has serious side effects such as decrease of Q × f value. Therefore many researches[6-10] have been focused on dopants for compensation of Q × f value decreased by sintering aids. However, the role of ZnO in altering Q × f value of ZST still has some controversy. In the present paper, by means of post-annealing in the oxygen atmosphere ZST and preparing ZST via a alternative route from conventional procedure, obtaining of ZST with high Q × f value was tried by the control of Zn content in the lattice .

EXPERIMENT

Sample preparation

In the method designated by process A, the $(Zr_{0.8}Sn_{0.2})TiO_4$ powders were prepared by conventional solid state synthesis from the simple oxides. The starting materials were reagent-grade ZrO_2, SnO_2, TiO_2, and ZnO. The mixtures were ball-milled for 24 hr in distilled water, filter pressed, dried and grounded. After calcined at 1000℃ for 2hr, 2 mol% ZnO was added as sintering aid. ZST powders prepared were pressed at 1000kg/cm² pressure into disks (8 mm diameter and ～4 mm thickness), then sintered at 1350℃ for 2 hr in static air atmosphere.

Different from process A, host materials were weighed according to the composition and at the same time, 2 mol% ZnO was added before calcination. The mixtures containing ZnO were ball-milled for 24 hr in distilled water, filter pressed, dried and grounded. Prepared powders were pre-pressed, then put into calcination(1000℃ for 2 hr) and subsequent sintering (1250℃ for 2 hr) [11]. This method is designated by process B. After sintered by both process A and B, samples were annealed at 900 ~ 1100℃ for 5 hr in the oxygen atmosphere.

293

Properties characterization

The sintered densities of the samples were measured by the Archimedes method. Dielectric properties in the microwave frequency range were measured, using Hakki- colemann method. A network analyzer (HP 8720C, Hewlett-Packard Co.) was used for the microwave measurement system. X-ray powder diffraction (M18XHF-SRA, McScience, Japan) was used to determine the crystalline phases, using Cu Kα radiation. The ZST ceramics was were processed by polishing, dimpling, and ion milling (Gattan, Model DuoMill 600) to observe by transmission electron microscope (CM 20, Philips Electronic Instruments, Inc., Amsterdam, Netherlands) and energy dispersive x-ray spectrometer for microstructure and composition, respectively.

RESULTS

Fig. 1. shows XRD patterns of ZST ceramics as-sintered at 1350℃ for 2h and annealed at 900 ~ 1100℃ for 5 h in the oxygen atmosphere. In the samples as-sintered as well as annealed at the all temperature, second phases and lattice constant as well as distortion were not observed.

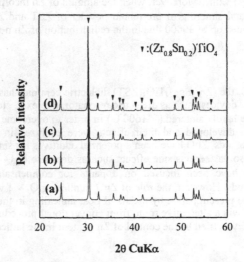

Fig. 1. XRD patterns of ZST ceramics (a) as-sintered, post-annealed at (b) 900℃, (c) 1000℃ and (d) 1100℃ for 2hr in the oxygen atmosphere, respectively.

Dielectric properties of post-annealed ZST ceramics at 900 ~ 1100℃ are shown in Fig. 2. Although samples are annealed, relative density, dielectric constant, and temperature coefficient of the resonant frequency remained virtually unchanged at as-sintered values; 97.5 %, 38 and -3ppm/℃, respectively. On the other hand, it was observed that among specimens, Q × f value of ZST post-annealed at 900℃ increased to 46000 from the as-sintered value of 40000. Therefore, it could be assumed that Q × f value in this case was affected by certain factors other than the existence of second phase and/or change of lattice parameter and distortion.

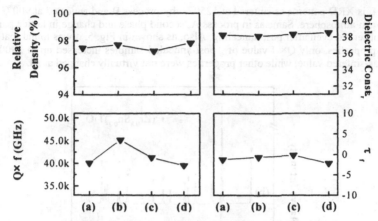

Fig. 2. Dielectric properties of ZST ceramics sintered at (a) 1350℃ and post-annealed at (b) 900℃, (c) 1000℃, (d) 1100℃ for 5 h in the oxygen atmosphere, respectively.

Fig. 3. shows TEM image and X-ray microanalysis of grain boundary phases and internal grain on the ZST ceramics prepared by process A. Comparing sample as-sintered at 1350℃ to annealed at 900℃ for 5h in the oxygen atmosphere, Zn content at certain depth from grain boundary remarkably decreased after post-annealing. That is, annealing treatment of as-sintered ZST ceramics can causes Zn to out-diffuse. Therefore, it is inferred that the incorporation of Zn into the grains are solely responsible for Q× f value of ZST in the absence of other factors mentioned above.

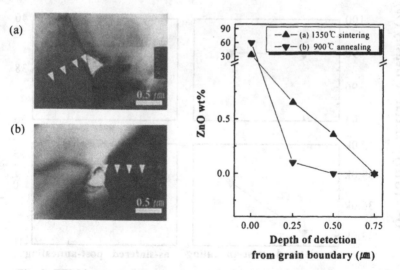

Fig. 3. TEM image and X-ray microanalysis of grain boundary area of ZST prepared by process A;(a) sintered at 1350℃ for 2h, (b) post-annealed at 900℃ for 5h in the oxygen atmosphere, ▽ detection point.

Fig. 4. is XRD patterns as-sintered at 1250℃ by process B and annealed at 900℃ for 5h in the oxygen atmosphere. Same as in process A, second phase and change in lattice parameter were not observed even after post-annealing. Also, as shown in Fig. 5, it was found that among dielectric properties, only Q× f value of post-annealed samples increased up to 39000 from 33000 of as-sintered value, while other properties were not virtually changed at all.

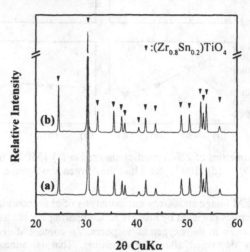

Fig. 4. XRD patterns of (a) as-sintered at 1250℃, process B and (b) post-annealed at 900℃ for 5h in the oxygen atmosphere.

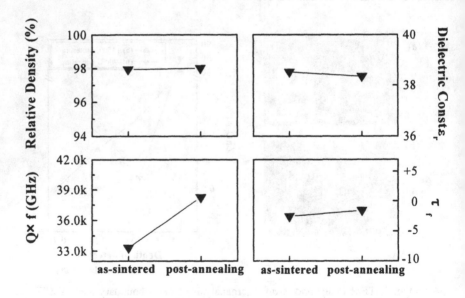

Fig. 5. Comparison of dielectric properties of ZST ceramics sintered at 1250℃ process B;before and after annealing at 900℃ for 5 h in the oxygen atmosphere.

Fig. 6 shows TEM image and X-ray microanalysis of grain of the ZST ceramics prepared by process B. In bright image shown in Fig. 6(a), at triple point of grains, Zn –rich phase were observed.(Zn content of as-sintered ZST at this point was 31.7wt%.) Because ZnO was added before calcination, Zn atoms in the process B could be incorporated into ZST ceramics during two separate heat treatment ; comparing with process A, both calcination and then sintering allow Zn to an opportunity of diffusion in grain or ZST lattice When the distance from grain boundary(thus triple point) increased to 0.25, 0.50 and 0.75 μm, Zn contents decreased gradually down to 12.9, 3.8 and 1.2 wt%, respectively. Moreover, compared with the result (Fig. 3) of sample prepared by conventional technique, in the case of as-sintered ZST ceramic made in process B, the more Zn quantity diffuses into the grain. This higher incorporation of Zn into the grain would be reason that samples by process B had the lower 33,000 of as-sintered Q × f value than 40000 from the process A.

After annealed at 900℃ for 5h in the oxygen atmosphere, Zn content of triple point enriched up to 69.5%. Also, Zn diffused from the internal grain into grain boundary. ; at distances from 0.25, 0.50 and 0.75 μm from the grain boundary, Zn contents reduced to 2.9, 1.5 and 0.0 wt%, respectively. Therefore, same as discussed above, it could be concluded that the recovery of Q × f value after post-annealing resulted from Zn out-diffusion toward grain boundary.

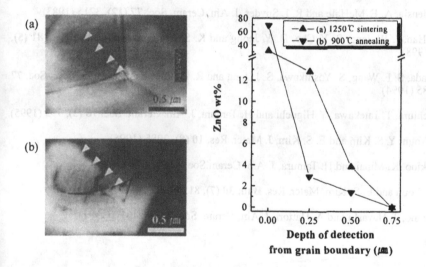

Fig. 6. TEM image and X-ray microanalysis of grain boundary phases of ZST prepared by process B;(a) sintered at 1250℃ for 2h, (b) post-annealed at 900℃ for 5h in the oxygen atmosphere.

CONCLUSIONS

Redistribution of Zn in the $(Zr_{0.8}Sn_{0.2})TiO_4$ ceramics via post-annealing could improve Q × f value without any sacrifice of other dielectric properties such as dielectric constant. After sintering via conventional route, a significant amount of Zn can diffused into the grain as well as segregated at grain boundary of ZST . On the other hand, after annealing at 900℃ in the oxygen atmosphere, it was confirmed that Zn diffuse out toward grain boundary resulting in increase of Q × f value up to 46000. Similarly, when the addition of ZnO before calcination was adopted as an alternative route, the post annealing step could also cure the losses in Q × f value up to 39000 by causing out-diffusion of Zn from grain inside. Considering both case, when ZnO is employed to low sintering temperature for ZST, Zn atom incorporated inside grain should maintained as low as possible in order to prevent decrease of Q × f value.

REFERENCES

1. W. Wersing, in *Electronic Ceramics*, edited by B. C. H. Steele (Elsevier A. A., London and New York, 1991), pp. 76-83.

2. K. Wakino, T. Nishikawa, Y. Ishikawa and H. Tamura, Br. Ceram. Trans. **89**, 39 (1990).

3. A. E. McHale and R. S. Roth, J. Am. Ceram. Soc. **66 (2)**, C-18 (1983).

4. R. Kudensia, A. E. McHale and R. L Snyder, J. Am. Ceram. Soc. **77 (12)**, 3215 (1983).

5. K. R. Han, J. W. Jang, S. Y. Cho, D. Y. Jeong and K. S. Hong, J. Am. Ceram. Soc. **81 (5)**, 1209 (1998).

6. T. Takada, S. F. Wang, S. Yoshikawa, S. J. Jang and R. E. Newnham, J. Am. Ceram. Soc. **77 (9)**, 2485 (1994).

7. N. Michiura, T. Tatekawa, Y. Higuchi and H. Tamura, J. Am. Ceram. Soc. **78 (3)**, 793 (1995).

8. K. H. Youn, Y. S. Kim and E. S. Kim, J. Mater. Res. **10 (8)**, 2085 (1995).

9. K. Wakino, K. Minai and H. Tamura, J. Am. Ceram.Soc. **67 (4)**, 278 (1984).

10. K. H. Youn and E. S. Kim, Mater. Res. Bull. **30 (7)**, 813 (1995).

11. S. Hirano, T. Hayashi and A. Hattori, J. Am. Ceram. Soc. **74 (6)**, 1320 (1991).

STEREOCHEMICAL STRUCTURE FOR SODIUM
IN NATIVE AND THERMAL SILICA LAYERS

A.-M. FLANK[*], F. TENEGAL[*], P. LAGARDE[*], C. MAZZARA[**], J. JUPILLE[**]

[*] LURE, Bât 209D, B.P. 34, Centre Universitaire Paris-Sud, F-91898 Orsay Cedex
[**] Laboratoire CNRS/Saint-Gobain " Surface du Verre et Interfaces ", BP 135, F-93303 Aubervilliers, France

ABSTRACT

Sodium-covered silica films formed on silicon substrates have been examined by X-ray photoemission spectroscopy (XPS) and X-ray absorption spectroscopy (EXAFS) in ultra-high vacuum conditions at 300K. The results show that sodium diffuses into the silica layer on a reversible manner and that it modifies the silica network in order to create its own site. Sodium atoms are surrounded by oxygen atoms at an average distance of 2.3 Å and by a second shell which is assigned to silicon atoms located at 3.8 Å. At high Na concentrations, sodium atoms are also present in the close environment of one sodium atom.

Keywords : Soda-silicate glasses, X-ray Absorption Spectroscopy, Sodium Local Order, X-ray Photoemission

INTRODUCTION

The addition of alkali-metal oxides to the pure silica strongly reduces its melting temperature and thus facilitate its manufacturing. Indeed, the understanding of the complex environment of alkali silicate glasses is an important issue regarding the industrial importance of these materials. The basic chemistry of commercial silica glasses involves network-forming elements whose role is to built chains linked by strong chemical bonds, mostly covalent, and network-modifying elements which reduces the connectivity of the glass former network [1] .

The neighbourhood of alkali ions in silica glasses is more puzzling than that of silicon. In the Zachariasen-Warren's view [2], alkali ions are not supposed to meet any particular coordination shell, but to simply occupy the available sites of the silica network. More recent studies have provided evidences for some ordering in the chemical environment of the network-modifying ions [3]. On the basis of EXAFS analysis at the sodium K-edge of silica glasses, Greaves et. al. [4] have developed a 'Modified Random Network' model, in which alkali ions and non-bridging oxygens (NBO) to which alkali atoms are mostly linked, concentrate into channels surrounding zones dominated by pure silica. Nevertheless, all these studies have been conducted on bulk materials manufactured with the usual methods. At the contrary, we present here results on thin silica films prepared using 'surface' techniques. Moreover, since pure silica is believed to mostly involve bridging oxygen atoms, such samples offer the opportunity to examine the behaviour of sodium atoms inserted in an environment which consists almost exclusively of bridging oxygen atoms.

EXPERIMENT

Photoemission experiments (XPS) have been performed in an ultra-high vacuum system equipped with an Mg Kα X-ray photoemission spectrometer VG CLAM II. X-ray absorption data on the sodium K-edge have been recorded on the SA32 soft X-ray beam line of Super-ACO equipped with a toroidal focusing mirror, and a two-crystal monochromator using beryl (10$\bar{1}$0) crystals. At the sodium K-edge energy the resolution is close to 0.5 eV. The data have been collected in the total electron yield mode (TEY) from the sample. Model compounds have also been examined, namely Na_2O and Na_2SiO_3 powders (Strem Chemicals). Moreover, we have run the sodium K-edge of two soda-silicate glasses with 10% and 25% of Na_2O.

Native and thermal silica films have been used in the present work for XPS and EXAFS experiments. Their thickness have been evaluated by microprobe analysis to 26 and 53 Å respectively, within an accuracy of 3 Å. Once in vacuum, samples are outgassed at 870 K for several hours in vacuum and then, while kept at the same temperature, they are exposed to 10^{-6} mbar of oxygen for 30 mn to remove the carbon contamination.

Sodium has been evaporated on the silica films at room temperature from a carefully outgassed SAES getter. Microprobe analysis has also been used to obtain the amount of sodium deposited onto the silica layers.

RESULTS

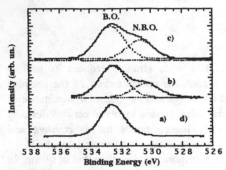 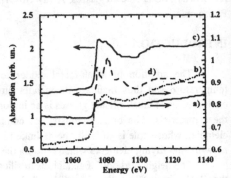

Fig. 1. Oxygen 1s photoemission spectra of a glass (top), and of a silica layer with (middle) and without (bottom) sodium.

Fig. 2. Sodium 1s near edge structure of two silica layers (a , b), a Na_2O-SiO_2 glass (c) and Na_2SiO_3 (d).

The photoemission O 1s chemical shift

Photoelectron spectrum of the O1s level from silicate glasses and from native silica films are shown in Fig. 1. Fig. 1a) is the spectrum of a pure silica glass. Fig. 1b) has been recorded after deposition of sodium at room temperature, up to a content, measured afterwards, of 18 at. % of Na. As expected, due to the presence of sodium, a strong shoulder appears at – 2 eV on the lower binding energy side of the core level spectrum of pure silica. Such a binding energy shift closely compares to that observed between bridging and non-bridging oxygen in the case of silica glass [5] . For comparison, the spectrum 1c) has been collected on a commercial soda-lime sample which has been exposed for hours in vacuum to the X-rays source of our photoemission system so as to enhance the surface sodium concentration. We then have here a first indication that the oxygen behavior in the film is identical to that in the bulk material. Spectrum 1d) has been taken on sample b after annealing at 870K. Within the detection limit of the XPS analyser, the two spectra a and d are identical, and no Na1s line could be detected in sample d. Therefore, it appears that the sodium dissolved in the bulk of the silica film has been removed by this annealing.

EXAFS data

The absorption edges corresponding to the two sodium-covered silica samples with 6.5 (sample S3, thermal silica, curve a) and 20 at % Na (sample S1, native silica, curve b), are shown in Fig. 2. They present both a double peak just after the edge, and a broad resonance some 20 eV beyond. The near edge structures of one glass (c) and of Na_2SiO_3 (d) present also the same features although that the intensity ratio between the two first peaks beyond the edge is slightly different and that, in the case of Na_2SiO_3 a more fine structure comes from the crystalline character of this compound. Therefore, the local structure around the sodium atom in both the silica and the glass samples appears close to that of Na_2SiO_3, but affected by a topological disorder.

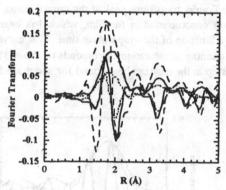

Fig. 3. Modulus and Imaginary parts of the Fourier transforms of sodium EXAFS in different samples (see text).

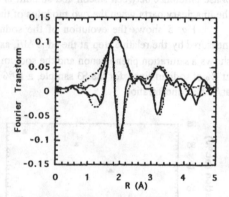

Fig. 4. Fourier transforms of the experimental data (solid line) and the Feff6 simulation (dots) for 6.5 at.% Na.

EXAFS data have been analyzed in a now classical way and the backscattering parameters have been checked on model compounds Na_2O and Na_2SiO_3 using the Feff6 code [6]. Fig. 3 compares the modulus and imaginary parts of the Fourier transforms for sodium-covered samples S2 (11 at.% Na on native silica, dots), S3 (solid line) and for the 25% Na_2O soda-silica glass (dashs). The similarity between all spectra is consistent with the previously demonstrated diffusion of sodium into the silica network and supports the suggestion which has been made by examining the profiles of the sodium edges that sodium occupies sites which are close to those encountered in a Na_2O-SiO_2 glass. Beyond a first shell made of oxygen atoms, we observe a second contribution at about 3.2 Å (uncorrected from phase shifts) which is a definite signature of a medium range order around the sodium atoms. It has also to be pointed out that the Fourier transform of the glass exhibits a peak around 2.5Å not seen in the films. To achieve a more quantitative description of the sodium-covered silica films, we have built a cluster which represents this local coordination by taking the five-fold oxygen local environment of sodium in Na_2SiO_3 as a basis and given some inputs [7]:

1) the interatomic distances Si-O and Na-O must be respectively close to 1.61 and 2.4 Å. These values are those found in silica and in soda-silicate glasses.

2) the valence bond angle of oxygen will be close to its average value of 150° found for bridging oxygens in silicon oxide. This results in a second shell distance of about 3.8 Å and then the above cluster of oxygen has been surrounded by a second shell of silicon atoms at 3.8 Å. A Feff6 calculation, which has first been checked on model compounds Na_2O and Na_2SiO_3, has then been performed and the results are shown for the sample S3 in Fig. 4 , where the dots are the theoretical functions. The quality of the fit is very good and the numerical values resulting of these different fits are gathered in Table I. In order to check the unicity of this structural solution, we have then tried different types of surroundings, keaping the first shell of oxygen constant and allowing the nature and the interatomic distance of the second shell to vary. It appears that no valid structural solution could be obtained with another structural hypothesis, in particular when a sodium atom is supposed to occupy the second shell. The backscattering phase difference between silicon and sodium is high enough to destroy the agreement between the imaginary parts when the two modulus of the Fourier transforms peak at the same distance.

Fig. 5 shows the evolution of the sodium concentration in the film, which has been measured by the relative step at the edge $\Delta I/I$, as a function of the evaporation time. This curve shows a saturation phenomenon and the sodium quantity at saturation corresponds to about 35 at. % of sodium since, for the S3 sample, a 10% step at the edge has been found for about a 6.5 at.% Na concentration.

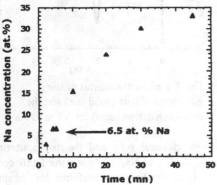

Fig. 5. Evolution of the Na concentration with the evaporation time.

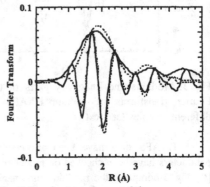

Fig. 6. Fourier transforms of the experimental data (solid line) and the Feff6 simulation (dots) for 35 at.% Na.

The Fourier Transform of the last data (a 45 mn evaporation time) is shown in Fig. 6 with a solid line. Compared to the same analysis done on samples S1, S2 and S3 (Fig. 3) differences appear in the 2-2.5 Å domain, differences that we have tentatively attributed to the presence of sodium atoms in the second shell of one given sodium. We therefore have added this contribution to the cluster defined above and the fitting procedure has only allowed the structural parameters of this new shell to vary. The result is shown in the same figure in dotted line, and the main numerical value is the Na-Na distance which is found at 3.6 Å. Assuming that two sodium atoms are linked through an oxygen bridge, the bond angle Na-O-Na of this oxygen is 103°. Finally this sodium-sodium distance appears slightly larger than the one found by EXAFS on bulk soda-silicate glasses [4].

	sodium - oxygen			sodium - silicon		
	R (Å)	N	σ^2 (Å^2)	R (Å)	N	σ^2 (Å^2)
Sample S3 (ther. sil.) (6.5 at. %)	2.30 (±0.04)	3.7 (±0.5)	0.006 (±0.003)	3.80 (±0.05)	3.7 (±0.5)	0.007 (±0.005)
Sample S2 (nat. sil.) (20 at. %)	2.30 (±0.04)	3.7 (±0.5)	0.009 (±0.003)	3.80 (±0.05)	3.7 (±0.5)	0.015 (±0.005)
Sample S1 (nat. sil.) (11 at. %)	2.30 (±0.04)	1.85 (±0.5)	0.0045 (±0.003)	3.80 (±0.05)	2 (±0.5)	0.007 (±0.005)

Table I. Local environment of the sodium atoms for the different samples

CONCLUSION

Although the present work is dealing with sodium-silica films which are chemically different from soda-silicate glasses, the sodium site defined here can be compared to the models describing the local order in a glass. The present data confirm that sodium diffuses within the bulk of the silica at room temperature, a behaviour which has been predicted by molecular dynamic calculations for lithium and, in smaller extent, for potassium [8]. The EXAFS demonstrates that, once inserted in the silica network, sodium atoms become surrounded by a first shell of oxygen atoms at a well defined distance. The presence of a second shell of atoms reveals that the local order around sodium extends up to the neighbouring silica tetrahedra. The EXAFS technique thus demonstrates the existence of a defined stereochemical order around sodium diffusing in an amorphous silica network. This means that in this thermodynamically favourable configuration, sodium creates its own site via a relaxation of the surrounding arrangement of silica tetrahedra, at variance with the " continuous random network " model in which, beyond their direct action in breaking the bridging bonds, network modifiers are supposed to only occupy available sites of the glass former network. On the contrary, the present data at high Na content is still compatible with the 'modified random network' [4] which postulates a local order around sodium for silica glasses and with the finding by Gaskell et al. [3] of a medium-range order in the cation network.

Nevertheless there are a few pending questions. The experimentally observed narrow distribution of sodium-oxygen distances can only be achieved via relaxations of the silica network around the sodium atoms. Under the assumption that, as predicted by theory for silicate glass, the sodium-BO distance should be higher by a quarter of an angstrom than the sodium-NBO distance, which is far beyond the error attached to the determination of distances by means of the EXAFS calculation, the above result indicates that sodium diffusing in silica becomes surrounded by non-bridging oxygen atoms. Consistently, the – 2 eV XPS O 1s core level shift associated to the O1s component which compares to the BO-NBO shift observed in soda-silica glass, also favours the formation of non-bridging oxygen. Since pure silica dominantly involves bridging oxygen, this means that the introduction of sodium breaks some Si-O bonds and that Si-O distances might also relax around dissolved sodium atoms. However, to account for the almost perfect recovery of the initial XPS O 1s spectrum - including lineshape and intensity – arising after the removal of the sodium dissolved in the silica film by annealing at 870 K (Fig. 1), the silica network should only undergo local relaxations.

While sodium atoms evaporated from a SAES Getter cell is supposed to be neutral, we find here a behaviour of this metal typical of Na^+, as it is in the glasses where sodium is introduced via Na_2O. Then the question arises about the electric neutrality of the thin film after sodium deposition, since Si2p photoemission results, non shown here, exhibit only, on the pure native silica layer, the signatures of Si^0 from the silicon substrate and Si^{4+} from the SiO_2, without any evidence of sub-oxide components.

REFERENCES

1. P.H. Gaskell, Models for the Structure of Amorphous Solids in *Glasses and Amorphous Materials, Materials Science and Technology*, edited by J. Zarzycki (VCH, Weinheim, 1991).

2. B.E. Warren, J. Am. Ceram. Soc., **24**, 256 (1941)

3. P.H. Gaskell, M.C. Eckersley, A.C. Barnes, P. Chieux, Nature **350**, 675 (1991)

4. G.N. Greaves and K.L. Ngai Phys. Rev. **B52**, 6358 (1995) and references therein

5. J.S. Jen and M.R. Kalinowski J. Non-Cryst. Solids **38-39**, 21 (1989)

6. J.J. Rehr Jpn. J. Appl. Phys. **32**, 8 (1993)

7. C. Mazzara, J. Jupille, A.M. Flank, P. Lagarde to be published in J. Phys. Chem.

8. R. Carracciolo and S.H. Garofalini J. Am. Ceram. Soc.**71**, C-346 (1988)

AUTHOR INDEX

305

SUBJECT INDEX

307

oxide(s), 223
    layers, 33

PECVD, 57, 63
phase shifter, 175, 217
photovoltaics, 155
piezoreflectance, 115
plasma, 63
PMN, 287
post-annealing, 293
precursor(s), 51, 133
    sol aging, 211
preferred orientation, 281
pulsed laser deposition, 275
pyrite, 133
pyrolysis, 281
PZT, 187

Raman scattering, 63
RBS, 115
rechargeable lithium microbattery, 205
recycling, 199
resistivity, 211
ruthenium oxide, 211

scandium oxide, 51
SiH_4, 97
silica, 251
silicon, 45, 245
    carbide, 281
    carbo-nitride, 115
    nitride, 109, 121
    oxide, 121
silsesquioxane, 251
sintering aids, 293
SiO_2:F:C, 57
soda-silicate glasses, 299
sodium local order, 299
soft solution processing, 205
solar cells, 147, 199
sol-coating, 287
sol-gel, 169, 175, 211, 217

process, 181
spin-on-glass (SOG), 251
stability of properties, 57
stress, 187
strontium barium niobate, 181
surface exchange, 275
superlattices, 33
synthesis, 223, 237

tantalum
    nitride, 13
    oxide, 13, 51
Ta_2O_5, 257
TDMAT, 103
TEOS, 63
tetraiodosilane, 109
thermal CVD
    low temperature, 109
thin film(s), 45, 69, 75, 181, 211, 275
$Ti(NMe_2)_4$, 97
Ti-Si-N, 97
titanium
    nitride, 91
    oxide, 163

ultrafine, 193
UV effect, 257

volatile precursors, 139

wafer cleaner, 245
water absorption, 57

x-ray
    absorption spectroscopy, 299
    diffraction, 281
    photoemission, 299

zirconium oxide, 69
ZnO, 155, 293
ZrO_2, 263
ZST, 281

Printed in the United States
By Bookmasters